Texts in Applied Mathematics 55

Texts in Applied Mathematics

(continued after index)

Grégoire Allaire   Sidi Mahmoud Kaber

Numerical Linear Algebra

Translated by Karim Trabelsi

 Springer

Grégoire Allaire
CMAP Ecole Polytechnique
91128 Palaiseau
France
allaire@cmapx.polytechnique.fr

Sidi Mahmoud Kaber
Université Pierre et Marie Curie
75252 Paris
France
kaber@ann.jussieu.fr

Series Editors

J.E. Marsden
Control and Dynamic Systems, 107-81
California Institute of Technology
Pasadena, CA 91125
USA

L. Sirovich
Division of Applied Mathematics
Brown University
Providence, RI 02912
USA

S.S. Antman

Department of Mathematics
and
Institute for Physical Science
  and Technology
University of Maryland
College Park, MD 20742-4015
USA
ssa@math.umd.edu

ISBN 978-1-4899-9741-8   ISBN 978-0-387-68918-0 (eBook)
DOI: 10.1007/978-0-387-68918-0

Mathematics Subject Classification (2000): 15-01, 54F05, 65F10, 65F15, 65F35

Original French edition: Introduction à Scilab-exercices pratiques corrigés d'algebre linéaire & Algebre linéaire numérique. Cours et exercices, 2002

Printed on acid-free paper

9 8 7 6 5 4 3 2 1

springer.com

Series Preface

Mathematics is playing an ever more important role in the physical and biological sciences, provoking a blurring of boundaries between scientific disciplines and a resurgence of interest in the modern as well as the classical techniques of applied mathematics. This renewal of interest, both in research and teaching, has led to the establishment of the series Texts in Applied Mathematics (TAM).

The development of new courses is a natural consequence of a high level of excitement on the research frontier as newer techniques, such as numerical and symbolic computer systems, dynamical systems, and chaos, mix with and reinforce the traditional methods of applied mathematics. Thus, the purpose of this textbook series is to meet the current and future needs of these advances and to encourage the teaching of new courses.

TAM will publish textbooks suitable for use in advanced undergraduate and beginning graduate courses, and will complement the Applied Mathematical Sciences (AMS) series, which will focus on advanced textbooks and research-level monographs.

Pasadena, California   J.E. Marsden
Providence, Rhode Island   L. Sirovich
College Park, Maryland   S.S. Antman

Preface

The origin of this textbook is a course on numerical linear algebra that we taught to third-year undergraduate students at Université Pierre et Marie Curie (Paris 6 University). Numerical linear algebra is the intersection of numerical analysis and linear algebra and, more precisely, focuses on practical algorithms for solving on a computer problems of linear algebra.

Indeed, most numerical computations in all fields of applications, such as physics, mechanics, chemistry, and finance. involve numerical linear algebra. All in all, these numerical simulations boil down to a series of matrix computations. There are mainly two types of matrix computations: solving linear systems of equations and computing eigenvalues and eigenvectors of matrices. Of course, there are other important problems in linear algebra, but these two are predominant and will be studied in great detail in this book.

From a theoretical point of view, these two questions are by now completely understood and solved. Necessary and sufficient conditions for the existence and/or uniqueness of solutions to linear systems are well known, as well as criteria for diagonalizing matrices. However, the steady and impressive progress of computer power has changed those theoretical questions into practical issues. An applied mathematician cannot be satisfied by a mere existence theorem and rather asks for an algorithm, i.e., a method for computing unknown solutions. Such an algorithm must be efficient: it must not take too long to run and too much memory on a computer. It must also be stable, that is, small errors in the data should produce similarly small errors in the output. Recall that errors cannot be avoided, because of rounding off in the computer. These two requirements, efficiency and stability, are key issues in numerical analysis. Many apparently simple algorithms are rejected because of them.

This book is intended for advanced undergraduate students who have already been exposed to linear algebra (for instance, [9], [10], [16]). Nevertheless, to be as self-contained as possible, its second chapter recalls the necessary definitions and results of linear algebra that will be used in the sequel. On the other hand, our purpose is to be introductory concerning numerical analysis, for which we do not ask for any prerequisite. Therefore, we do not pretend to be exhaustive nor to systematically give the most efficient or recent algorithms if they are too complicated. We leave this task to other books at the graduate level, such as [2], [7], [11], [12], [14], [17], [18]. For pedagogical reasons we satisfy ourselves in giving the simplest and most illustrative algorithms.

Since the inception of computers and, all the more, the development of simple and user-friendly software such as Maple, Mathematica, Matlab, Octave, and Scilab, mathematics has become a truly experimental science like physics or mechanics. It is now possible and very easy to perform numerical experiments on a computer that help in increasing intuition, checking conjectures or theorems, and quantifying the effectiveness of a method. One original feature of this book is to follow an experimental approach in all exercises. The reader should use Matlab for solving these exercises, which are given at the end of each chapter. For some of them, marked by a (*), complete solutions, including Matlab scripts, are given in the last chapter. The solutions of the other exercises are available in a solution manual available for teachers and professors on request to Springer. The original french version of this book (see our web page http://www.ann.jussieu.fr/numalgebra) used Scilab, which is probably less popular than Matlab but has the advantage of being free software (see http://www.scilab.org). Finally we thank Karim Trabelsi for translating a large part of this book from a French previous version of it.

We hope the reader will enjoy more mathematics by seeing it "in practice."

G.A., S.M.K.
Paris

Contents

1

Introduction

As we said in the preface, linear algebra is everywhere in numerical simulations, often well hidden for the average user, but always crucial in terms of performance and efficiency. Almost all numerical computations in physics, mechanics, chemistry, engineering, economics, finance, etc., involve numerical linear algebra, i.e., computations involving matrices. The purpose of this introduction is to give a few examples of applications of the two main types of matrix computations: solving linear systems of equations on the one hand, and computing eigenvalues and eigenvectors on the other hand. The following examples serve as a motivation for the main notions, methods, and algorithms discussed in this book.

1.1 Discretization of a Differential Equation

We first give a typical example of a mathematical problem whose solution is determined by solving a linear system of large size. This example (which can be generalized to many problems all of which have extremely important applications) is linked to the approximate numerical solution of partial differential equations. A partial differential equation is a differential equation in several variables (hence the use of partial derivatives). However, for simplicity, we shall confine our exposition to the case of a single space variable x, and to real-valued functions.

A great number of physical phenomena are modeled by the following equation, the so-called Laplace, Poisson, or conduction equation (it is also a time-independent version of the heat or wave equation; for more details, see [3]):

$$\begin{cases} -u''(x) + c(x)u(x) = f(x) \text{ for all } x \in (0,1), \\ u(0) = \alpha, \quad u(1) = \beta, \end{cases} \tag{1.1}$$

where α, β are two real numbers, $c(x)$ is a nonnegative and continuous function on $[0,1]$, and $f(x)$ is a continuous function on $[0,1]$. This second-order differential equation with its boundary conditions "at both ends" is called a boundary value problem. We shall admit the existence and uniqueness of a solution $u(x)$ of class C^2 on $[0, 1]$ of the boundary value problem (1.1). If $c(x)$ is constant, one can find an explicit formula for the solution of (1.1). However, in higher spatial dimensions or for more complex boundary value problems (varying $c(x)$, nonlinear equations, system of equations, etc.), there is usually no explicit solution. Therefore, the only possibility is to approximate the solution numerically. The aim of example (1.1) is precisely to show a method of discretization and computation of approximate solutions that is called the method of *finite difference*. We call discretization of a differential equation the formulation of an approximate problem for which the unknown is no longer a function but a finite (discrete) collection of approximate values of this function. The finite difference method, which is very simple in dimension 1, can easily be generalized, at least in principle, to a wide class of boundary value problems.

In order to compute numerically an approximation of the solution of (1.1), we divide the interval $[0, 1]$ into n equal subintervals (i.e., of size $1/n$), where n is an integer chosen according to the required accuracy (the larger the value of n, the "closer" the approximate solution will be to the exact one). We denote by x_i the $(n+1)$ endpoints of these intervals:

$$x_i = \frac{i}{n}, \quad 0 \le i \le n.$$

We call c_i the value of $c(x_i)$, f_i the value of $f(x_i)$, and u_i the approximate value of the solution $u(x_i)$. To compute these approximate values $(u_i)_{0<i<n}$, we substitute the differential equation (1.1) with a system of $(n-1)$ algebraic equations. The main idea is to write the differential equation at each point x_i and to replace the second derivative by an appropriate linear combination of the unknowns u_i. To do so, we use Taylor's formula by assuming that $u(x)$ is four times continuously differentiable:

$$\begin{cases} u(x_{i+1}) = u(x_i) + \frac{1}{n}u'(x_i) + \frac{1}{2n^2}u''(x_i) + \frac{1}{6n^3}u'''(x_i) + \frac{1}{24n^4}u''''(x_i + \frac{\theta^+}{n}), \\ u(x_{i-1}) = u(x_i) - \frac{1}{n}u'(x_i) + \frac{1}{2n^2}u''(x_i) - \frac{1}{6n^3}u'''(x_i) + \frac{1}{24n^4}u''''(x_i - \frac{\theta^-}{n}), \end{cases}$$

where $\theta^-, \theta^+ \in (0, 1)$. Adding these two equations, we get

$$-u''(x_i) = \frac{2u(x_i) - u(x_{i-1}) - u(x_{i+1})}{n^{-2}} + \mathcal{O}(n^{-2}).$$

Neglecting the lowest-order term n^{-2} yields a "finite difference" formula, or *discrete derivative*

$$-u''(x_i) \approx \frac{2u(x_i) - u(x_{i-1}) - u(x_{i+1})}{n^{-2}}.$$

Substituting $-u''(x_i)$ with its discrete approximation in the partial differential equation (1.1), we get $(n-1)$ algebraic equations

$$\frac{2u_i - u_{i-1} - u_{i+1}}{n^{-2}} + c_i u_i = f_i, \quad 1 \le i \le n - 1,$$

completed by the two boundary conditions

$$u_0 = \alpha, \quad u_n = \beta.$$

Since the dependence in u_i is linear in these equations, we obtain a so-called linear system (see Chapter 5) of size $(n-1)$:

$$A_n u^{(n)} = b^{(n)}, \tag{1.2}$$

where $u^{(n)}$ is the vector of entries $(u_1, \ldots, u_{n-1})$, while $b^{(n)}$ is a vector and A_n a matrix defined by

$$A_n = n^2 \begin{pmatrix} 2 + \frac{c_1}{n^2} & -1 & 0 & \cdots & & 0 \\ -1 & 2 + \frac{c_2}{n^2} & \ddots & \ddots & & \vdots \\ 0 & \ddots & \ddots & -1 & & 0 \\ \vdots & & \ddots & -1 & 2 + \frac{c_{n-2}}{n^2} & -1 \\ 0 & & \cdots & 0 & -1 & 2 + \frac{c_{n-1}}{n^2} \end{pmatrix}, \quad b^{(n)} = \begin{pmatrix} f_1 + \alpha n^2 \\ f_2 \\ \vdots \\ f_{n-2} \\ f_{n-1} + \beta n^2 \end{pmatrix}.$$

The matrix A_n is said to be tridiagonal, since it has nonzero entries only on its main diagonal and on its two closest diagonals (the subdiagonal and the superdiagonal).

One can prove (see Lemma 5.3.2) that the matrix A_n is invertible, so that there exists a unique solution $u^{(n)}$ of the linear system (1.2). Even more, it is possible to prove that the solution of the linear system (1.2) is a correct approximation of the exact solution of the boundary value problem (1.1). We said that the previous finite difference method converges as the number of intervals n increases. This is actually a delicate result (see, e.g., [3] for a proof), and we content ourselves in stating it without proof.

Theorem 1.1.1. *Assume that the solution $u(x)$ of (1.1) is of class C^4 on $[0, 1]$. Then the finite difference method converges in the sense that*

$$\max_{0 \le i \le n} |u_i^{(n)} - u(x_i)| \le \frac{1}{96n^2} \sup_{0 \le x \le 1} |u''''(x)|.$$

Figure 1.1 shows the exact and approximate ($n = 20$) solutions of equation (1.1) on $]0, 1[$, where the functions c and f are chosen so that the exact solution is $u(x) = x \sin(2\pi x)$:

$$c(x) = 4, \quad f(x) = 4(\pi^2 + 1)u(x) - 4\pi \cos(2\pi x), \quad \alpha = \beta = 0.$$

The problem of solving the differential equation (1.1) has thus been reduced to solving a linear system. In practice, these linear systems are very large.

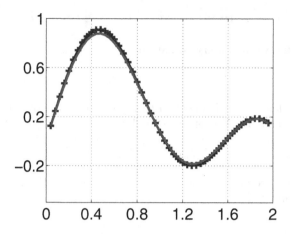

Fig. 1.1. Computation of an approximate solution of (1.1) and comparison with the exact solution.

Indeed, most physical phenomena are three-dimensional (unlike our simplified example in one spatial dimension). A numerical simulation requires the discretization of a three-dimensional domain. For instance, if we decide to place 100 discretization points in each spatial direction, the total number of points, or unknowns, is 1 million (100^3), and hence the linear system to be solved is of size 1 million. This is a typical size for such a system even if some are smaller... or larger. In practice, one needs to have at one's disposal high-performance algorithms for solving such linear systems, that is, fast algorithms that require little memory storage and feature the highest accuracy possible. This last point is a delicate challenge because of the inevitable rounding errors (an issue that will be discussed in Section 5.3.1). Solving linear systems efficiently is the topic of Chapters 5 to 9.

1.2 Least Squares Fitting

We now consider a data analysis problem. Assume that during a physical experiment, we measure a magnitude or quantity y that depends on a real parameter t. We carry out m experiments and different measurements by varying the parameter t. The problem is to find a way of deducing from these m measurements a very simple experimental law that enables one to approximate as well as possible the studied quantity as an $(n-1)$th-degree polynomial (at most) of the parameter. Let us remark that the form of the experimental law is imposed (it is polynomial); but the coefficients of this polynomial are unknown. In general, the number of measurements m is very large with respect to the degree n of the sought-after polynomial. In other words, given m values of the parameter $(t_i)_{i=1}^m$ and m corresponding values of the measures $(y_i)_{i=1}^m$, we look for a polynomial $p \in \mathbb{P}_{n-1}$, the set of polynomials of one variable t of degree less than or equal to $n-1$, that minimizes the error between the experimental value y_i and the predicted theoretical value $p(t_i)$. Here, the error is measured in the sense of "least squares fitting," that is, we minimize the sum of the squares of the individual errors, namely

$$E = \sum_{i=1}^m |y_i - p(t_i)|^2. \tag{1.3}$$

We write p in a basis $(\varphi_j)_{j=0}^{n-1}$ of $\mathbb{P}_{n-1}$:

$$p(t) = \sum_{j=0}^{n-1} a_j \varphi_j(t). \tag{1.4}$$

The quantity E defined by (1.3) is a quadratic function of the n coefficients a_i, since $p(t_i)$ depends linearly on these coefficients. In conclusion, we have to minimize the function E with respect to the n variables $(a_0, a_1, \ldots, a_{n-1})$.

Linear regression. Let us first study the case $n = 2$, which comes down to looking for a straight line to approximate the experimental values. This line is called the "least squares fitting" line, or linear regression line. In this case, we choose a basis $\varphi_0(t) = 1, \varphi_1(t) = t$, and we set $p(t) = a_0 + a_1 t$. The quantity to minimize is reduced to

$$E(a_0, a_1) = \sum_{i=1}^m |y_i - (a_0 + a_1 t_i)|^2$$
$$= A a_1^2 + B a_0^2 + 2C a_0 a_1 + 2D a_1 + 2E a_0 + F$$

with

$$A = \sum_{i=1}^m t_i^2, \qquad B = m, \qquad C = \sum_{i=1}^m t_i,$$
$$D = -\sum_{i=1}^m t_i y_i, \quad E = -\sum_{i=1}^m y_i, \quad F = \sum_{i=1}^m y_i^2.$$

Noting that $A > 0$, we can factor $E(a_0, a_1)$ as

$$E(a_0, a_1) = A\left(a_1 + \frac{C}{A}a_0 + \frac{D}{A}\right)^2 + \left(B - \frac{C^2}{A}\right)a_0^2 + 2\left(E - \frac{CD}{A}\right)a_0 + F - \frac{D^2}{A}.$$

The coefficient of a_0^2 is also positive if the values of the parameter t_i are not all equal, which is assumed henceforth:

$$AB - C^2 = m\left(\sum_{i=1}^m t_i^2\right) - \left(\sum_{i=1}^m t_i\right)^2 = m\sum_{i=1}^m \left(t_i - \frac{1}{m}\sum_{j=1}^m t_j\right)^2.$$

We can thus rewrite $E(a_0, a_1)$ as follows:

$$E(a_0, a_1) = A\left(a_1 + \frac{C}{A}a_0 + \frac{D}{A}\right)^2 + \Delta\left(a_0 + \frac{E - \frac{CD}{A}}{\Delta}\right)^2 + G, \qquad (1.5)$$

where $\Delta = B - \frac{C^2}{A}$ and $G = F - \frac{D^2}{A} - \frac{(E-\frac{CD}{A})^2}{\Delta}$. The function E has then a unique minimum point given by

$$a_0 = -\frac{E - \frac{CD}{A}}{\Delta}, \qquad a_1 = -\frac{C}{A}a_0 - \frac{D}{A}.$$

Example 1.2.1. Table 1.1 (source I.N.S.E.E.) gives data on the evolution of the cost construction index taken in the first trimester of every year, from 1990 to 2000.

1990	1991	1992	1993	1994	1995	1996	1997	1998	1999	2000
939	972	1006	1022	1016	1011	1038	1047	1058	1071	1083

Table 1.1. Evolution of the cost construction index.

Figure 1.2 displays the least squares fitting line of best approximation for Table 1.1, that is, a line "deviating the least" from the cloud of given points. This example ($n = 2$) is simple enough that we can solve it exactly "by hand."

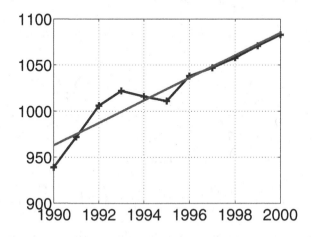

Fig. 1.2. Approximation of the data of Table 1.1 by a line.

Polynomial regression. We return now to the general case $n > 2$. We denote by $b \in \mathbb{R}^m$ the vector of measurements, by $c \in \mathbb{R}^m$ that of parameters, by

$x \in \mathbb{R}^n$ that of unknowns, and by $q(t) \in \mathbb{R}^m$ that of the predictions by the polynomial p:

$$b = \begin{pmatrix} y_1 \\ y_2 \\ \vdots \\ y_m \end{pmatrix}, \quad c = \begin{pmatrix} t_1 \\ t_2 \\ \vdots \\ t_m \end{pmatrix}, \quad x = \begin{pmatrix} a_0 \\ a_1 \\ \vdots \\ a_{n-1} \end{pmatrix}, \quad q(t) = \begin{pmatrix} p(t_1) \\ p(t_2) \\ \vdots \\ p(t_m) \end{pmatrix}.$$

As already observed, the polynomial p, and accordingly the vector $q(t)$, depend linearly on x (its coefficients in the basis of $\varphi_i(t)$). Namely $q(t) = Ax$ and $E = \|Ax - b\|^2$ with

$$A = \begin{pmatrix} \varphi_0(t_1) & \varphi_1(t_1) & \cdots & \varphi_{n-1}(t_1) \\ \varphi_0(t_2) & \varphi_1(t_2) & \cdots & \varphi_{n-1}(t_2) \\ \vdots & \vdots & & \vdots \\ \varphi_0(t_m) & \varphi_1(t_m) & \cdots & \varphi_{n-1}(t_m) \end{pmatrix} \in \mathcal{M}_{m,n}(\mathbb{R}),$$

where $\|.\|$ denotes here the Euclidean norm of $\mathbb{R}^m$. The least squares fitting problem reads then, find $x \in \mathbb{R}^n$ such that

$$\|Ax - b\| = \inf_{u \in \mathbb{R}^n} \|Au - b\|. \tag{1.6}$$

We shall prove (see Chapter 7) that $x \in \mathbb{R}^n$ is a solution of the minimization problem (1.6) if and only if x is solution of the so-called normal equations

$$A^t Ax = A^t b. \tag{1.7}$$

The solutions of the least squares fitting problem are therefore given by solving either the linear system (1.7) or the minimization problem (1.6). Figure 1.3 shows the solution of this equation for $m = 11$, $n = 5$, and the data y_i of Table 1.1.

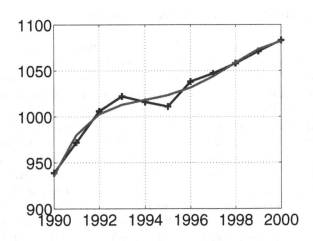

Fig. 1.3. Approximation of the data of Table 1.1 by a fourth-degree polynomial.

Multiple variables regression. So far we have assumed that the measured physical quantity y depended on a single real parameter t. We now consider the case that the experiments and the measured quantity depends on n parameters. We still have m experiments and each of them yields a possibly different measure. The goal is to deduce from these m measurements a very simple experimental law that approximates the physical quantity as a linear combination of the n parameters. Note that the form of this experimental law is imposed (it is linear), whereas the coefficients of this linear form are unknown.

In other words, let $a_i \in \mathbb{R}^n$ be a vector, the entries of which are the values of the parameters for the ith experiment, and let $f(a)$ be the unknown function from $\mathbb{R}^n$ into $\mathbb{R}$ that gives the measured quantity in terms of the parameters. The linear regression problem consists in finding a vector $x \in \mathbb{R}^n$ (the entries of which are the coefficients of this experimental law or linear regression) satisfying

$$\sum_{i=1}^{m} |f(a_i) - \langle a_i, x \rangle_n|^2 = \min_{y \in \mathbb{R}^n} \sum_{i=1}^{m} |f(a_i) - \langle a_i, y \rangle_n|^2,$$

where $\langle .,. \rangle_n$ denotes the scalar product in $\mathbb{R}^n$. Hence, we attempt to best approximate the unknown function f by a linear form. Here "best" means "in the least squares sense," that is, we minimize the error in the Euclidean norm of $\mathbb{R}^m$. Once again this problem is equivalent to solving the so-called normal equations (1.7), where b is the vector of $\mathbb{R}^n$ whose entries are the $f(a_i)$ and A is the matrix of $\mathcal{M}_{m,n}(\mathbb{R})$ whose m rows are the vectors a_i. The solution $x \in \mathbb{R}^n$ of (1.7) is the coefficients of the experimental law.

Least squares problems are discussed at length in Chapter 7.

1.3 Vibrations of a Mechanical System

The computation of the eigenvalues and eigenvectors of a matrix is a fundamental mathematical tool for the study of the vibrations of mechanical structures. In this context, the eigenvalues are the squares of the frequencies, and the eigenvectors are the modes of vibration of the studied system. Consider, for instance, the computation of the vibration frequencies of a building, which is an important problem in order to determine its strength, for example, against earthquakes. To simplify the exposition we focus on a toy model, but the main ideas are the same for more complex and more realistic models.

We consider a two-story building whose sufficiently rigid ceilings are assumed to be point masses m_1, m_2, m_3. The walls are of negligible mass but their elasticity is modeled, as that of a spring, by stiffness coefficients k_1, k_2, k_3 (the larger is k_i, the more rigid or "stiff" is the wall). The horizontal displacements of the ceilings are denoted by y_1, y_2, y_3, whereas the base of the building

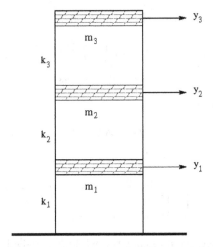

Fig. 1.4. A two-story building model.

is clamped on the ground; see Figure 1.4. In other words, this two-story building is represented as a system of three masses linked by springs to a fixed support; see Figure 1.5. We write the fundamental equation of mechanics,

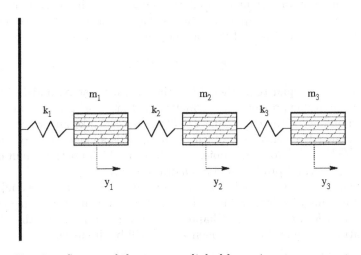

Fig. 1.5. System of three masses linked by springs to a support.

which asserts that mass multiplied by acceleration is equal to the sum of the applied forces. The only forces here are return forces exerted by the springs. They are equal to the product of stiffness and elongation of the spring. The displacements y_1, y_2, y_3 (with respect to the equilibrium) are functions of time t. Their first derivatives, denoted by $\dot{y}_1, \dot{y}_2, \dot{y}_3$, are velocities, and their second derivatives $\ddot{y}_1, \ddot{y}_2, \ddot{y}_3$ are accelerations of the masses m_1, m_2, m_3, respectively.

Thus, we deduce the following three equations:

$$\begin{cases} m_1\ddot{y}_1 + k_1 y_1 + k_2(y_1 - y_2) = 0, \\ m_2\ddot{y}_2 + k_2(y_2 - y_1) + k_3(y_2 - y_3) = 0, \\ m_3\ddot{y}_3 + k_3(y_3 - y_2) = 0, \end{cases} \tag{1.8}$$

which read, in matrix form,

$$M\ddot{y} + Ky = 0, \tag{1.9}$$

where

$$y = \begin{pmatrix} y_1 \\ y_2 \\ y_3 \end{pmatrix}, \quad M = \begin{pmatrix} m_1 & 0 & 0 \\ 0 & m_2 & 0 \\ 0 & 0 & m_3 \end{pmatrix}, \quad K = \begin{pmatrix} k_1 + k_2 & -k_2 & 0 \\ -k_2 & k_2 + k_3 & -k_3 \\ 0 & -k_3 & k_3 \end{pmatrix}.$$

The matrix M is called the mass matrix, while K is called the stiffness matrix. They are both symmetric. We look for particular solutions of equations (1.8) that are periodic (or harmonic) in time in order to represent the vibrations of the system. Accordingly, we set

$$y(t) = y^0 e^{i\omega t},$$

where i is the basis of imaginary numbers, and ω is the vibration frequency of the solution. A simple computation shows that in this case, the acceleration is $\ddot{y}(t) = -\omega^2 y(t)$, and that (1.9) simplifies to

$$Ky^0 = \omega^2 My^0. \tag{1.10}$$

If all masses are equal to 1, then M is the identity matrix, and (1.10) is a standard eigenvalue problem for the matrix K, that is, y^0 is an eigenvector of K corresponding to the eigenvalue ω^2. If the masses take any values, (1.10) is a "generalized" eigenvalue problem, that is, ω^2 is an eigenvalue of the matrix $M^{-1/2}KM^{-1/2}$ (on this topic, see Theorem 2.5.3 on the simultaneous reduction of a scalar product and a quadratic form).

Of course, had we considered a building with n floors, we would have obtained a similar matrix problem of size $n + 1$, the matrix M being always diagonal, and K tridiagonal. In Chapter 10, several algorithms for the efficient computation of eigenvalues and eigenvectors will be discussed.

1.4 The Vibrating String

We generalize the example of Section 1.3 to the case of an infinite number of masses and springs. More precisely, we pass from a discrete model to a continuous one. We consider the vibrating string equation that is the generalization of (1.8) to an infinite number of masses; for more details, see [3]. This equation is again a partial differential equation, very similar to that introduced in

Section 1.1. Upon discretization by finite differences, an approximate solution of the vibrating string equation is obtained by solving an eigenvalue problem for a large matrix.

The curvilinear abscissa along the string is denoted by x, and time is t. The deflection of the string with respect to its horizontal equilibrium position is therefore a real-valued function $u(t, x)$. We call $\ddot{u}(t, x)$ its second derivative with respect to time t, and $u''(t, x)$ its second derivative with respect to x. With no exterior forces, the vibrating string equation reads

$$\begin{cases} m\ddot{u}(t, x) - ku''(t, x) = 0, & \text{for all } x \in (0, 1), \ t > 0, \\ u(t, 0) = 0, \quad u(t, 1) = 0, \end{cases} \tag{1.11}$$

where m and k are the mass and stiffness per unit length of the string. The boundary conditions "at both ends" $u(t, 0) = u(t, 1) = 0$ specify that at any time t, the string is fixed at its endpoints; see Figure 1.6. We look for special

Fig. 1.6. Vibrating string problem.

"vibrating" solutions of (1.11) that are periodic in time and of the form

$$u(t, x) = v(x)e^{i\omega t},$$

where ω is the vibration frequency of the string. A simple computation shows that $v(x)$ is a solution to

$$\begin{cases} -v''(x) = \frac{m\omega^2}{k}v(x), & \text{for all } x \in (0, 1), \\ v(0) = 0, \quad v(1) = 0. \end{cases} \tag{1.12}$$

We say that $m\omega^2/k$ is an eigenvalue, and $v(x)$ is an eigenfunction of problem (1.12). In the particular case studied here, solutions of (1.12) can be computed explicitly; they are sine functions of period linked to the eigenvalue. However, in higher space dimensions, or if the linear mass or the stiffness varies with the point x, there is in general no explicit solution of this boundary value problem, in which case solutions $(\omega, v(x))$ must be determined numerically.

As in Section 1.1, we compute approximate solutions by a *finite difference* method. Let us recall that this method consists in dividing the interval $[0, 1]$ into n subintervals of equal size $1/n$, where n is an integer chosen according to the desired accuracy (the larger n is, the closer the approximate solution will be to the exact solution). We denote by $x_i = i/n$, $0 \leq i \leq n$, the $n+1$ limit points of the intervals. We call v_i the approximated value of the solution $v(x)$ at point x_i, and λ an approximation of $m\omega^2/k$. The idea is to write equation (1.12) at each point x_i, and substitute the second derivative by an appropriate linear combination of the unknowns v_i using Taylor's formula:

$$-v''(x_i) = \frac{2v(x_i) - v(x_{i-1}) - v(x_{i+1})}{n^{-2}} + \mathcal{O}(n^{-2}).$$

Hence, we obtain a system of $(n-1)$ equations

$$\frac{2v_i - v_{i-1} - v_{i+1}}{n^{-2}} = \lambda v_i, \quad 1 \leq i \leq n-1,$$

supplemented by the two boundary conditions $v_0 = v_n = 0$. We can rewrite the system in matrix form:

$$A_n v = \lambda v, \tag{1.13}$$

where v is the vector whose entries are $(v_1, \ldots, v_{n-1})$, and

$$A_n = n^2 \begin{pmatrix} 2 & -1 & 0 & \cdots & 0 \\ -1 & 2 & \ddots & \ddots & \vdots \\ 0 & \ddots & \ddots & -1 & 0 \\ \vdots & \ddots & -1 & 2 & -1 \\ 0 & \cdots & 0 & -1 & 2 \end{pmatrix}.$$

In other words, the pair (λ, v) are an eigenvalue and eigenvector of the tridiagonal symmetric matrix A_n. Since A_n is real symmetric, it is diagonalizable (see Theorem 2.5.2), so (1.13) admits $n-1$ linearly independent solutions. Therefore, it is possible to approximately compute the vibrating motion of a string by solving a matrix eigenvalue problem. In the case at hand, we can compute explicitly the eigenvalues and eigenvectors of matrix A_n; see Exercise 5.16. More generally, one has to resort to numerical algorithms for approximating eigenvalues and eigenvectors of a matrix; see Chapter 10.

1.5 Image Compression by the SVD Factorization

A black-and-white image can be identified with a rectangular matrix A the size of which is equal to the number of pixels of the image and with entries $a_{i,j}$ belonging to the range $[0, 1]$, where 0 corresponds to a white pixel and 1 to a black pixel. Intermediate values $0 < a_{i,j} < 1$ correspond to different levels of gray. We assume that the size of the image is very large, so it cannot reasonably be stored on a computer (not enough disk space) or sent by email (network saturation risk). Let us show how the SVD (singular value decomposition) is useful for the *compression* of images, i.e., for minimizing the storage size of an image by replacing it by an approximation that is visually equivalent.

As we shall see in Section 2.7 the SVD factorization of a matrix $A \in \mathcal{M}_{m,n}(\mathbb{C})$ of rank r is

$$A = V \tilde{\Sigma} U^* \quad \text{and} \quad \tilde{\Sigma} = \begin{pmatrix} \Sigma & 0 \\ 0 & 0 \end{pmatrix},$$

where $U \in \mathcal{M}_n(\mathbb{C})$ and $V \in \mathcal{M}_m(\mathbb{C})$ are two unitary matrices and Σ is the diagonal matrix equal to $\mathrm{diag}(\mu_1, \ldots, \mu_r)$, where $\mu_1 \geq \mu_2 \geq \cdots \geq \mu_r > 0$ are the positive square roots of the eigenvalues of $A^* A$ (where A^* denotes the adjoint matrix of A), called the "singular values" of A. Therefore, computing the SVD factorization is a type of eigenvalue problem. Denoting by u_i and v_i the columns of U and V, the SVD factorization of A can also be written

$$A = V \tilde{\Sigma} U^* = \sum_{i=1}^{r} \mu_i v_i u_i^*. \tag{1.14}$$

Since the singular values $\mu_1 \geq \cdots \geq \mu_r > 0$ are arranged in decreasing order, an approximation of A can easily be obtained by keeping only the $k \leq r$ first terms in (1.14),

$$A_k = \sum_{i=1}^{k} \mu_i v_i u_i^*.$$

Actually, Proposition 3.2.1 will prove that A_k is the best approximation (in some sense) of A among matrices of rank k. Of course, if k is much smaller than r (which is less than n and m), approximating A by A_k yields a big saving in terms of memory requirement.

Indeed, the storage of the whole matrix $A \in \mathcal{M}_{m,n}(\mathbb{R})$ requires a priori $m \times n$ scalars. To store the approximation A_k, it suffices, after having performed the SVD factorization of A, to store k vectors $\mu_i v_i \in \mathbb{C}^m$ and k vectors $u_i \in \mathbb{C}^n$, i.e., a total of $k(m+n)$ scalars. This is worthwhile if k is small and if we are satisfied with such an approximation of A. In Figure 1.7, the original image is a grid of 500×752 pixels, the corresponding matrix A is thus of size 500×752. We display the original image as well as three images corresponding to three approximations A_k of A. For $k = 10$, the image is very blurred, but for $k = 20$, the subject is recognizable. There does not seem to be any differences between the image obtained with $k = 60$ and the original image, even though the storage space is divided by 5:

$$\frac{k(m+n)}{m\,n} = \frac{60(500+752)}{500 \times 752} \approx 20\%.$$

The main computational cost of this method of image compression is the SVD factorization of the matrix A. Recall that the singular values of A are the positive square roots of the eigenvalues of $A^* A$. Thus, the SVD factorization

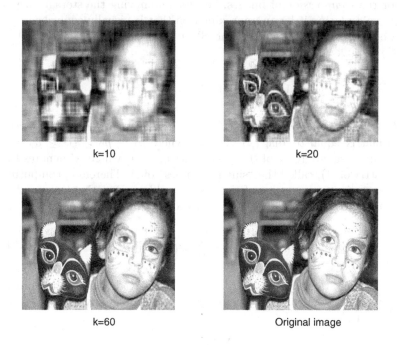

k=10   k=20

k=60   Original image

Fig. 1.7. An application of the SVD factorization: image compression.

is a variant of the problem of determining the eigenvalues and eigenvectors of a matrix which, we shall study in Chapter 10.

Finally, let us mention that there exist other algorithms that are more efficient and cheaper than the SVD factorization for image processing. Their analysis is beyond the scope of this course.

2

Definition and Properties of Matrices

Throughout this book we consider matrices with real or complex entries. Most of the results are valid for real and complex matrices (but not all of them!). That is, in order to avoid tedious repetition, we denote by $\mathbb{K}$ a field that is either the field of all real numbers $\mathbb{R}$ or the field of all complex numbers $\mathbb{C}$, i.e., $\mathbb{K} = \mathbb{R}, \mathbb{C}$.

The goal of this chapter is to recall basic results and definitions that are useful in the sequel. Therefore, many statements are given without proofs. We refer to classical courses on linear algebra for further details (see, e.g., [10], [16]).

2.1 Gram–Schmidt Orthonormalization Process

We consider the vector space $\mathbb{K}^d$ with the scalar product $\langle x, y \rangle = \sum_{i=1}^{n} x_i y_i$ if $\mathbb{K} = \mathbb{R}$, or the Hermitian product $\langle x, y \rangle = \sum_{i=1}^{n} x_i \overline{y_i}$ if $\mathbb{K} = \mathbb{C}$. We describe a constructive process for building an orthonormal family out of a family of linearly independent vectors in $\mathbb{K}^d$, known as the Gram–Schmidt orthonormalization process. This algorithm is often used in numerical linear algebra. In the sequel, the notation $span \{\ldots\}$ is used for the subspace spanned by the vectors between parentheses.

Theorem 2.1.1 (Gram–Schmidt). *Let $(x_1, \ldots, x_n)$ be a linearly independent family in $\mathbb{K}^d$. There exists an orthonormal family $(y_1, \ldots, y_n)$ such that*

$$span \{y_1, \ldots, y_p\} = span \{x_1, \ldots, x_p\}, \text{for any index } p \text{ in the range } 1 \leq p \leq n.$$

If $\mathbb{K} = \mathbb{R}$, this family is unique up to a change of sign of each vector y_p. If $\mathbb{K} = \mathbb{C}$, this family is unique up to a multiplicative factor of unit modulus for each vector y_p.

Proof. We proceed by induction on n. For $n = 1$, we define

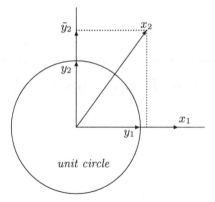

Fig. 2.1. Gram–Schmidt orthonormalization: x_1 and x_2 are linearly independent vectors, y_1 and y_2 are orthonormal.

$$y_1 = \frac{x_1}{\|x_1\|},$$

which is the unique vector (up to a change of sign in the real case, and to multiplication by a complex number of unit modulus in the complex case) to satisfy the desired property. Assume that the result holds up to order $n-1$. Let $(y_1, \ldots, y_{n-1})$ be the unique orthonormal family such that

$$span\,\{y_1, \ldots, y_p\} = span\,\{x_1, \ldots, x_p\},$$

for each index p in the range $1 \le p \le n-1$. If y_n together with the previous $(y_1, \ldots, y_{n-1})$ satisfy the recurrence property, then, since $span\,\{y_1, \ldots, y_n\} = span\,\{x_1, \ldots, x_n\}$, we necessarily have

$$y_n = \sum_{i=1}^{n} \alpha_i x_i,$$

where the α_i coefficients belong to $\mathbb{K}$. By assumption at order $n-1$, we have $span\,\{y_1, \ldots, y_{n-1}\} = span\,\{x_1, \ldots, x_{n-1}\}$, so the linear combination of $(x_1, \ldots, x_{n-1})$ can be replaced by that of $(y_1, \ldots, y_{n-1})$ to yield

$$y_n = \alpha_n x_n + \sum_{i=1}^{n-1} \beta_i y_i$$

with some other coefficients β_i. Since y_n must be orthogonal to all previous y_i for $1 \le i \le n-1$, we deduce

$$\langle y_n, y_i \rangle = 0 = \alpha_n \langle x_n, y_i \rangle + \beta_i,$$

which implies

$$y_n = \alpha_n \tilde{y}_n \quad \text{with} \quad \tilde{y}_n = x_n - \sum_{i=1}^{n-1} \langle x_n, y_i \rangle y_i.$$

Since the family $(x_1, \ldots, x_n)$ is linearly independent and $span\,\{y_1, \ldots, y_{n-1}\} = span\,\{x_1, \ldots, x_{n-1}\}$, $\tilde{y}_n$ cannot vanish; see Figure 2.1. Now, y_n must have a unit norm, it yields

$$|\alpha_n| = \frac{1}{\|\tilde{y}_n\|}.$$

In the real case, we deduce $\alpha_n = \pm 1/\|\tilde{y}_n\|$. In the complex case, α_n is equal to $1/\|\tilde{y}_n\|$ up to a multiplicative factor that is a complex number of modulus 1. Such a choice of y_n clearly satisfies the recurrence property which concludes the proof. $\qquad\square$

Remark 2.1.1. If the family $(x_1, \ldots, x_n)$ is not linearly independent, and rather generates a linear subspace of dimension $r < n$, then the Gram–Schmidt orthonormalization process yields an orthonormal family $(y_1, \ldots, y_r)$ of only r vectors.

2.2 Matrices

Definitions

Definition 2.2.1. *A matrix A is a rectangular array $(a_{i,j})_{1 \leq i \leq n,\, 1 \leq j \leq p}$, where $a_{i,j} \in \mathbb{K}$ is the entry in row i and column j, i.e.,*

$$A = \begin{pmatrix} a_{1,1} & \cdots & a_{1,p} \\ \vdots & & \vdots \\ a_{n,1} & \cdots & a_{n,p} \end{pmatrix}.$$

The set of all matrices of size $n \times p$ (n rows and p columns) is denoted by $\mathcal{M}_{n,p}(\mathbb{K})$.

Definition 2.2.2. *Let A and B be two matrices in $\mathcal{M}_{n,p}(\mathbb{K})$ defined by*

$$A = (a_{i,j})_{1 \leq i \leq n,\, 1 \leq j \leq p} \quad \text{and} \quad B = (b_{i,j})_{1 \leq i \leq n,\, 1 \leq j \leq p}.$$

The sum $A + B$ is the matrix in $\mathcal{M}_{n,p}(\mathbb{K})$ defined by

$$A + B = (a_{i,j} + b_{i,j})_{1 \leq i \leq n,\, 1 \leq j \leq p}.$$

Let $\lambda \in \mathbb{K}$. The scalar multiplication of A by λ is defined by

$$\lambda A = (\lambda a_{i,j})_{1 \leq i \leq n,\, 1 \leq j \leq p}.$$

Definition 2.2.3. *The product of two matrices* $A \in \mathcal{M}_{n,p}(\mathbb{K})$ *and* $B \in \mathcal{M}_{p,q}(\mathbb{K})$, *defined by*

$$A = (a_{i,j})_{1 \leq i \leq n,\, 1 \leq j \leq p} \quad and \quad B = (b_{i,j})_{1 \leq i \leq p,\, 1 \leq j \leq q},$$

is a matrix $C = AB$ *in* $\mathcal{M}_{n,q}(\mathbb{K})$ *defined by*

$$C = \left(c_{i,j} = \sum_{k=1}^{p} a_{i,k} b_{k,j} \right)_{1 \leq i \leq n,\, 1 \leq j \leq q}.$$

Remark 2.2.1. The number of columns of A and that of rows of B must be equal in order to define their product AB. Otherwise, they need not have the same dimensions and belong to the same matrix space. It is an easy exercise to check that the matrix multiplication is associative, i.e., $(MN)P = M(NP)$. However, even for $n = p = q$, the multiplication is usually not commutative, i.e., $AB \neq BA$.

Definition 2.2.4. *For any matrix* $A = (a_{i,j})_{1 \leq i \leq n,\, 1 \leq j \leq p} \in \mathcal{M}_{n,p}(\mathbb{K})$, *its transpose matrix* $A^t \in \mathcal{M}_{p,n}(\mathbb{K})$ *is defined by*

$$A^t = (a_{j,i})_{1 \leq j \leq p,\, 1 \leq i \leq n}.$$

In other words, the rows of A^t *are the columns of* A, *and the columns of* A^t *are the rows of* A:

$$A = \begin{pmatrix} a_{1,1} & \cdots & a_{1,p} \\ \vdots & & \vdots \\ a_{n,1} & \cdots & a_{n,p} \end{pmatrix}, \quad A^t = \begin{pmatrix} a_{1,1} & \cdots & a_{n,1} \\ \vdots & & \vdots \\ a_{1,p} & \cdots & a_{n,p} \end{pmatrix}.$$

If $A^t = A$ *(which can happen only if* A *is a square matrix, i.e., if* $n = p$*), then* A *is said to be symmetric.*

The notation A^t for the transpose matrix of A is not universal. Some authors prefer to denote it by A^T, or put the exponent t before the matrix, as in $^t A$.

When the number of rows is equal to the number of columns, the matrix is said to be a square matrix, the set of which is denoted by $\mathcal{M}_n(\mathbb{K}) \equiv \mathcal{M}_{n,n}(\mathbb{K})$, where n is the size of the matrix. The set $\mathcal{M}_n(\mathbb{K})$ is thus a noncommutative algebra for the multiplication. Its neutral element is the identity matrix, denoted by I (or I_n if one wants to give a precise indication of the dimension) and defined by its entries $(\delta_{i,j})_{1 \leq i,j \leq n}$, where $\delta_{i,j}$ is the Kronecker symbol taking the values $\delta_{i,i} = 1$ and $\delta_{i,j} = 0$ if $i \neq j$.

Definition 2.2.5. *A matrix* $A \in \mathcal{M}_n(\mathbb{K})$ *is said to be invertible (or nonsingular) if there exists a matrix* $B \in \mathcal{M}_n(\mathbb{K})$ *such that* $AB = BA = I_n$. *This matrix* B *is denoted by* A^{-1} *and is called the inverse matrix of* A.

A noninvertible matrix is said to be singular. The kernel, or null space, of a matrix $A \in \mathcal{M}_{n,p}(\mathbb{K})$ is the set of vectors $x \in \mathbb{K}^p$ such that $Ax = 0$; it is denoted by $\mathrm{Ker}\, A$. The image, or range, of A is the set of vectors $y \in \mathbb{K}^n$ such that $y = Ax$, with $x \in \mathbb{K}^p$; it is denoted by $\mathrm{Im}\, A$. The dimension of the linear space $\mathrm{Im}\, A$ is called the rank of A; it is denoted by $\mathrm{rk}\, A$.

Lemma 2.2.1. *For any $A \in \mathcal{M}_n(\mathbb{K})$ the following statements are equivalent:*

1. *A is invertible;*
2. *$\mathrm{Ker}\, A = \{0\}$;*
3. *$\mathrm{Im}\, A = \mathbb{K}^n$;*
4. *there exists $B \in \mathcal{M}_n(\mathbb{K})$ such that $AB = I_n$;*
5. *there exists $B \in \mathcal{M}_n(\mathbb{K})$ such that $BA = I_n$.*

In the last two cases, the matrix B is precisely equal to the inverse A^{-1}.

Lemma 2.2.2. *Let A and B be two invertible matrices in $\mathcal{M}_n(\mathbb{K})$. Then*

$$(AB)^{-1} = B^{-1}A^{-1}.$$

2.2.1 Trace and Determinant

In this section we consider only square matrices in $\mathcal{M}_n(\mathbb{K})$.

Definition 2.2.6. *The trace of a matrix $A = (a_{i,j})_{1 \leq i,j \leq n}$ is the sum of its diagonal elements*

$$\mathrm{tr}\, A = \sum_{i=1}^{n} a_{i,i}.$$

Lemma 2.2.3. *If A and B are two matrices in $\mathcal{M}_n(\mathbb{K})$, then*

$$\mathrm{tr}\,(AB) = \mathrm{tr}\,(BA).$$

Definition 2.2.7. *A permutation of order n is a one-to-one mapping from the set $\{1, 2, \ldots, n\}$ into itself. We denote by $\mathcal{S}_n$ the set of all permutations of order n. The signature of a permutation σ is the number $\varepsilon(\sigma)$, equal to $+1$ or -1, defined by*

$$\varepsilon(\sigma) = (-1)^{p(\sigma)} \quad \text{with} \quad p(\sigma) = \sum_{1 \leq i \leq j \leq n} \mathrm{Inv}_\sigma(i, j),$$

where the number $\mathrm{Inv}_\sigma(i, j)$ indicates whether the order between i and j is inverted or not by the permutation σ, and is defined, for $i \leq j$, by

$$\mathrm{Inv}_\sigma(i, j) = \begin{cases} 0 \text{ if } \sigma(i) \leq \sigma(j), \\ 1 \text{ if } \sigma(i) > \sigma(j). \end{cases}$$

Definition 2.2.8. *The determinant of a square matrix* $A = (a_{i,j})_{1 \leq i,j \leq n} \in \mathcal{M}_n(\mathbb{K})$ *is*

$$\det A = \sum_{\sigma \in \mathcal{S}_n} \varepsilon(\sigma) \prod_{i=1}^{n} a_{i,\sigma(i)}.$$

Lemma 2.2.4. *Let A and B be two square matrices in $\mathcal{M}_n(\mathbb{K})$. Then*

1. $\det(AB) = (\det A)(\det B) = \det(BA)$;
2. $\det(A^t) = \det(A)$;
3. A *is invertible if and only if* $\det A \neq 0$.

2.2.2 Special Matrices

Definition 2.2.9. *A matrix $A = (a_{i,j})_{1 \leq i,j \leq n} \in \mathcal{M}_n(\mathbb{K})$ is said to be diagonal if its entries satisfy $a_{i,j} = 0$ for $i \neq j$. A diagonal matrix is often denoted by $A = \text{diag}(a_{1,1}, \ldots, a_{n,n})$.*

Definition 2.2.10. *Let $T = (t_{i,j})_{1 \leq i \leq n,\ 1 \leq j \leq p}$ be a matrix in $\mathcal{M}_{n,p}(\mathbb{K})$. It is said to be an upper triangular matrix if $t_{i,j} = 0$ for all indices (i,j) such that $i > j$. It is said to be a lower triangular matrix if $t_{i,j} = 0$ for all indices (i,j) such that $i < j$.*

Lemma 2.2.5. *Let T be a lower triangular matrix (respectively, upper triangular) in $\mathcal{M}_n(\mathbb{K})$. Its inverse (when it exists) is also a lower triangular matrix (respectively, upper triangular) with diagonal entries equal to the inverse of the diagonal entries of T. Let T' be another lower triangular matrix (respectively, upper triangular) in $\mathcal{M}_n(\mathbb{K})$. The product TT' is also a lower triangular matrix (respectively, upper triangular) with diagonal entries equal to the product of the diagonal entries of T and T'.*

Definition 2.2.11. *Let $A = (a_{i,j})_{1 \leq i,j \leq n}$ be a complex square matrix in $\mathcal{M}_n(\mathbb{C})$. The matrix $A^* \in \mathcal{M}_n(\mathbb{C})$, defined by $A^* = \overline{A}^t = (\overline{a}_{j,i})_{1 \leq i,j \leq n}$, is the adjoint matrix of A.*

Definition 2.2.12. *Let A be a complex square matrix in $\mathcal{M}_n(\mathbb{C})$.*

1. A *is self-adjoint or Hermitian if* $A = A^*$;
2. A *is unitary if* $A^{-1} = A^*$;
3. A *is normal if* $AA^* = A^*A$.

Definition 2.2.13. *Let A be a real square matrix in $\mathcal{M}_n(\mathbb{R})$.*

1. A *is symmetric or self-adjoint if* $A = A^t$ *(or equivalently $A = A^*$)*;
2. A *is orthogonal or unitary if* $A^{-1} = A^t$ *(or equivalently $A^{-1} = A^*$)*;
3. A *is normal if* $AA^t = A^t A$ *(or equivalently $AA^* = A^*A$)*.

2.2.3 Rows and Columns

A matrix $A \in \mathcal{M}_{n,p}(\mathbb{C})$ may be defined by its columns $c_j \in \mathbb{C}^n$ as

$$A = [c_1 | \ldots | c_p]$$

or by its rows $\ell_i \in \mathcal{M}_{1,p}(\mathbb{C})$ as

$$A = \begin{bmatrix} \ell_1 \\ \hline \vdots \\ \hline \ell_n \end{bmatrix}.$$

Recalling that $c_j^* \in \mathcal{M}_{1,n}(\mathbb{C})$ (the adjoint of c_j) is a row vector and $\ell_i^* \in \mathbb{C}^p$ (the adjoint of ℓ_i) is a column vector, we have

$$A^* = \begin{bmatrix} c_1^* \\ \hline \vdots \\ \hline c_p^* \end{bmatrix} = [\ell_1^* | \ldots | \ell_n^*],$$

and for any $x \in \mathbb{C}^p$,

$$Ax = \begin{bmatrix} \ell_1 x \\ \hline \vdots \\ \hline \ell_n x \end{bmatrix} = \sum_{i=1}^p x_i c_i.$$

For $X \in \mathcal{M}_{m,n}(\mathbb{C})$, we have

$$XA = X[c_1 | \ldots | c_p] = [Xc_1 | \ldots | Xc_p].$$

Similarly for $X \in \mathcal{M}_{p,m}(\mathbb{C})$, we have

$$AX = \begin{bmatrix} \ell_1 \\ \hline \vdots \\ \hline \ell_n \end{bmatrix} X = \begin{bmatrix} \ell_1 X \\ \hline \vdots \\ \hline \ell_n X \end{bmatrix}.$$

By the same token, given $u_1, \ldots, u_m$, vectors in $\mathbb{C}^n$, and $v_1, \ldots, v_m$, vectors in $\mathbb{C}^p$, one can define a product matrix in $\mathcal{M}_{n,p}(\mathbb{C})$ by

$$\sum_{i=1}^{m} u_i v_i^* = [u_1|\ldots|u_m] \begin{bmatrix} v_1^* \\ \hline \vdots \\ \hline v_m^* \end{bmatrix}.$$

2.2.4 Row and Column Permutation

Let A be a matrix in $\mathcal{M}_{n,p}(\mathbb{K})$. We interpret some usual operations on A as multiplying A by some other matrices.

- ✗ Multiplying each row i of A by a scalar $\alpha_i \in \mathbb{K}$ is done by left-multiplying the matrix A by a diagonal matrix, $\mathrm{diag}\,(\alpha_1,\ldots,\alpha_n)A$.
- ✗ Multiplying each column j of A by $\beta_j \in \mathbb{K}$ is done by right-multiplying the matrix A by a diagonal matrix, $A\,\mathrm{diag}\,(\beta_1,\ldots,\beta_p)$.
- ✗ To exchange rows l_1 and $l_2 \neq l_1$, we multiply A on the left by an elementary permutation matrix $P(l_1,l_2)$, which is a square matrix of size n defined by its entries:

$$p_{i,i} = \begin{cases} 0 & \text{if } i \in \{l_1,l_2\}, \\ 1 & \text{otherwise;} \end{cases}$$

and for $i \neq j$,

$$p_{i,j} = \begin{cases} 1 & \text{if } (i,j) \in \{(l_1,l_2),(l_2,l_1)\}, \\ 0 & \text{otherwise.} \end{cases}$$

The matrix $P(l_1,l_2)$ is nothing but the identity matrix whose rows l_1 and l_2 are permuted. The resulting matrix $P(l_1,l_2)A$ has exchanged rows l_1 and l_2 of the initial matrix A.

- ✗ To exchange columns c_1 and c_2, we multiply A on the right by an elementary permutation matrix $P(c_1,c_2)$ of size p.
- ✗ A general permutation matrix is any matrix obtained from the identity matrix by permuting its rows (not necessarily only two). Such a permutation matrix is actually a product of elementary permutation matrices, and its inverse is just its transpose. Therefore its determinant is equal to ± 1.

2.2.5 Block Matrices

So far, we have discussed matrices with entries belonging to a field $\mathbb{K}$ equal to $\mathbb{R}$ or $\mathbb{C}$. Actually, one can define matrices with entries in a noncommutative ring and still keep the same matrix addition and multiplication as introduced in Definitions 2.2.2 and 2.2.3. This is of particular interest for the so-called "block matrices" that we now describe.

Let A be a square matrix in $\mathcal{M}_n(\mathbb{C})$. Let $(n_I)_{1 \leq I \leq p}$ be a family of positive integers such that $\sum_{I=1}^{p} n_I = n$. Let $(e_1, \ldots, e_n)$ be the canonical basis of $\mathbb{C}^n$. We call V_1 the subspace of $\mathbb{C}^n$ spanned by the first n_1 basis vectors, V_2 the subspace of $\mathbb{C}^n$ spanned by the next n_2 basis vectors, and so on up to V_p spanned by the last n_p basis vectors. The dimension of each V_I is n_I. Let $A_{I,J}$ be the submatrix (or block) of size $n_I \times n_J$ defined as the restriction of A on the domain space V_J into the target space V_I. We write

$$A = \begin{pmatrix} A_{1,1} & \cdots & A_{1,p} \\ \vdots & \ddots & \vdots \\ A_{p,1} & \cdots & A_{p,p} \end{pmatrix}.$$

In other words, A has been partitioned into p horizontal and vertical strips of unequal sizes $(n_I)_{1 \leq I \leq p}$. The diagonal blocks $A_{I,I}$ are square matrices, but usually not the off-diagonal blocks $A_{I,J}$, $I \neq J$.

Lemma 2.2.6. *Let $A = (A_{I,J})_{1 \leq I,J \leq p}$ and $B = (B_{I,J})_{1 \leq I,J \leq p}$ be two block matrices with entries $A_{I,J}$ and $B_{I,J}$ belonging to $\mathcal{M}_{n_I, n_J}(\mathbb{K})$. Let $C = AB$ be the usual matrix product of A and B. Then C can also be written as a block matrix with entries $C = (C_{I,J})_{1 \leq I,J \leq p}$ given by the following block multiplication rule:*

$$C_{I,J} = \sum_{K=1}^{p} A_{I,K} B_{K,J}, \quad \text{for all } 1 \leq I, J \leq p,$$

where $C_{I,J}$ has the same size as $A_{I,J}$ and $B_{I,J}$.

It is essential that A and B share the same block partitioning, i.e., $A_{I,J}$ and $B_{I,J}$ have equal sizes, in order to correctly define the product $A_{I,K}$ times $B_{K,J}$ in the above lemma. One must also keep in mind that the matrix multiplication is not commutative, so the order of multiplication in the above block multiplication rule is important. Although block matrices are very handy (and we shall use them frequently in the sequel), not all matrix operations can be generalized to block matrices. In particular, there is no block determinant rule.

2.3 Spectral Theory of Matrices

In the sequel and unless mentioned otherwise, we assume that all matrices are square and complex (which encompasses the real case as well).

Definition 2.3.1. *Let $A \in \mathcal{M}_n(\mathbb{C})$. The characteristic polynomial of A is the polynomial $P_A(\lambda)$ defined on $\mathbb{C}$ by*

$$P_A(\lambda) = \det(A - \lambda I).$$

It is a polynomial of degree equal to n. It has thus n roots in $\mathbb{C}$ (for a proof, see for instance [10]), which we call the eigenvalues of A. The algebraic multiplicity of an eigenvalue is its multiplicity as a root of $P_A(\lambda)$. An eigenvalue whose algebraic multiplicity is equal to one is said to be a simple eigenvalue, otherwise, it is called a multiple eigenvalue.

We call a nonzero vector $x \in \mathbb{C}^n$ such that $Ax = \lambda x$ the eigenvector of A associated with the eigenvalue λ.

We shall sometimes denote by $\lambda(A)$ an eigenvalue of A. The set of eigenvalues of a matrix A is called spectrum of A and is denoted by $\sigma(A)$.

Definition 2.3.2. *We call the maximum of the moduli of the eigenvalues of a matrix $A \in \mathcal{M}_n(\mathbb{C})$ the spectral radius of A, and we denote it by $\varrho(A)$.*

Let us make some remarks.

1. If λ is an eigenvalue of A, then there always exists a corresponding eigenvector x, namely a vector $x \neq 0$ in $\mathbb{C}^n$ such that $Ax = \lambda x$ (x is not unique). Indeed, $P_A(\lambda) = 0$ implies that the matrix $(A - \lambda I)$ is singular. In particular, its kernel is not reduced to the zero vector. Conversely, if there exists $x \neq 0$ such that $Ax = \lambda x$, then λ is an eigenvalue of A.
2. If A is real, there may exist complex eigenvalues of A.
3. The characteristic polynomial (and accordingly the eigenvalues) is invariant under basis change, since

$$\det \left(Q^{-1}AQ - \lambda I \right) = \det \left(A - \lambda I \right),$$

   for any invertible matrix Q.
4. There exist two distinct ways of enumerating the eigenvalues of a matrix of size n. Either they are denoted by $(\lambda_1, \ldots, \lambda_n)$ (i.e., we list the roots of the characteristic polynomial repeating a root as many times as its multiplicity) and we say "eigenvalues repeated with multiplicity," or we denote them by $(\lambda_1, \ldots, \lambda_p)$ with $1 \leq p \leq n$ keeping only the distinct roots of the characteristic polynomial (i.e., an eigenvalue appears only once in this list no matter its algebraic multiplicity) and we say "distinct eigenvalues."

Remark 2.3.1. The eigenvalues of a Hermitian matrix are real. Actually, if λ is an eigenvalue of a Hermitian matrix A, and $u \neq 0$ a corresponding eigenvector, we have

$$\lambda \|u\|^2 = \langle \lambda u, u \rangle = \langle Au, u \rangle = \langle u, A^*u \rangle = \langle u, Au \rangle = \langle u, \lambda u \rangle = \bar{\lambda} \|u\|^2,$$

which shows that $\lambda = \bar{\lambda}$, i.e., $\lambda \in \mathbb{R}$.

Definition 2.3.3. *Let λ be an eigenvalue of A. We call the vector subspace defined by*

$$E_\lambda = \operatorname{Ker} \left(A - \lambda I \right)$$

the eigensubspace associated with the eigenvalue λ. We call the vector subspace defined by

$$F_\lambda = \bigcup_{k \geq 1} \mathrm{Ker}\,(A - \lambda I)^k$$

the generalized eigenspace associated with λ.

Remark 2.3.2. In the definition of the generalized eigenspace F_λ, the union of the kernels of $(A - \lambda I)^k$ is finite, i.e., there exists an integer k_0 such that

$$F_\lambda = \bigcup_{1 \leq k \leq k_0} \mathrm{Ker}\,(A - \lambda I)^k.$$

Indeed, the sequence of vector subspaces $\mathrm{Ker}\,(A - \lambda I)^k$ is an increasing nested sequence in a space of finite dimension. For k larger than an integer k_0 the sequence of dimensions is stationary; otherwise, this would contradict the finiteness of the dimension of the space $\mathbb{C}^n$. Consequently, for $k \geq k_0$, all spaces $\mathrm{Ker}\,(A - \lambda I)^k$ are equal to $\mathrm{Ker}\,(A - \lambda I)^{k_0}$.

Definition 2.3.4. Let $P(X) = \sum_{i=0}^d a_i X^i$ be a polynomial on $\mathbb{C}$ and A a matrix of $\mathcal{M}_n(\mathbb{C})$. The corresponding matrix polynomial $P(A)$ is defined as $P(A) = \sum_{i=0}^d a_i A^i$.

Lemma 2.3.1. If $Ax = \lambda x$ with $x \neq 0$, then $P(A)x = P(\lambda)x$ for all polynomials $P(X)$. In other words, if λ is an eigenvalue of A, then $P(\lambda)$ is an eigenvalue of $P(A)$.

Theorem 2.3.1 (Cayley–Hamilton). Let $P_A(\lambda) = \det(A - \lambda I)$ be the characteristic polynomial of A. We have

$$P_A(A) = 0.$$

Remark 2.3.3. The Cayley–Hamilton theorem shows that the smallest-degree possible for a polynomial that vanishes at A is less than or equal to n. This smallest degree may be strictly less than n. We call the smallest-degree polynomial that vanishes at A and whose highest-degree term has coefficient 1 the minimal polynomial of A

Theorem 2.3.2 (Spectral decomposition). Consider a matrix $A \in \mathcal{M}_n(\mathbb{C})$ that has p distinct eigenvalues $(\lambda_1, \ldots, \lambda_p)$, with $1 \leq p \leq n$, of algebraic multiplicity $n_1, \ldots, n_p$ with $1 \leq n_i \leq n$ and $\sum_{i=1}^p n_i = n$. Then its generalized eigenspaces satisfy

$$\mathbb{C}^n = \oplus_{i=1}^p F_{\lambda_i}, \quad F_{\lambda_i} = \mathrm{Ker}\,(A - \lambda_i I)^{n_i}, \quad and \quad \dim F_{\lambda_i} = n_i.$$

We recall that $\oplus$ denotes the direct sum of subspaces. More precisely, $\mathbb{C}^n = \oplus_{i=1}^p F_{\lambda_i}$ means that any vector $x \in \mathbb{C}^n$ can be uniquely decomposed as $x = \sum_{i=1}^p x^i$ with $x^i \in F_{\lambda_i}$.

Remark 2.3.4. Theorem 2.3.2 can be interpreted as follows. Let B_i be a basis of the generalized eigenspace F_{λ_i}. The union of all $(B_i)_{1 \leq i \leq p}$ form a basis B of $\mathbb{C}^n$. Let P be the change of basis matrix from the canonical basis to B. Since each F_{λ_i} is stable by A, we obtain a new matrix that is diagonal by blocks in the basis B, that is,

$$P^{-1}AP = \begin{pmatrix} A_1 & & 0 \\ & \ddots & \\ 0 & & A_p \end{pmatrix},$$

where each A_i is a square matrix of size n_i. We shall see in the next section that by a suitable choice of the basis B_i, each block A_i can be written as an upper triangular matrix with the eigenvalue λ_i on its diagonal. The Jordan form (cf. [9], [10]) allows us to simplify further the structure of this triangular matrix.

2.4 Matrix Triangularization

There exist classes of particularly simple matrices. For instance, diagonal matrices are matrices $A = (a_{i,j})_{1 \leq i,j \leq n}$ such that $a_{i,j} = 0$ if $i \neq j$, and upper (respectively, lower) triangular matrices are matrices such that $a_{i,j} = 0$ if $i > j$ (respectively, if $i < j$). Reducing a matrix is the process of transforming it by a change of basis into one of these particular forms.

Definition 2.4.1. *A matrix $A \in \mathcal{M}_n(\mathbb{C})$ can be reduced to triangular form (respectively, to diagonal form) if there exists a nonsingular matrix P and a triangular matrix T (respectively, diagonal matrix D) such that*

$$A = PTP^{-1} \quad \text{(respectively, } A = PDP^{-1}\text{).}$$

Remark 2.4.1. The matrices A and T (or D) are similar: they correspond to the same linear transformation expressed in two different bases, and P is the matrix of this change of basis. More precisely, if this linear transformation has A for its matrix in the basis $B = (e_i)_{1 \leq i \leq n}$, and T (or D) in the basis $B' = (f_i)_{1 \leq i \leq n}$, then P is the matrix for passing from B to B', and we have $P = (p_{i,j})_{1 \leq i,j \leq n}$ with $p_{i,j} = f_j^* e_i$. Furthermore, when A can be diagonalized, the column vectors of P are eigenvectors of A.

If A can be reduced to diagonal or triangular form, then the eigenvalues of A, repeated with their algebraic multiplicities $(\lambda_1, \ldots, \lambda_n)$, appear on the diagonal of D, or of T. In other words, we have

$$D = \begin{pmatrix} \lambda_1 & & 0 \\ & \ddots & \\ 0 & & \lambda_n \end{pmatrix} \quad \text{or} \quad T = \begin{pmatrix} \lambda_1 & \cdots & x \\ & \ddots & \vdots \\ 0 & & \lambda_n \end{pmatrix}.$$

In all cases, the characteristic polynomial of A is

$$P(\lambda) = \det(A - \lambda I) = \prod_{i=1}^{n}(\lambda_i - \lambda).$$

Proposition 2.4.1. *Any matrix $A \in \mathcal{M}_n(\mathbb{C})$ can be reduced to triangular form.*

Proof. We proceed by induction on the dimension n. The proposition is obviously true for $n = 1$. We assume that it holds up to order $n - 1$. For $A \in \mathcal{M}_n(\mathbb{C})$, its characteristic polynomial $\det(A - \lambda I)$ has at least one root $\lambda_1 \in \mathbb{C}$ with a corresponding eigenvector $e_1 \neq 0$ such that $Ae_1 = \lambda_1 e_1$. We complement e_1 with other vectors $(e_2, \ldots, e_n)$ to obtain a basis of $\mathbb{C}^n$. For $2 \leq j \leq n$, there exist coefficients α_j and $b_{i,j}$ such that

$$Ae_j = \alpha_j e_1 + \sum_{i=2}^{n} b_{i,j} e_i. \qquad (2.1)$$

We denote by B the matrix of size $n - 1$ defined by its entries $(b_{i,j})_{2 \leq i,j \leq n}$. Introducing the change of basis matrix P_1 for passing from the canonical basis to $(e_1, \ldots, e_n)$, identity (2.1) is equivalent to

$$P_1^{-1} A P_1 = \begin{pmatrix} \lambda_1 & \alpha_2 \ldots \alpha_n \\ 0 & \\ \vdots & B \\ 0 & \end{pmatrix}.$$

Applying the induction assumption, there exists a nonsingular matrix P_2 of size $n - 1$ such that $P_2^{-1} B P_2 = T_2$, where T_2 is an upper triangular matrix of order $n - 1$. From P_2 we create a matrix P_3 of size n defined by

$$P_3 = \begin{pmatrix} 1 & 0 \ldots 0 \\ 0 & \\ \vdots & P_2 \\ 0 & \end{pmatrix}.$$

Then, setting $P = P_1 P_3$ yields

$$P^{-1} A P = \begin{pmatrix} \lambda_1 & \beta_2 & \cdots & \beta_n \\ 0 & & & \\ \vdots & & P_2^{-1} B P_2 & \\ 0 & & & \end{pmatrix} = \begin{pmatrix} \lambda_1 & \beta_2 \ldots \beta_n \\ 0 & \\ \vdots & T_2 \\ 0 & \end{pmatrix} = T,$$

where T is an upper triangular matrix and $(\beta_2, \ldots, \beta_n) = (\alpha_2, \ldots, \alpha_n)P_2$. $\square$

Remark 2.4.2. If A is real, the result still applies, but T and P may be complex! For instance, the matrix

$$A = \begin{pmatrix} 0 & -1 \\ 1 & 0 \end{pmatrix}$$

has complex eigenvalues and eigenvectors.

We already know that all matrices can be reduced to triangular form. The purpose of the next theorem is to prove furthermore that this may be performed through a change of orthonormal basis.

Theorem 2.4.1 (Schur Factorization). *For any matrix $A \in \mathcal{M}_n(\mathbb{C})$ there exists a unitary matrix U (i.e., $U^{-1} = U^*$) such that $U^{-1}AU$ is triangular.*

Proof. Let $(e_i)_{i=1}^n$ be the canonical basis and let $(f_i)_{i=1}^n$ be the basis in which A is triangular. We call P the corresponding change of basis matrix, that is, the matrix whose columns are the vectors $(f_i)_{i=1}^n$. Proposition 2.4.1 tells us that $A = PTP^{-1}$. We apply the Gram–Schmidt orthonormalization process (see Theorem 2.1.1) to the basis $(f_i)_{i=1}^n$, which yields an orthonormal basis $(g_i)_{i=1}^n$ such that for any $1 \le i \le n$,

$$span\,\{g_1, \ldots, g_i\} = span\,\{f_1, \ldots, f_i\}.$$

Since $AP = PT$ with T upper triangular, keeping only the first i columns of this equality gives

$$span\,\{Af_1, \ldots, Af_i\} \subset span\,\{f_1, \ldots, f_i\}, \quad \text{for all } 1 \le i \le n. \tag{2.2}$$

Thus, we deduce that

$$span\,\{Ag_1, \ldots, Ag_i\} \subset span\,\{g_1, \ldots, g_i\}. \tag{2.3}$$

Conversely, (2.3) implies that there exists an upper triangular matrix R such that $AU = UR$, where U is the unitary matrix whose columns are the orthonormal vectors $(g_i)_{i=1}^n$. $\qquad\square$

2.5 Matrix Diagonalization

Proposition 2.5.1. *Let $A \in \mathcal{M}_n(\mathbb{C})$ with distinct eigenvalues $(\lambda_1, \ldots, \lambda_p)$, $1 \le p \le n$. The matrix A is diagonalizable if and only if*

$$\mathbb{C}^n = \bigoplus_{i=1}^p E_{\lambda_i},$$

or, equivalently, if and only if $F_{\lambda_i} = E_{\lambda_i}$ for any $1 \le i \le p$.

Proof. If $\mathbb{C}^n = \oplus_{i=1}^{p} E_{\lambda_i}$, then A is diagonal in a basis obtained as the union of bases for the subspaces E_{λ_i}. Conversely, if there exists a nonsingular matrix P such that $P^{-1}AP$ is diagonal, it is clear that $\mathbb{C}^n = \oplus_{i=1}^{p} E_{\lambda_i}$.

What is more, we always have $E_{\lambda_i} \subset F_{\lambda_i}$ and $\mathbb{C}^n = \oplus_{i=1}^{p} F_{\lambda_i}$ by virtue of Theorem 2.3.2. Hence, the identities $F_{\lambda_i} = E_{\lambda_i}$ for all $1 \leq i \leq p$ are equivalent to requiring that A be diagonalizable. $\qquad\square$

In general, not every matrix is diagonalizable. Moreover, there is no simple characterization of the set of diagonalizable matrices. However, if we restrict ourselves to matrices that are diagonalizable in an orthonormal basis of eigenvectors, then such matrices have an elementary characterization. Namely, the set of diagonalizable matrices in an orthonormal basis coincides with the set of normal matrices, i.e., satisfying $AA^* = A^*A$.

Theorem 2.5.1 (Diagonalization). *A matrix $A \in \mathcal{M}_n(\mathbb{C})$ is normal (i.e., $AA^* = A^*A$) if and only if there exists a unitary matrix U such that*

$$A = U \operatorname{diag}(\lambda_1, \ldots, \lambda_n)U^{-1},$$

where $(\lambda_1, \ldots, \lambda_n)$ are the eigenvalues of A.

Remark 2.5.1. There are diagonalizable matrices in a nonorthonormal basis that are not normal. For instance, the matrix

$$A = \begin{pmatrix} -1 & 2 \\ 0 & 1 \end{pmatrix}$$

is not normal, because

$$AA^t = \begin{pmatrix} 5 & 2 \\ 2 & 1 \end{pmatrix} \neq A^tA = \begin{pmatrix} 1 & -2 \\ -2 & 5 \end{pmatrix}.$$

Nevertheless A is diagonalizable in a basis of eigenvectors, but these vectors are not orthogonal:

$$A = PDP^{-1} \equiv \begin{pmatrix} 1 & 1/\sqrt{2} \\ 0 & 1/\sqrt{2} \end{pmatrix} \begin{pmatrix} -1 & 0 \\ 0 & 1 \end{pmatrix} \begin{pmatrix} 1 & -1 \\ 0 & \sqrt{2} \end{pmatrix}.$$

Proof of Theorem 2.5.1. Clearly, a matrix $A = UDU^*$, with U unitary and D diagonal, is normal. Conversely, we already know by Theorem 2.4.1 that any matrix A can be reduced to triangular form in an orthonormal basis. In other words, there exists a unitary matrix U and an upper triangular matrix T such that $A = UTU^*$. Now, $AA^* = A^*A$ implies that $TT^* = T^*T$, i.e., the matrix T is normal. Let us show that a matrix that is both triangular and normal is diagonal. By definition we have $T = (t_{i,j})_{1 \leq i,j \leq n}$ with $t_{i,j} = 0$ if $i > j$. Identifying the entry in the first row and first column of the product $T^*T = TT^*$, we deduce that

$$|t_{1,1}|^2 = \sum_{k=1}^{n} |t_{1,k}|^2,$$

which yields $t_{1,k} = 0$ for all $2 \leq k \leq n$, i.e., the first row of T has only zero entries, except for the diagonal entry. By induction, we assume that the first $(i-1)$ rows of T have only zeros, except for the diagonal entries. Identifying the entry in the ith row and ith column of the product $T^*T = TT^*$ yields

$$|t_{i,i}|^2 = \sum_{k=i}^{n} |t_{i,k}|^2,$$

so that $t_{i,k} = 0$ for all $i+1 \leq k \leq n$, which means that the ith row of T also has only zeros off the diagonal. Hence T is diagonal. $\quad\square$

Remark 2.5.2. Take a matrix $A \in \mathcal{M}_n(\mathbb{C})$ that is diagonalizable in an orthonormal basis, i.e., $A = U \operatorname{diag}(\lambda_1, \ldots, \lambda_n)U^*$ with U unitary. Another way of writing A is to introduce the columns $(u_i)_{1 \leq i \leq n}$ of U (which are also the eigenvectors of A), and to decompose $A = \sum_{i=1}^{n} \lambda_i u_i u_i^*$.

Theorem 2.5.2. *A matrix $A \in \mathcal{M}_n(\mathbb{C})$ is self-adjoint (or Hermitian, i.e., $A = A^*$) if and only if it is diagonalizable in an orthonormal basis with real eigenvalues, in other words, if there exists a unitary matrix U such that*

$$A = U \operatorname{diag}(\lambda_1, \ldots, \lambda_n)U^{-1} \quad \text{with} \quad \lambda_i \in \mathbb{R}.$$

Proof. If $A = U \operatorname{diag}(\lambda_1, \ldots, \lambda_n)U^{-1}$, then

$$A^* = (U^{-1})^* \operatorname{diag}(\overline{\lambda}_1, \ldots, \overline{\lambda}_n)U^*,$$

and since U is unitary and the eigenvalues are real, we have $A = A^*$. Reciprocally, we assume that $A = A^*$. In particular, A is normal, hence diagonalizable in an orthonormal basis of eigenvectors. Then, according to Remark 2.3.1, its eigenvalues are real. $\quad\square$

We can improve the previous theorem in the case of a real symmetric matrix A (which is a special case of a self-adjoint matrix) by asserting that the unitary matrix U is also real.

Corollary 2.5.1. *A matrix $A \in \mathcal{M}_n(\mathbb{R})$ is real symmetric if and only if there exist a real unitary matrix Q (also called orthogonal, $Q^{-1} = Q^t$) and real eigenvalues $\lambda_1, \ldots, \lambda_n \in \mathbb{R}$ such that*

$$A = Q \operatorname{diag}(\lambda_1, \ldots, \lambda_n)Q^{-1}.$$

A self-adjoint matrix $A \in \mathcal{M}_n(\mathbb{C})$ (i.e., $A^* = A$) is said to be positive definite if all of its eigenvalues are strictly positive. It is said to be nonnegative definite (or positive semidefinite) if all of its eigenvalues are nonnegative.

Theorem 2.5.3. *Let A be a self-adjoint matrix and B a positive definite self-adjoint matrix in $\mathcal{M}_n(\mathbb{C})$. There exists a basis of $\mathbb{C}^n$ that is both orthonormal for the Hermitian product $\overline{y}^t B x$ and orthogonal for the Hermitian form $\overline{x}^t A x$. In other words, there exists a nonsingular matrix M such that*

$$B = M^*M \quad and \quad A = M^* \operatorname{diag}(\mu_1, \ldots, \mu_n) M$$

with $\mu_i \in \mathbb{R}$.

2.6 Min–Max Principle

We establish a variational principle, known as the min–max principle, or the Courant–Fisher principle, which gives the eigenvalues of a Hermitian matrix as the result of a simple optimization process.

Definition 2.6.1. *Let A be a self-adjoint or Hermitian matrix in $\mathcal{M}_n(\mathbb{C})$, i.e., $A^* = A$. The function from $\mathbb{C}^n \setminus \{0\}$ into $\mathbb{R}$ defined by*

$$R_A(x) = \frac{\langle Ax, x \rangle}{\langle x, x \rangle}$$

is called the Rayleigh quotient of A.

By virtue of Remark 2.3.1, the Rayleigh quotient of a Hermitian matrix is always real. Indeed, for any vector $x \in \mathbb{C}^n$, we have $\langle Ax, x \rangle = \langle x, A^*x \rangle = \langle x, Ax \rangle = \overline{\langle Ax, x \rangle}$, which thus belongs to $\mathbb{R}$.

Theorem 2.6.1. *Let A be a Hermitian matrix of $\mathcal{M}_n(\mathbb{C})$. Its smallest eigenvalue, denoted by λ_1, satisfies*

$$\lambda_1 = \min_{x \in \mathbb{C}^n, x \neq 0} R_A(x) = \min_{x \in \mathbb{C}^n, \|x\|=1} \langle Ax, x \rangle,$$

and both minima are attained for at least one eigenvector $e_1 \neq 0$ satisfying $Ae_1 = \lambda_1 e_1$.

Remark 2.6.1. The same kind of result is true for the largest eigenvalue λ_n of A, namely,

$$\lambda_n = \max_{x \in \mathbb{C}^n, x \neq 0} R_A(x) = \max_{x \in \mathbb{C}^n, \|x\|=1} \langle Ax, x \rangle,$$

and both maxima are attained for at least one eigenvector $e_n \neq 0$ satisfying $Ae_n = \lambda_n e_n$.

Proof of Theorem 2.6.1. Since A is Hermitian, it is diagonalizable in an orthonormal basis of eigenvectors $(e_1, \ldots, e_n)$. We call $\lambda_1 \leq \cdots \leq \lambda_n$ its real eigenvalues (see Remark 2.3.1) sorted in increasing order. Let $(x_1, \ldots, x_n)$ be the coordinates of the vector x in this basis. We have

$$\langle Ax, x \rangle = \sum_{i=1}^{n} \lambda_i |x_i|^2 \geq \lambda_1 \sum_{i=1}^{n} |x_i|^2 = \lambda_1 \langle x, x \rangle.$$

As a consequence, both minima are larger than λ_1. Moreover, we have $Ae_1 = \lambda_1 e_1$ and $\|e_1\| = 1$, by definition. Therefore, we get

$$\langle Ae_1, e_1 \rangle = \lambda_1 \langle e_1, e_1 \rangle,$$

so both minima are attained at this vector e_1. $\square$

Theorem 2.6.1 and Remark 2.6.1 can be generalized to all other intermediate eigenvalues as follows.

Proposition 2.6.1. *Let A be a Hermitian matrix with eigenvalues $\lambda_1 \leq \cdots \leq \lambda_n$. For each index $i \in \{1, \ldots, n\}$, we have*

$$\lambda_i = \min_{x \perp span\{e_1, \ldots, e_{i-1}\}} R_A(x) = \max_{x \perp span\{e_{i+1}, \ldots, e_n\}} R_A(x), \qquad (2.4)$$

where $(e_1, \ldots, e_n)$ are the eigenvectors of A associated with $(\lambda_1, \ldots, \lambda_n)$.

Remark 2.6.2. In formula (2.4), it should be understood that for $i = 1$, the minimization is carried without any orthogonality constraint on the vector x, so we recover the statement of Theorem 2.6.1. Similarly, it should be understood that for $i = n$, the maximization is carried without any orthogonality constraint on the vector x.

Proof of Proposition 2.6.1. With the previous notation, we have $x = \sum_{i=1}^{n} x_i e_i$. Consequently, $x \perp span\{e_1, \ldots, e_{i-1}\}$ implies that $x = \sum_{j=i}^{n} x_j e_j$, hence

$$\frac{\langle Ax, x \rangle}{\langle x, x \rangle} \geq \lambda_i \quad \text{and} \quad \frac{\langle Ae_i, e_i \rangle}{\langle e_i, e_i \rangle} = \lambda_i,$$

which completes the proof for the minimization (same argument for the maximum). $\square$

Now we can prove the main result of this section, namely the "min–max" principle, also called Courant–Fisher theorem.

Theorem 2.6.2 (Courant–Fisher or min–max). *Let A be a Hermitian matrix with eigenvalues $\lambda_1 \leq \cdots \leq \lambda_n$. For each index $i \in \{1, \ldots, n\}$, we have*

$$\lambda_i = \min_{(a_1, \ldots, a_{n-i}) \in \mathbb{C}^n} \max_{x \perp span\{a_1, \ldots, a_{n-i}\}} R_A(x) \qquad (2.5)$$

$$= \max_{(a_1, \ldots, a_{i-1}) \in \mathbb{C}^n} \min_{x \perp span\{a_1, \ldots, a_{i-1}\}} R_A(x). \qquad (2.6)$$

Remark 2.6.3. In the "min–max" formula of (2.5), we first start by carrying out, for some fixed family of vectors $(a_1, \ldots, a_{n-i})$ of $\mathbb{C}^n$, the maximization of the Rayleigh quotient on the subspace orthogonal to $span\{a_1, \ldots, a_{n-i}\}$. The next stage consists in minimizing the latter result as the family $(a_1, \ldots, a_{n-i})$ varies in $\mathbb{C}^n$. The "max–min" formula is interpreted in a similar fashion. As already mentioned in Remark 2.6.2, in the "max–min" formula, for $i = 1$, minimization in x is done all over $\mathbb{C}^n \setminus \{0\}$, and there is no maximization on the (empty) family $(a_1, \ldots, a_{i-1})$. Hence we retrieve Theorem 2.6.1. The same remark applies to the "min–max" formula for $i = n$ too.

Proof of the Courant–Fisher theorem. We prove only the max–min formula (the min–max formula is proved in the same way). First of all, for the particular choice of the family $(e_1, \ldots, e_{i-1})$, we have

$$\max_{(a_1, \ldots, a_{i-1}) \in \mathbb{C}^n} \min_{x \perp span\{a_1, \ldots, a_{i-1}\}} R_A(x) \geq \min_{x \perp span\{e_1, \ldots, e_{i-1}\}} R_A(x) = \lambda_i.$$

Furthermore, for all choices of $(a_1, \ldots, a_{i-1})$, we have

$$\dim span \{a_1, \ldots, a_{i-1}\}^\perp \geq n - i + 1.$$

On the other hand, $\dim span \{e_1, \ldots, e_i\} = i$, so the subspace

$$span \{a_1, \ldots, a_{i-1}\}^\perp \cap span \{e_1, \ldots, e_i\}$$

is nonempty since its dimension is necessarily larger than 1. Therefore, we can restrict the minimization space to obtain an upper bound:

$$\min_{x \perp span\{a_1, \ldots, a_{i-1}\}} R_A(x) \leq \min_{x \in span\{a_1, \ldots, a_{i-1}\}^\perp \cap span\{e_1, \ldots, e_i\}} R_A(x)$$

$$\leq \max_{x \in span\{a_1, \ldots, a_{i-1}\}^\perp \cap span\{e_1, \ldots, e_i\}} R_A(x)$$

$$\leq \max_{x \in span\{e_1, \ldots, e_i\}} R_A(x) = \lambda_i,$$

where the last inequality is a consequence of the fact that enlarging the maximization space does not decrease the value of the maximum. Taking the maximum with respect to all families $(a_1, \ldots, a_{i-1})$ in the above equation, we conclude that

$$\max_{(a_1, \ldots, a_{i-1}) \in \mathbb{C}^n} \min_{x \perp span\{a_1, \ldots, a_{i-1}\}} R_A(x) \leq \lambda_i,$$

which ends the proof. □

2.7 Singular Values of a Matrix

Throughout this section, we consider matrices in $\mathcal{M}_{m,n}(\mathbb{C})$ that are not necessarily square. To define the singular values of a matrix, we first need a technical lemma.

Lemma 2.7.1. *For any $A \in \mathcal{M}_{m,n}(\mathbb{C})$, the matrix A^*A is Hermitian and has real, nonnegative eigenvalues.*

Proof. Obviously A^*A is a square Hermitian matrix of size n. We deduce from Remark 2.3.1 that its eigenvalues are real. It remains to show that they are nonnegative. Let λ be an eigenvalue of A^*A, and let $x \neq 0$ be a corresponding eigenvector such that $A^*Ax = \lambda x$. Taking the Hermitian product of this equality with x, we obtain

$$\lambda = \frac{\langle A^*Ax, x \rangle}{\langle x, x \rangle} = \frac{\langle Ax, Ax \rangle}{\langle x, x \rangle} = \frac{\|Ax\|^2}{\|x\|^2} \in \mathbb{R}^+,$$

and the result is proved. $\qquad\qquad\square$

Definition 2.7.1. *The singular values of a matrix $A \in \mathcal{M}_{m,n}(\mathbb{C})$ are the nonnegative square roots of the n eigenvalues of A^*A.*

This definition makes sense thanks to Lemma 2.7.1, which proves that the eigenvalues of A^*A are real nonnegative, so their square roots are real. The next lemma shows that the singular values of A could equally be defined as the nonnegative square roots of the m eigenvalues of AA^*, since both matrices A^*A and AA^* share the same nonzero eigenvalues.

Lemma 2.7.2. *Let $A \in \mathcal{M}_{m,n}(\mathbb{C})$ and $B \in \mathcal{M}_{n,m}(\mathbb{C})$. The nonzero eigenvalues of the matrices AB and BA are the same.*

Proof. Take λ an eigenvalue of AB, and $u \in \mathbb{C}^m$ a corresponding nonzero eigenvector such that $ABu = \lambda u$. If $u \in \mathrm{Ker}\,(B)$, we deduce that $\lambda u = 0$, and since $u \neq 0$, $\lambda = 0$. Consequently, if $\lambda \neq 0$, u does not belong to $\mathrm{Ker}\,(B)$. Multiplying the equality by B, we obtain $BA(Bu) = \lambda(Bu)$, where $Bu \neq 0$, which proves that λ is also an eigenvalue of BA. $\qquad\square$

Remark 2.7.1. A square matrix is nonsingular if and only if its singular values are positive. Clearly, if A is nonsingular, so is A^*A, and thus its eigenvalues (the squared singular values of A) are nonzero. Reciprocally, if A is singular, there exists a nonzero vector u such that $Au = 0$. For this same vector, we have $A^*Au = 0$, and therefore A^*A is singular too.

Owing to Theorem 2.5.1, we can characterize the singular values of a normal matrix.

Proposition 2.7.1. *The singular values of a normal matrix are the moduli of its eigenvalues.*

Proof. Indeed, if a matrix A is normal, there exists a unitary matrix U such that $A = U^*DU$ with $D = \mathrm{diag}\,(\lambda_i)$, and so $A^*A = (U^*DU)^*(U^*DU) = U^*(D^*D)U$. We deduce that the matrices A^*A and $D^*D = \mathrm{diag}\,(|\lambda_i|^2)$ are similar, and have accordingly the same eigenvalues. $\qquad\square$

Remark 2.7.2. As a consequence of Proposition 2.7.1, the spectral radius of a normal matrix is equal to its largest singular value.

We finish this chapter by introducing the "singular value decomposition" of a matrix, in short the SVD factorization.

Theorem 2.7.1 (SVD factorization). *Let $A \in \mathcal{M}_{m,n}(\mathbb{C})$ be a matrix having r positive singular values. There exist two unitary matrices $U \in \mathcal{M}_n(\mathbb{C})$, $V \in \mathcal{M}_m(\mathbb{C})$, and a diagonal matrix $\tilde{\Sigma} \in \mathcal{M}_{m,n}(\mathbb{R})$ such that*

$$A = V \tilde{\Sigma} U^* \quad and \quad \tilde{\Sigma} = \begin{pmatrix} \Sigma & 0 \\ 0 & 0 \end{pmatrix}, \tag{2.7}$$

where $\Sigma = \mathrm{diag}\,(\mu_1, \ldots, \mu_r)$, and $\mu_1 \geq \mu_2 \geq \cdots \geq \mu_r > 0$ are the positive singular values of A.

Before proving this result, let us make some remarks. Without loss of generality we shall assume that $m \geq n$, since for $m < n$ we can apply the SVD factorization to the matrix A^* and deduce the result for A by taking the adjoint of (2.7).

Remark 2.7.3. For $n = m$, the SVD factorization of Theorem 2.7.1 has nothing to do with the usual diagonalization of a matrix because, in general, V is different from U so U^* is different from V^{-1} (see Definition 2.4.1). In other words, Theorem 2.7.1 involves two unitary changes of basis (associated with U and V) while Definition 2.4.1 relies on a single change of basis.

Remark 2.7.4. As a byproduct of Theorem 2.7.1, we obtain that the rank of A is equal to r, i.e., the number of nonzero singular values of A. In particular, it satisfies $r \leq \min(m, n)$.

Remark 2.7.5. We have $A^*A = U \tilde{\Sigma}^t \tilde{\Sigma} U^*$. The columns of matrix U are thus the eigenvectors of the Hermitian matrix A^*A, and the diagonal entries of $\tilde{\Sigma}^t \tilde{\Sigma} \in \mathcal{M}_n(\mathbb{R})$ are the eigenvalues of A^*A, i.e., the squares of the singular values of A. On the other hand, $AA^* = V \tilde{\Sigma} \tilde{\Sigma}^t V^*$, so the columns of V are the eigenvectors of the other Hermitian matrix AA^* and the diagonal entries of $\tilde{\Sigma} \tilde{\Sigma}^t \in \mathcal{M}_m(\mathbb{R})$ are the eigenvalues of AA^* too.

Remark 2.7.6. Theorem 2.7.1 can be refined in the case of a real matrix A, by showing that both matrices U and V are real too.

Proof of Theorem 2.7.1.   We denote by u_i the eigenvectors of A^*A corresponding to the eigenvalues μ_i^2 (see Lemma 2.7.1), $A^*Au_i = \mu_i^2 u_i$, and U is the unitary matrix defined by $U = [u_1 | \ldots | u_n]$. We have

$$A^*AU = [A^*Au_1 | \ldots | A^*Au_n] = [\mu_1^2 u_1 | \ldots | \mu_n^2 u_n] = U \, \mathrm{diag}\,(\mu_1^2, \ldots, \mu_n^2),$$

so $U^*A^*AU = \tilde{\Sigma}^t \tilde{\Sigma}$, setting

$$\tilde{\Sigma} = \begin{pmatrix} \mu_1 & 0 & \cdots & 0 \\ 0 & \mu_2 & & \vdots \\ \vdots & & \ddots & 0 \\ \vdots & & \mu_n & \\ \vdots & & & 0 \\ \vdots & & & \vdots \\ 0 & \cdots & \cdots & 0 \end{pmatrix} \in \mathcal{M}_{m,n}(\mathbb{R}).$$

We arrange in decreasing order the singular values $\mu_1 \geq \cdots \geq \mu_r > \mu_{r+1} = \cdots = \mu_n = 0$, of which only the first r are nonzero. We also notice that

$$(Au_i, Au_j) = (A^* Au_i, u_j) = \mu_i^2(u_i, u_j) = \mu_i^2 \delta_{i,j},$$

and in particular, $Au_i = 0$ if $r < i \leq n$. For $1 \leq i \leq r$, $\mu_i \neq 0$, so we can define unit vectors $v_i \in \mathbb{C}^m$ by $v_i = Au_i/\mu_i$. These vectors, complemented in order to obtain an orthonormal basis $v_1, \ldots, v_m$ of $\mathbb{C}^m$, yield a matrix $V = [v_1| \ldots |v_m]$. Let us check equality (2.7):

$$V \tilde{\Sigma} U^* = [v_1| \ldots |v_m] \tilde{\Sigma} U^* = [\mu_1 v_1| \ldots |\mu_n v_n] U^* = [Au_1| \ldots |Au_r|0| \ldots |0] U^*.$$

Since $Au_i = 0$ for $r < i \leq n$, we deduce $V \tilde{\Sigma} U^* = AUU^* = A$.   □

Geometrical interpretation. The SVD factorization shows that the image of the unit sphere S^{n-1} by a nonsingular matrix A is an ellipsoid. For instance, Figure 2.2 displays the image of the unit circle of $\mathbb{R}^2$ by the 2×2 matrix of Exercise 2.25. We recall that $S^{n-1} = \{(x_1, \ldots, x_n)^t \in \mathbb{R}^n, \sum_i x_i^2 = 1\}$. Let A be a nonsingular matrix of $\mathcal{M}_n(\mathbb{R})$ whose SVD factorization is $A = V \Sigma U^t$. We wish to characterize the set $V \Sigma U^t S^{n-1}$. Since the matrix U^t is orthogonal, it transforms any orthonormal basis into another orthonormal basis. Hence $U^t S^{n-1} = S^{n-1}$. Then we clearly have $\Sigma S^{n-1} = \{(x'_1, \ldots, x'_n)^t \in \mathbb{R}^n, \sum_i (x'_i/\mu_i)^2 = 1\}$, which is precisely the definition of an ellipsoid E^{n-1} of semiaxes $\mu_i e_i$. Finally, since V is still a matrix of change of orthonormal basis, $AS^{n-1} = VE^{n-1}$ is just a rotation of E^{n-1}. To sum up, AS^{n-1} is an ellipsoid, of semiaxes $\mu_i v_i$, where v_i is the ith column of V.

Pseudoinverse. The SVD factorization allows us to introduce the so-called pseudoinverse of a matrix, which generalizes the notion of inverse for rectangular matrices or square matrices that are singular.

Definition 2.7.2. *Let $A = V \tilde{\Sigma} U^*$ be the SVD factorization of some matrix $A \in \mathcal{M}_{m,n}(\mathbb{C})$ having r nonzero singular values. We call the matrix $A^\dagger \in \mathcal{M}_{n,m}(\mathbb{C})$ defined by $A^\dagger = U \tilde{\Sigma}^\dagger V^*$ with*

$$\tilde{\Sigma}^\dagger = \begin{pmatrix} \Sigma^{-1} & 0 \\ 0 & 0 \end{pmatrix} \in \mathcal{M}_{n,m}(\mathbb{R})$$

the pseudoinverse matrix of A.

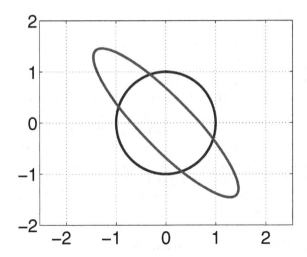

Fig. 2.2. Image of the unit circle of the plane by the matrix defined in Exercise 2.25.

The reader will easily check the following relations:

$$A^\dagger A = U \tilde{\Sigma}^\dagger \tilde{\Sigma} U^* = U \begin{pmatrix} I_r & 0 \\ 0 & 0 \end{pmatrix} U^* = \sum_{i=1}^{r} u_i u_i^*, \tag{2.8}$$

$$AA^\dagger = V \tilde{\Sigma} \tilde{\Sigma}^\dagger V^* = V \begin{pmatrix} I_r & 0 \\ 0 & 0 \end{pmatrix} V^* = \sum_{i=1}^{r} v_i v_i^*, \tag{2.9}$$

$$A = \sum_{i=1}^{r} \mu_i v_i u_i^*, \quad \text{and} \quad A^\dagger = \sum_{i=1}^{r} \frac{1}{\mu_i} u_i v_i^*. \tag{2.10}$$

In addition, if A has maximal rank ($r = n \leq m$), its pseudoinverse is given by

$$A^\dagger = (A^*A)^{-1} A^*.$$

In particular, if A is a nonsingular square matrix ($r = n = m$), we obtain $A^\dagger A = AA^\dagger = I_n$. Thus $A^\dagger = A^{-1}$. In this sense, the pseudoinverse is indeed the generalization of the inverse.

Remark 2.7.7. It can be shown that the pseudoinverse matrix is the only matrix X satisfying the following conditions (known as the Moore–Penrose conditions):

$$AXA = A, \quad XAX = X, \quad XA = (XA)^*, \quad AX = (AX)^*.$$

2.8 Exercises

2.1 (∗). We fix the dimension $n \geq 2$.

1. What is the vector u in terms of the matrix a defined by the instructions
   `a=eye(n,n);u=a(:,i)` for an integer i between 1 and n?
2. We recall that the `Matlab` instruction `rand(m,n)` returns a matrix of size
   $m \times n$ whose entries are random real numbers in the range $[0, 1]$. For
   n fixed, we define two vectors `u=rand(n,1)` and `v=rand(n,1)`. Compute
   using `Matlab` the vector $w = v - \frac{\langle v,u \rangle}{\|u\|_2^2} u$ and the scalar product $\langle w, u \rangle$.
3. Let A be a real square matrix (initialized by `rand`), and define two matrices
   B and C by `B=0.5*(A+A')` and `C=0.5*(A-A')`
   (a) Compute the scalar product $\langle Cx, x \rangle$ for various vectors x. Justify the
   observed result.
   (b) Compute the scalar product $\langle Bx, x \rangle$ for various vectors x. Check that
   it is equal to $\langle Ax, x \rangle$ and explain why.

2.2 (∗). The goal of this exercise is to define `Matlab` functions returning matrices having special properties that we shall exploit in the upcoming exercises.
All variables will be initialized by `rand`. Note that there are several possible
answers, as it is often the case for computer programs.

1. Write a function (called `SymmetricMat(n)`) returning a real symmetric
   matrix of size $n \times n$.
2. Write a function (called `NonsingularMat(n)`) returning a real nonsingular
   matrix of size $n \times n$.
3. Write a function (called `LowNonsingularMat(n)`) returning a real nonsingular lower triangular matrix of size $n \times n$.
4. Write a function (called `UpNonsingularMat(n)`) returning a real nonsingular upper triangular matrix of size $n \times n$.
5. Write a function (called `ChanceMat(m,n,p)`) returning a real matrix of
   size $m \times n$ whose entries are chosen randomly between the values $-p$ and
   p.
6. Write a function (called `BinChanceMat(m,n)`) returning a real matrix of
   size $m \times n$ whose entries are chosen randomly equal to 0 or 1.
7. Write a function (called `HilbertMat(m,n)`) returning the so-called Hilbert
   matrix $H \in \mathcal{M}_{m,n}(\mathbb{R})$ defined by its entries:

$$H_{i,j} = \frac{1}{i + j - 1}.$$

2.3. Define a matrix A by the `Matlab` instructions
    `p=NonsingularMat(n);A=p*diag([ones(n-1,1); e])*inv(p),`
where e is a real number. What is the determinant of A? (Do not use `Matlab`
to answer.) Take $e = 10^{-20}$, $n = 5$, and compute by `Matlab` the determinant
of A. What do you notice?

2.4. Set A=[1:3; 4:6; 7:9; 10:12; 13:15]. What is the rank of A? (use the function **rank**). Let B and C be two nonsingular matrices respectively of size 3×3 and 5×5 given by the function **NonsingularMat** defined in Exercise 2.2. Compare the ranks of the products CA and AB. Justify your experimental results.

2.5. Vary n from 1 to 10 and

1. determine the rank of a matrix A defined by A=rand(8,n)*rand(n,6). What is going on?
2. Same question for A=BinChanceMat(8,n)*BinChanceMat(n,6).
3. Justify your observations.

2.6. We fix the dimension $n = 5$.

1. For any integer r between 1 and 5, initialize (with **rand**) r vectors u_i and define a square matrix $A = \sum_{i=1}^{r} u_i u_i^t$. Compare the rank of A with r.
2. Same question for vectors generated by **BinChanceMat**.
3. Justify your observations.

2.7 (*). Write a function (called MatRank(m,n,r)) returning a real matrix of size $m \times n$ and of fixed rank r.

2.8. Define rectangular matrices
     A=[1:3;4:6;7:9;10:12] and B=[-1 2 3;4:6;7:9;10:12].

1. Are the square matrices $A^t A$ and $B^t B$ nonsingular?
2. Determine the rank of each of the two matrices.
3. Justify the answers to the first question.
4. Same questions for A^t and B^t.

2.9. Let A be a matrix defined by A=MatRank(n,n,r) with $r \leq n$ and Q a matrix defined by Q=null(A'), that is, a matrix whose columns form a basis of the null space of A^t. Let u be a column of Q, compute the rank of $A + uu^t$. Prove the observed result.

2.10 (*). Let $a_1, \ldots, a_n$ be a family of n vectors of $\mathbb{R}^m$, and $A \in \mathcal{M}_{m,n}(\mathbb{R})$ the matrix whose columns are the vectors $(a_j)_{1 \leq j \leq n}$. We denote by r the rank of A. The goal is to write a program that delivers an orthonormal family of r vectors $u_1, \ldots, u_r$ of $\mathbb{R}^m$ by applying the Gram–Schmidt orthogonalization procedure to A. We consider the following algorithm, written here in pseudolanguage (see Chapter 4):

$$
\begin{aligned}
&\text{For } p = 1 \nearrow n \\
&\quad s = 0 \\
&\quad \text{For } k = 1 \nearrow p - 1 \\
&\quad\quad\quad s = s + \langle a_p, u_k \rangle u_k \\
&\quad \text{End} \\
&\quad s = a_p - s
\end{aligned}
$$

$$\text{If } \|s\| \neq 0 \text{ then}$$
$$u_p = s/\|s\|$$
$$\text{Else}$$
$$u_p = 0$$
$$\text{End}$$

End

Gram–Schmidt Algorithm.

1. Determine the computational complexity (number of multiplications and divisions for n large) of this algorithm.
2. Write a program GramSchmidt whose input argument is A and whose output argument is a matrix the pth column of which is the vector u_p, if it exists, and zero otherwise. Test this program with $A \in \mathcal{M}_{10,5}(\mathbb{R})$ defined by
   ```
   n=5;u=1:n; u=u'; c2=cos(2*u); c=cos(u); s=sin(u);
   A=[u c2 ones(n,1) rand()*c.*c exp(u) s.*s];
   ```
   We denote by U the matrix obtained by applying the orthonormalization procedure to A.
   (a) Compute UU^t and U^tU. Comment.
   (b) Apply the GramSchmidt algorithm to U. What do you notice?
3. Change the program GramSchmidt into a program GramSchmidt1 that returns a matrix whose first r columns are the r vectors u_k, and whose last $n - r$ columns are zero.

2.11 (∗). The goal of this exercise is to study a modified Gram–Schmidt algorithm. With the notation of the previous exercise, each time a new vector u_p is found, we may subtract from each a_k $(k > p)$ its component along u_p. Assume now that the vectors a_k are linearly independent in $\mathbb{R}^m$.

- Set $u_1 = a_1/\|a_1\|$ and replace each a_k with $a_k - \langle u_1, a_k \rangle u_1$ for all $k > 1$.
- If the first $p - 1$ vectors $u_1, \ldots, u_{p-1}$ are known, set $u_p = a_p/\|a_p\|$ and replace each a_k with $a_k - \langle u_p, a_k \rangle u_p$ for all $k > p$.

1. Write a function MGramSchmidt coding this algorithm.
2. Fix $m = n = 10$. Compare both the Gram–Schmidt and modified Gram–Schmidt algorithms (by checking the orthonormality of the vectors) for a matrix whose entries are chosen by the function rand, then for a Hilbert matrix $H \in \mathcal{M}_{m,n}(\mathbb{R})$.
3. If we have at our disposal only the Gram–Schmidt algorithm, explain how to improve the computation of u_p.

2.12 (∗). The goal of the following exercises is to compare different definitions of matrices. We use the Matlab functions tic and toc to estimate the running time of Matlab. The function toc returns the time elapsed since the last call tic. Run the following instructions plus2pt

```
n=400;tic;
for j=1:n for i=1:n,a(i,j)=cos(i)*sin(j);end;end;t1=toc;clear a;

tic;a=zeros(n,n);
for~j=1:n~for i=1:n,~a(i,j)=cos(i)*sin(j);end;end;t2=toc;clear~a;

tic;a=zeros(n,n);
for~i=1:n~for j=1:n,~a(i,j)=cos(i)*sin(j);end;end;t3=toc;clear a;

tica=zeros(n,n);a=cos(1:n)'*sin(1:n);t4=toc;
```

Display the variables t1, t2, t3, and t4. Explain.

2.13. Define a matrix A of size $n \times n$ (vary n) by the instructions A=rand(n,n);A=triu(A)-diag(diag(A)). What is the purpose of triu? Compute the powers of A. Justify the observed result.

2.14. Let H be the Hilbert matrix of size 6×6.

1. Compute the eigenvalues λ of H (use the function eig).
2. Compute the eigenvectors u of H.
3. Verify the relations $Hu = \lambda u$.

2.15. Define a matrix A by the instructions
```
P=[1 2 2 1; 2 3 3 2; -1 1 2 -2; 1 3 2 1];
D=[2 1 0 0; 0 2 1 0; 0 0 3 0; 0 0 0 4];
A=P*D*inv(P);
```

1. Without using Matlab, give the eigenvalues of A.
2. Compute the eigenvalues by Matlab. What do you notice?
3. For $n = 3$ and $n = 10$, compute the eigenvalues of A^n with Matlab, and compare with their exact values. What do you observe?
4. Diagonalize A using the function eig. Comment.

2.16. For various values of n, compare the spectra of the matrices A and A^t with A=rand(n,n). Justify the answer.

2.17. Fix the dimension n. For u and v two vectors of $\mathbb{R}^n$ chosen randomly by rand, determine the spectrum of $I_n + uv^t$. What are your experimental observations? Rigorously prove the observed result.

2.18 (*). Define a matrix A=[10 2;2 4]. Plot the curve of the Rayleigh quotient $x \mapsto x^t A x$, where x spans the unit circle of the plane. What are the maximal and minimal attained values? Compare with the eigenvalues of the matrix A. Explain.
Hint: use the Matlab function plot3.

2.19. For various values of the integers m and n, compare the spectra of AA^t and $A^t A$, where A=rand(n,m). Justify the observations.

2.20 (∗). Show that the following function PdSMat returns a positive definite symmetric matrix of size $n \times n$

```
function A=PdSMat(n)
A=SymmetricMat(n);    // defined in Exercise 2.2
[P,D]=eig(A);D=abs(D);
D=D+norm(D)*eye(size(D));
A=P*D*inv(P);
```

1. For different values of n, compute the determinant of A=PdSMat(n). What do you observe? Justify.
2. Fix $n = 10$. For k varying from 1 to n, define a matrix A_k of size $k \times k$ by Ak = A(1:k,1:k). Check that the determinants of all the matrices A_k are positive. Prove this result.
3. Are the eigenvalues of A_k eigenvalues of A?

2.21.

1. Let A=SymmetricMat(n). Compute the eigenvectors u_i and the eigenvalues λ_i of A. Compute $\sum_{k=1}^{n} \lambda_i u_i u_i^t$ and compare this matrix with A. What do you observe?
2. Denote by D and P the matrices defined by [P,D]=eig(A). We modify an entry of the diagonal matrix D by setting $D(1,2) = 1$, and we define a matrix B by B=P*D*inv(P). Compute the eigenvectors v_i and eigenvalues μ_i of B. Compute $\sum_{k=1}^{n} \mu_i v_i v_i^t$ and compare this matrix with B. What do you observe?
3. Justify.

2.22. For various values of n, compute the rank of the matrix defined by the instruction rank(rand(n,1)*rand(1,n)). What do you notice? The goal is to prove the observed result.

1. Let u and v be two nonzero vectors of $\mathbb{R}^n$. What is the rank of the matrix $A = vu^t$?
2. Let $A \in \mathcal{M}_n(\mathbb{R})$ be a rank-one matrix. Show that there exist two vectors u and v of $\mathbb{R}^n$ such that $A = vu^t$.

2.23 (∗). Define a square matrix by A=rand(n,n).

1. Compute the spectrum of A.
2. For $1 \leq i \leq n$, define $\gamma_i = \sum_{j\neq i, j=1}^{n} |a_{i,j}|$ and denote by D_i the (so-called Gershgorin) disk of radius γ_i and center $a_{i,i}$:

$$D_i = \left\{ z \in \mathbb{C}, \quad |z - a_{i,i}| \leq \gamma_i \right\}.$$

(a) Compute the γ_i's with the Matlab function sum.
(b) Let λ be an eigenvalue of A. Check with Matlab that there exists (at least) one index i such that $\lambda \in D_i$.

(c) Rigorously prove this result.

3. A matrix A is said to be strictly diagonally dominant if

$$|a_{i,i}| > \sum_{j \neq i} |a_{i,j}| \qquad (1 \leq i \leq n).$$

(a) Write a function DiagDomMat(n) returning an $n \times n$ diagonally dominant matrix.

(b) For various values of n, compute the determinant of A=DiagDomMat(n). What do you notice?

(c) Justify your answer.

4. Write a program PlotGersh that plots the Gershgorin disks for a given matrix.

Application:

$$A = \begin{pmatrix} 1 & 0 & 1 \\ -2 & 6 & 1 \\ 1 & -1 & -3 \end{pmatrix}.$$

2.24. For each of the matrices

$$A_1 = \begin{pmatrix} 1 & 2 & 3 \\ 3 & 2 & 1 \\ 4 & 2 & 1 \end{pmatrix}, \quad A_2 = \begin{pmatrix} .75 & 0. & .25 \\ 0. & 1. & 0. \\ .25 & 0. & .75 \end{pmatrix},$$

$$A_3 = \begin{pmatrix} .375 & 0 & -.125 \\ 0 & .5 & 0 \\ -.125 & 0 & .375 \end{pmatrix}, \quad A_4 = \begin{pmatrix} -.25 & 0. & -.75 \\ 0. & 1. & 0. \\ -.75 & 0. & -.25 \end{pmatrix},$$

compute A_i^n for $n = 1, 2, 3, \ldots$ In your opinion, what is the limit of A_i^n as n goes to infinity? Justify the observed results.

2.25. Plot the image of the unit circle of $\mathbb{R}^2$ by the matrix

$$A = \begin{pmatrix} -1.25 & 0.75 \\ 0.75 & -1.25 \end{pmatrix} \tag{2.11}$$

to reproduce Figure 2.2. Use the Matlab function svd.

2.26. For different choices of m and n, compare the singular values of a matrix A=rand(m,n) and the eigenvalues of the block matrix $B = \begin{pmatrix} 0 & A \\ A^t & 0 \end{pmatrix}$. Justify.

2.27. Compute the pseudoinverse $A^\dagger$ (function pinv) of the matrix

$$A = \begin{pmatrix} 1 & -1 & 4 \\ 2 & -2 & 0 \\ 3 & -3 & 5 \\ -1 & -1 & 0 \end{pmatrix}.$$

Compute $A^\dagger A$, $AA^\dagger$, $AA^\dagger A$, and $A^\dagger AA^\dagger$. What do you observe? Justify.

2.28. Fix $n = 100$. For different values of $r \leq n$, compare the rank of A=MatRank(n,n,r) and the trace of $AA^\dagger$. Justify.

2.29 ($*$). The goal of this exercise is to investigate another definition of the pseudoinverse matrix. Fix $m = 10, n = 7$. Let A be a matrix defined by A=MatRank(m,n,5). We denote by P the orthogonal projection onto $(\operatorname{Ker} A)^\perp$, and by Q the orthogonal projection onto $\operatorname{Im} A$.

1. Compute a basis of $(\operatorname{Ker} A)^\perp$, then the matrix P.
2. Compute a basis of $\operatorname{Im} A$, then the matrix Q.
3. Compare on the one hand $A^\dagger A$ with P, and on the other hand, $AA^\dagger$ with Q. What do you notice? Justify your answer.
4. Let $y \in \mathbb{C}^m$ and define $x_1 = Px$, where $x \in \mathbb{C}^n$ is such that $Ax = Qy$. Prove (without using Matlab) that there exists a unique such x_1. Consider the linear map $\varphi : \mathbb{C}^m \to \mathbb{C}^n$ by $\varphi(y) = x_1$. Show (without using Matlab) that the matrix corresponding to this map (in the canonical basis) is $A^\dagger$.

3

Matrix Norms, Sequences, and Series

3.1 Matrix Norms and Subordinate Norms

We recall the definition of a norm on the vector space $\mathbb{K}^n$ (with $\mathbb{K} = \mathbb{R}$ or $\mathbb{C}$.)

Definition 3.1.1. *We call a mapping denoted by $\| \cdot \|$, from $\mathbb{K}^n$ into $\mathbb{R}^+$ satisfying the following properties a norm on $\mathbb{K}^n$*

1. $\forall x \in \mathbb{K}^n, \quad \|x\| = 0 \Longrightarrow x = 0;$
2. $\forall x \in \mathbb{K}^n, \forall \lambda \in \mathbb{K}, \quad \|\lambda x\| = |\lambda| \|x\|;$
3. $\forall x \in \mathbb{K}^n, \forall y \in \mathbb{K}^n, \quad \|x + y\| \leq \|x\| + \|y\|$.

If $\mathbb{K}^n$ is endowed with a scalar (or Hermitian) product $\langle \cdot, \cdot \rangle$, then the mapping $x \to \langle x, x \rangle^{1/2}$ defines a norm on $\mathbb{K}^n$. The converse is not true in general: we cannot deduce from any norm a scalar product. The most common norms on $\mathbb{K}^n$ are (x_i denotes the coordinates of a vector x in the canonical basis of $\mathbb{K}^n$):

- the Euclidean norm, $\|x\|_2 = \left(\sum_{i=1}^n |x_i|^2 \right)^{1/2}$;
- the ℓ^p-norm, $\|x\|_p = \left(\sum_{i=1}^n |x_i|^p \right)^{1/p}$ for $p \geq 1$;
- the ℓ^∞-norm, $\|x\|_\infty = \max_{1 \leq i \leq n} |x_i|$.

Figure 3.1 shows the unit circles of $\mathbb{R}^2$ defined by each of these norms l^1, l^2, and l^∞. We recall the following important result.

Theorem 3.1.1. *If E is a vector space of finite dimension, then all norms are equivalent on E. That is, for all pairs of norms $\| \cdot \|, \| \cdot \|'$ there exist two constants c and C such that $0 < c \leq C$, and for all $x \in E$, we have*

$$c\|x\| \leq \|x\|' \leq C\|x\|.$$

Let us recall the equivalence constants between some vector norms ℓ_p (here $x \in \mathbb{K}^n$ and $p \geq 1$ is an integer):

$$\|x\|_\infty \leq \|x\|_p \leq n^{1/p} \|x\|_\infty,$$
$$\|x\|_2 \leq \|x\|_1 \leq \sqrt{n} \|x\|_2. \tag{3.1}$$

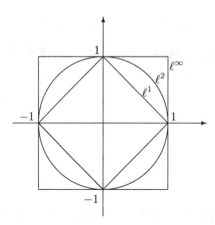

Fig. 3.1. Unit circles of $\mathbb{R}^2$ for the norms ℓ^1, ℓ^2, and ℓ^∞.

In this section, we confine ourselves to the case of square matrices. Since the space $\mathcal{M}_n(\mathbb{K})$ of square matrices of size n with entries in $\mathbb{K}$ is a vector space on $\mathbb{K}$, isomorphic to $\mathbb{K}^{n^2}$, we can define the following norms:

- Frobenius (or Schur or Euclidean) norm, $\|A\|_F = \left(\sum_{i=1}^n \sum_{j=1}^n |a_{i,j}|^2\right)^{1/2}$;
- l^q norm, $\|A\| = \left(\sum_{i=1}^n \sum_{j=1}^n |a_{i,j}|^q\right)^{1/q}$ for $q \geq 1$;
- l^∞ norm, $\|A\| = \max_{1 \leq i \leq n, 1 \leq j \leq n} |a_{i,j}|$.

There are particular norms on $\mathcal{M}_n(\mathbb{K})$ that satisfy an additional inequality on the product of two matrices.

Definition 3.1.2. *A norm* $\|\cdot\|$ *defined on* $\mathcal{M}_n(\mathbb{K})$ *is a matrix norm if for all matrices* $A, B \in \mathcal{M}_n(\mathbb{K})$,

$$\|AB\| \leq \|A\| \|B\|.$$

Example 3.1.1. The Frobenius norm is a matrix norm. Indeed, for any matrices A and B the Cauchy–Schwarz inequality yields

$$\sum_{i=1}^n \sum_{j=1}^n \left|\sum_{k=1}^n A_{i,k} B_{k,j}\right|^2 \leq \sum_{i=1}^n \sum_{j=1}^n \left(\sum_{k=1}^n |A_{i,k}|^2\right) \left(\sum_{k=1}^n |B_{k,j}|^2\right),$$

which is the desired inequality.

Example 3.1.2. Not all norms defined on $\mathcal{M}_n(\mathbb{K})$ are matrix norms. For instance, the norm defined by

$$\|A\| = \max_{1 \le i,j \le n} |a_{i,j}| \qquad (3.2)$$

is not a matrix norm when $n > 1$, because for the matrix A whose entries are all equal to 1, we have

$$1 = \|A\|^2 < \|A^2\| = n.$$

We shall consider norms on $\mathcal{M}_n(\mathbb{K})$ that are yet more particular and that are said to be subordinate to a vector norm on $\mathbb{K}^n$.

Definition 3.1.3. *Let $\|\cdot\|$ be a vector norm on $\mathbb{K}^n$. It induces a matrix norm defined by*

$$\|A\| = \sup_{x \in \mathbb{K}^n, x \ne 0} \frac{\|Ax\|}{\|x\|}, \qquad (3.3)$$

which is said to be subordinate to this vector norm.

The reader can easily check that indeed, (3.3) is a matrix norm on $\mathcal{M}_n(\mathbb{K})$. For convenience, the vector and matrix norms are denoted in the same way.

Remark 3.1.1. Let us recall briefly the difference between a maximum and a supremum. The supremum, $\sup_{i \in I} x_i$, of a family $(x_i)_{i \in I}$ of real numbers is the smallest constant C that bounds from above all the x_i:

$$\sup_{i \in I} x_i = \min \mathbf{C} \quad \text{with} \quad \mathbf{C} = \{C \in \mathbb{R} \text{ such that } x_i \le C, \ \forall i \in I\}.$$

(This smallest constant exists, possibly equal to $+\infty$, since $\mathbf{C}$ is a closed set of $\mathbb{R}$ as the intersection of the closed sets $\{C \in \mathbb{R} \text{ such that } C \ge x_i\}$.) If there exists x_{i_0} such that $\sup_{i \in I} x_i = x_{i_0}$, then the supremum is said to be attained and, by convention, we denote it by $\max_{i \in I} x_i$. Of course, this last notation, which specifies to the reader that the supremum is attained, coincides with the usual maximal value for a finite family $(x_i)_{i \in I}$.

Proposition 3.1.1. *Let $\|\cdot\|$ be a subordinate matrix norm on $\mathcal{M}_n(\mathbb{K})$.*

1. For all matrices A, the norm $\|A\|$ is also defined by

$$\|A\| = \sup_{x \in \mathbb{K}^n, \|x\|=1} \|Ax\| = \sup_{x \in \mathbb{K}^n, \|x\| \le 1} \|Ax\|.$$

2. There exists $x_A \in \mathbb{K}^n, x_A \ne 0$, such that

$$\|A\| = \frac{\|Ax_A\|}{\|x_A\|}$$

and sup *can be replaced by* max *in the definitions of $\|A\|$.*
3. The identity matrix satisfies

$$\|I_n\| = 1. \qquad (3.4)$$

4. *A subordinate norm is indeed a matrix norm: for all matrices A and B, we have*
$$\|AB\| \leq \|A\| \, \|B\|.$$

Proof. The first point is obvious. The second point is proved by observing that the function $x \to \|Ax\|$ is continuous on the bounded, closed, and therefore compact set $\{x \in \mathbb{K}^n, \|x\| = 1\}$. Thus it attains its maximum. The third point is obvious, whereas the fourth is a consequence of the inequality $\|ABx\| \leq \|A\| \, \|Bx\|$.   □

Remark 3.1.2. There are matrix norms that are not subordinate to any vector norm. A well-known example is the Frobenius norm, for which the norm of the identity matrix is $\|I_n\|_F = \sqrt{n}$. This is not possible for a subordinate norm according to (3.4).

Note that the equivalences (3.1) between vector norms on K^n imply the corresponding equivalences between the subordinate matrix norms. Namely, for any matrix $A \in \mathcal{M}_n(\mathbb{K})$,

$$n^{-1/p}\|A\|_\infty \leq \|A\|_p \leq n^{1/p}\|A\|_\infty,$$
$$n^{-1/2}\|A\|_2 \leq \|A\|_1 \leq n^{1/2}\|A\|_2. \tag{3.5}$$

The computation of matrix norms by Definition 3.1.3 may be quite difficult. However, the usual norms $\|\cdot\|_1$ and $\|\cdot\|_\infty$ can be computed explicitly.

Proposition 3.1.2. *We consider matrices in* $\mathcal{M}_n(\mathbb{K})$.

1. *The matrix norm* $\|A\|_1$*, subordinate to the* l^1*-norm on* $\mathbb{K}^n$*, satisfies*

$$\|A\|_1 = \max_{1 \leq j \leq n} \left(\sum_{i=1}^n |a_{i,j}| \right).$$

2. *The matrix norm* $\|A\|_\infty$*, subordinate to the* l^∞*-norm on* $\mathbb{K}^n$*, satisfies*

$$\|A\|_\infty = \max_{1 \leq i \leq n} \left(\sum_{j=1}^n |a_{i,j}| \right).$$

Proof. We write

$$\|Ax\|_1 = \sum_{i=1}^n \left| \sum_{j=1}^n a_{i,j} x_j \right| \leq \sum_{j=1}^n |x_j| \sum_{i=1}^n |a_{i,j}| \leq \|x\|_1 \left(\max_{1 \leq j \leq n} \sum_{i=1}^n |a_{i,j}| \right),$$

from which we deduce the inequality

$$\|A\|_1 \leq \max_{1 \leq j \leq n} \sum_{i=1}^n |a_{i,j}|. \tag{3.6}$$

Let j_0 be the index satisfying

$$\max_{1\leq j\leq n} \sum_{i=1}^{n} |a_{i,j}| = \sum_{i=1}^{n} |a_{i,j_0}|.$$

Let x^0 be defined by $x_j^0 = 0$ if $j \neq j_0$, and $x_{j_0}^0 = 1$. We have

$$\|x^0\|_1 = 1 \quad \text{and} \quad \|Ax^0\|_1 = \max_{1\leq j\leq n} \sum_{i=1}^{n} |a_{i,j}|,$$

which implies that inequality (3.6) is actually an equality. Next, we write

$$\|Ax\|_\infty = \max_{1\leq i\leq n} \left| \sum_{j=1}^{n} a_{i,j}x_j \right| \leq \|x\|_\infty \left(\max_{1\leq i\leq n} \sum_{j=1}^{n} |a_{i,j}| \right),$$

from which we infer the inequality

$$\|A\|_\infty \leq \max_{1\leq i\leq n} \sum_{j=1}^{n} |a_{i,j}|. \tag{3.7}$$

Let i_0 be the index satisfying

$$\max_{1\leq i\leq n} \sum_{j=1}^{n} |a_{i,j}| = \sum_{j=1}^{n} |a_{i_0 j}|.$$

Let x^0 be defined by $x_j^0 = 0$ if $a_{i_0 j} = 0$, and $x_j^0 = \frac{\overline{a_{i_0 j}}}{|a_{i_0 j}|}$ if $a_{i_0 j} \neq 0$. If $A \neq 0$, then $x^0 \neq 0$, and therefore $\|x^0\|_\infty = 1$ (if $A = 0$, then there is nothing to prove). Furthermore,

$$\|Ax^0\|_\infty \geq \left| \sum_{j=1}^{n} a_{i_0 j} x_j^0 \right| = \sum_{j=1}^{n} |a_{i_0 j}| = \max_{1\leq i\leq n} \sum_{j=1}^{n} |a_{i,j}|,$$

which proves that inequality (3.6) is actually an equality.   □

We now proceed to the matrix norm subordinate to the Euclidean norm:

Proposition 3.1.3. *Let* $\|A\|_2$ *be the matrix norm subordinate to the Euclidean norm on* $\mathbb{K}^n$. *We have*

$$\|A\|_2 = \|A^*\|_2 = \text{largest singular value of } A.$$

Proof. First of all, we have

$$\|A\|_2^2 = \sup_{x\in\mathbb{K}^n, x\neq 0} \frac{\|Ax\|_2^2}{\|x\|_2^2} = \sup_{x\in\mathbb{K}^n, x\neq 0} \frac{\langle A^*Ax, x\rangle}{\langle x, x\rangle}.$$

By Lemma 2.7.1, A^*A is self-adjoint and positive. Hence it is diagonalizable and its eigenvalues $(\lambda_i(A^*A))_{1 \leq i \leq n}$ are nonnegative real numbers. In the orthonormal basis of its eigenvectors, we easily check that

$$\sup_{x \in \mathbb{K}^n, x \neq 0} \frac{\langle A^*Ax, x \rangle}{\langle x, x \rangle} = \max_{1 \leq i \leq n} \lambda_i(A^*A).$$

Since the singular values of A are the positive square roots of the eigenvalues of A^*A, we infer the desired result. Moreover, the Cauchy–Schwarz inequality yields

$$\frac{\langle A^*Ax, x \rangle}{\langle x, x \rangle} \leq \frac{\|A^*Ax\|_2 \|x\|_2}{\langle x, x \rangle} \leq \frac{\|A^*A\|_2 \|x\|_2^2}{\langle x, x \rangle} \leq \|A^*\|_2 \|A\|_2.$$

We deduce that $\|A\|_2 \leq \|A^*\|_2$. Applying this inequality to A^*, we obtain the desired inequality, which is to say $\|A\|_2 = \|A^*\|_2$. $\qquad \square$

Remark 3.1.3. A real matrix may be seen either as a matrix of $\mathcal{M}_n(\mathbb{R})$ or as a matrix of $\mathcal{M}_n(\mathbb{C})$, since $\mathbb{R} \subset \mathbb{C}$. If $\| \cdot \|_{\mathbb{C}}$ is a vector norm in $\mathbb{C}^n$, we can define its restriction $\| \cdot \|_{\mathbb{R}}$ to $\mathbb{R}^n$, which is also a vector norm in $\mathbb{R}^n$. For a real matrix $A \in \mathcal{M}_n(\mathbb{R})$, we can thus define two subordinate matrix norms $\|A\|_{\mathbb{C}}$ and $\|A\|_{\mathbb{R}}$ by

$$\|A\|_{\mathbb{C}} = \sup_{x \in \mathbb{C}^n, x \neq 0} \frac{\|Ax\|_{\mathbb{C}}}{\|x\|_{\mathbb{C}}} \quad \text{and} \quad \|A\|_{\mathbb{R}} = \sup_{x \in \mathbb{R}^n, x \neq 0} \frac{\|Ax\|_{\mathbb{R}}}{\|x\|_{\mathbb{R}}}.$$

At first glance, these two definitions seem to be distinct. Thanks to the explicit formulas of Proposition 3.1.2, we know that they coincide for the norms $\|x\|_1$, $\|x\|_2$, or $\|x\|_\infty$. However, for other vector norms we may have $\|A\|_{\mathbb{C}} > \|A\|_{\mathbb{R}}$. Since some fundamental results, like the Schur factorization theorem (Theorem 2.4.1), hold only for complex matrices, we shall assume henceforth that all subordinate matrix norms are valued in $\mathbb{C}^n$, even for real matrices (this is essential, in particular, for Proposition 3.1.4).

Remark 3.1.4. The spectral radius $\varrho(A)$ is not a norm on $\mathcal{M}_n(\mathbb{C})$. Indeed, we may have $\varrho(A) = 0$ with $A \neq 0$, for instance,

$$A = \begin{pmatrix} 0 & 1 \\ 0 & 0 \end{pmatrix}.$$

Nonetheless, the lemma below shows that $\varrho(A)$ is a norm on the set of normal matrices.

Lemma 3.1.1. *Let U be a unitary matrix $(U^* = U^{-1})$. We have*

$$\|UA\|_2 = \|AU\|_2 = \|A\|_2.$$

Consequently, if A is a normal matrix, then $\|A\|_2 = \varrho(A)$.

Proof. Since $U^*U = I$, we have

$$\|UA\|_2^2 = \sup_{x \in \mathbb{C}^n, x \neq 0} \frac{\|UAx\|_2^2}{\|x\|_2^2} = \sup_{x \in \mathbb{C}^n, x \neq 0} \frac{\langle U^*UAx, Ax \rangle}{\langle x, x \rangle} = \|A\|_2^2.$$

Moreover, the change of variable $y = Ux$ satisfies $\|x\|_2 = \|y\|_2$, hence

$$\|AU\|_2^2 = \sup_{x \in \mathbb{C}^n, x \neq 0} \frac{\|AUx\|_2^2}{\|x\|_2^2} = \sup_{y \in \mathbb{C}^n, y \neq 0} \frac{\|Ay\|_2^2}{\|U^{-1}y\|_2^2} = \sup_{y \in \mathbb{C}^n, y \neq 0} \frac{\|Ay\|_2^2}{\|y\|_2^2} = \|A\|_2^2.$$

If A is normal, it is diagonalizable in an orthonormal basis of eigenvectors $A = U \operatorname{diag}(\lambda_1, \ldots, \lambda_n)U^*$, and we have $\|A\|_2 = \|\operatorname{diag}(\lambda_i)\|_2 = \varrho(A)$. $\square$

Lemma 3.1.1 shows that $\varrho(A)$ and $\|A\|_2$ are equal for normal matrices. Actually, any matrix norm $\|A\|$ is larger than $\varrho(A)$, which is, in turn, always close to some subordinate matrix norm, as shown by the next proposition.

Proposition 3.1.4. *Let $\|\cdot\|$ be a matrix norm defined on $\mathcal{M}_n(\mathbb{C})$. It satisfies*

$$\varrho(A) \leq \|A\|.$$

Conversely, for any matrix A and for any real number $\varepsilon > 0$, there exists a subordinate norm $\|\cdot\|$ (which depends on A and ε) such that

$$\|A\| \leq \varrho(A) + \varepsilon.$$

Proof. Let $\lambda \in \mathbb{C}$ be an eigenvalue of A such that $\varrho(A) = |\lambda|$, and $x \neq 0$ a corresponding eigenvector. If the norm $\|.\|$ is subordinate to a vector norm, we write $\|\lambda x\| = \varrho(A)\|x\| = \|Ax\| \leq \|A\| \|x\|$ and therefore $\varrho(A) \leq \|A\|$. If $\|.\|$ is some matrix norm (not necessarily subordinate), we denote by $y \in \mathbb{C}^n$ a nonzero vector, so the matrix xy^* is nonzero, and we have $\lambda xy^* = Axy^*$. Then, taking the norm of this last equality yields $|\lambda| \|xy^*\| = \|Axy^*\| \leq \|A\| \|xy^*\|$, which implies $\varrho(A) \leq \|A\|$.

To prove the second inequality, we use the Schur factorization theorem (Theorem 2.4.1) which states that for any A, there exists a unitary matrix U such that $T = U^{-1}AU$ is triangular:

$$T = \begin{pmatrix} t_{1,1} & t_{1,2} & \cdots & t_{1,n} \\ 0 & \ddots & & \vdots \\ \vdots & \ddots & \ddots & \vdots \\ 0 & \cdots & 0 & t_{n,n} \end{pmatrix},$$

and the diagonal entries $t_{i,i}$ are the eigenvalues of A. For any $\delta > 0$ we introduce the diagonal matrix $D_\delta = \operatorname{diag}(1, \delta, \delta^2, \ldots, \delta^{n-1})$, and we define a matrix $T_\delta = (UD_\delta)^{-1}A(UD_\delta) = D_\delta^{-1}TD_\delta$ that satisfies

$$T_\delta = \begin{pmatrix} t_{1,1} & \delta t_{1,2} & \cdots & \delta^{n-1} t_{1,n} \\ 0 & \ddots & & \vdots \\ \vdots & \ddots & \ddots & \delta t_{n-1,n} \\ 0 & \cdots & 0 & t_{n,n} \end{pmatrix}.$$

Given $\varepsilon > 0$, we can choose δ sufficiently small that the off-diagonal entries of T_δ are also very small. Namely, they satisfy for all $1 \le i \le n-1$,

$$\sum_{j=i+1}^{n} \delta^{j-i} |t_{i,j}| \le \varepsilon.$$

Since the $t_{i,i}$ are the eigenvalues of T_δ, which is similar to A, we infer that $\|T_\delta\|_\infty \le \varrho(A) + \varepsilon$. Then, the mapping $B \to \|B\| = \|(UD_\delta)^{-1} B(UD_\delta)\|_\infty$ is a subordinate norm (that depends on A and ε) satisfying

$$\|A\| \le \varrho(A) + \varepsilon,$$

thereby yelding the result. $\square$

Remark 3.1.5. The second part of Proposition 3.1.4 may be false if the norm is not a matrix norm. For instance, the norm defined by (3.2) is not a matrix norm, and we have the counterexample

$$\varrho(A) = 2 > \|A\| = 1 \text{ for } A = \begin{pmatrix} 1 & -1 \\ -1 & 1 \end{pmatrix}.$$

In Exercise 3.8, we dwell on the link between the matrix norm and its spectral radius by showing that

$$\varrho(A) = \lim_{k \to \infty} \left(\|A^k\| \right)^{1/k}.$$

Remark 3.1.6. Propositions 3.1.2 and 3.1.4 provide an immediate upper bound for the spectral radius of a matrix:

$$\varrho(A) \le \min\left(\max_{1 \le j \le n} \sum_{i=1}^{n} |a_{i,j}| \, , \, \max_{1 \le i \le n} \sum_{j=1}^{n} |a_{i,j}| \right).$$

3.2 Subordinate Norms for Rectangular Matrices

Similar norms can be defined on the space $\mathcal{M}_{m,n}(\mathbb{K})$ of rectangular (or non-square) matrices of size $m \times n$ with entries in $\mathbb{K}$. For instance,

- the Frobenius (or Schur, or Euclidean) norm

$$\|A\|_F = \left(\sum_{i=1}^{m} \sum_{j=1}^{n} |a_{i,j}|^2 \right)^{1/2} ;$$

- the l^q-norm, $\|A\| = \left(\sum_{i=1}^{m} \sum_{j=1}^{n} |a_{i,j}|^q \right)^{1/q}$ for $q \geq 1$;
- the l^∞-norm, $\|A\| = \max_{1 \leq i \leq m, 1 \leq j \leq n} |a_{i,j}|$.

We may, of course, define a subordinate matrix norm in $\mathcal{M}_{m,n}(\mathbb{K})$ by

$$\|A\| = \sup_{x \in \mathbb{K}^n, x \neq 0} \frac{\|Ax\|_m}{\|x\|_n},$$

where $\| \cdot \|_n$ (respectively $\| \cdot \|_m$) is a vector norm on $\mathbb{K}^n$ (respectively $\mathbb{K}^n$).

We conclude this section by defining the best approximation of a given (not necessarily square) matrix by matrices of fixed rank. Recall that this property is important for the example of image compression described in Section 1.5.

Proposition 3.2.1. Let $A = V\tilde{\Sigma}U^*$ be the SVD factorization of some matrix $A \in \mathcal{M}_{m,n}(\mathbb{C})$ having r nonzero singular values arranged in decreasing order. For $1 \leq k < r$, the matrix $A_k = \sum_{i=1}^{k} \mu_i v_i u_i^*$ is the best approximation of A by matrices of rank k, in the following sense: for all matrices $X \in \mathcal{M}_{m,n}(\mathbb{C})$ of rank k, we have

$$\|A - A_k\|_2 \leq \|A - X\|_2. \tag{3.8}$$

Moreover, the error made in substituting A with A_k is $\|A - A_k\|_2 = \mu_{k+1}$.

Proof. According to (2.10), we have

$$A - A_k = \sum_{i=k+1}^{r} \mu_i v_i u_i^* = [v_{k+1}| \ldots |v_r] \operatorname{diag}(\mu_{k+1}, \ldots, \mu_r) \begin{bmatrix} u_{k+1}^* \\ \vdots \\ \hline u_r^* \end{bmatrix}.$$

Denoting by $D \in \mathcal{M}_{m,n}(\mathbb{R})$ the matrix $\operatorname{diag}(0, \ldots, 0, \mu_{k+1}, \ldots, \mu_r, 0, \ldots, 0)$, we have $A - A_k = VDU^*$, and since the Euclidean norm is invariant under unitary transformation, we have $\|A - A_k\|_2 = \|D\|_2 = \mu_{k+1}$. Let us now prove the approximation property. For all $x \in \mathbb{C}^n$, we have

$$\|Ax\|_2 = \|V\tilde{\Sigma}U^*x\|_2 = \|\tilde{\Sigma}U^*x\|_2. \tag{3.9}$$

Let E be the subspace of $\mathbb{C}^n$, of dimension $k + 1$, generated by the vectors $u_1, \ldots, u_{k+1}$. If $x \in E$, we have $x = \sum_{i=1}^{k+1} x_i u_i$ and

$$U^*x = U^* \sum_{i=1}^{k+1} x_i u_i = \sum_{i=1}^{k+1} x_i U^* u_i = \sum_{i=1}^{k+1} x_i e_i,$$

where e_i is the ith vector of the canonical basis of $\mathbb{C}^n$. Thus we have

$$\tilde{\Sigma} U^* x = (\mu_1 x_1, \ldots, \mu_{k+1} x_{k+1}, 0, \ldots, 0)^t.$$

So by (3.9) and the decreasing order of the singular values μ_i,

$$\|Ax\|_2 \geq \mu_{k+1} \|x\|_2, \qquad \forall x \in E. \tag{3.10}$$

If the matrix $X \in \mathcal{M}_{m,n}(\mathbb{C})$ is of rank $k < r$, its kernel is of dimension $n - k \geq r - k \geq 1$, and for all $x \in \mathrm{Ker}\,(X)$, we have

$$\|Ax\|_2 = \|(A - X)x\|_2 \leq \|A - X\|_2 \, \|x\|_2.$$

Assume that X contradicts (3.8):

$$\|A - X\|_2 < \|A - A_k\|_2.$$

Hence for all $x \in \mathrm{Ker}\,(X)$,

$$\|Ax\|_2 < \|A - A_k\|_2 \, \|x\|_2 = \mu_{k+1} \|x\|_2,$$

and therefore if $x \in E \cap \mathrm{Ker}\,(X)$ with $x \neq 0$, we end up with a contradiction to (3.10). Indeed, the two spaces have a nonempty intersection since $\dim E + \dim \mathrm{Ker}\,(X) > n$, so that (3.8) is finally satisfied. $\qquad \square$

3.3 Matrix Sequences and Series

In the sequel, we consider only square matrices.

Definition 3.3.1. *A sequence of matrices $(A_i)_{i \geq 1}$ converges to a limit A if for a matrix norm $\| \cdot \|$, we have*

$$\lim_{i \to +\infty} \|A_i - A\| = 0,$$

and we write $A = \lim_{i \to +\infty} A_i$.

The definition of convergence does not depend on the chosen norm, since $\mathcal{M}_n(\mathbb{C})$ is a vector space of finite dimension. Therefore Theorem 3.1.1, which asserts that all norms are equivalent, is applicable, and thus if a sequence converges for one norm, it converges for all norms.

Remark 3.3.1. Let us recall that $\mathcal{M}_n(\mathbb{C})$, having finite dimension, is a complete space, that is, every Cauchy sequence of elements of $\mathcal{M}_n(\mathbb{C})$ is a convergent sequence in $\mathcal{M}_n(\mathbb{C})$:

$$\lim_{i \to +\infty} \lim_{j \to +\infty} \|A_i - A_j\| = 0 \Rightarrow \exists A \in \mathcal{M}_n(\mathbb{C}) \text{ such that } \lim_{i \to +\infty} \|A_i - A\| = 0.$$

A matrix series is a sequence $(S_i)_{i \geq 0}$ defined by partial sums of another sequence of matrices $(A_i)_{i \geq 0}$: $S_i = \sum_{j=0}^{i} A_j$. A series is said to be convergent if the sequence of partial sums is convergent. Among all series, we shall be more particularly concerned with matrix power series defined by $(a_i A^i)_{i \geq 0}$, where each a_i is a scalar in $\mathbb{C}$ and A^i is the ith power of the matrix A. A necessary but not sufficient condition for the convergence of a power series is that the sequence $a_i A^i$ converge to 0. The following result provides a necessary and sufficient condition for the sequence of iterated powers of a matrix to converge to 0.

Lemma 3.3.1. *Let A be a matrix in $\mathcal{M}_n(\mathbb{C})$. The following four conditions are equivalent:*

1. $\lim_{i \to +\infty} A^i = 0$;
2. $\lim_{i \to +\infty} A^i x = 0$ *for all vectors $x \in \mathbb{C}^n$;*
3. $\varrho(A) < 1$;
4. *there exists at least one subordinate matrix norm such that $\|A\| < 1$.*

Proof. Let us first show that $(1) \Rightarrow (2)$. The inequality

$$\|A^i x\| \leq \|A^i\| \|x\|$$

implies $\lim_{i \to +\infty} A^i x = 0$. Next, $(2) \Rightarrow (3)$; otherwise, there would exist λ and $x \neq 0$ satisfying $Ax = \lambda x$ and $|\lambda| = \varrho(A)$, which would entail that the sequence $A^i x = \lambda^i x$ cannot converge to 0. Since $(3) \Rightarrow (4)$ is an immediate consequence of Proposition 3.1.4, it remains only to show that $(4) \Rightarrow (1)$. To this end, we consider the subordinate matrix norm such that $\|A\| < 1$, and accordingly,

$$\|A^i\| \leq \|A\|^i \to 0 \text{ when } i \to +\infty,$$

which proves that A^i tends to 0. $\qquad\square$

We now study some properties of matrix power series.

Theorem 3.3.1. *Consider a power series on $\mathbb{C}$ of positive radius of convergence R:*

$$\left| \sum_{i=0}^{+\infty} a_i z^i \right| < +\infty, \forall z \in \mathbb{C} \text{ such that } |z| < R.$$

For any matrix $A \in \mathcal{M}_n(\mathbb{C})$ such that $\varrho(A) < R$, the series $(a_i A^i)_{i \geq 0}$ is convergent, i.e., $\sum_{i=0}^{+\infty} a_i A^i$ is well defined in $\mathcal{M}_n(\mathbb{C})$.

Proof. Since $\varrho(A) < R$, thanks to Lemma 3.3.1, there exists a subordinate matrix norm for which we also have $\|A\| < R$. We check the Cauchy criterion for the sequence of partial sums:

$$\left\| \sum_{k=j+1}^{i} a_k A^k \right\| \leq \sum_{k=j+1}^{i} |a_k| \|A\|^k. \tag{3.11}$$

Now, a power series on $\mathbb{C}$ is absolutely convergent inside its disk of convergence. Hence $\|A\| < R$ implies that the right-hand term in (3.11) tends to 0 as j and i tend to $+\infty$. The convergence of the series is therefore established. $\square$

The previous notion of matrix power series generalizes, of course, the definition of matrix polynomial and may be extended to analytic functions.

Definition 3.3.2. *Let $f(z) : \mathbb{C} \mapsto \mathbb{C}$ be an analytic function defined in the disk of radius $R > 0$ written as a power series:*

$$f(z) = \sum_{i=0}^{+\infty} a_i z^i \quad \forall z \in \mathbb{C} \text{ such that } |z| < R.$$

By a slight abuse of notation, for any $A \in M_n(\mathbb{C})$ with $\varrho(A) < R$, we define the matrix $f(A)$ by

$$f(A) = \sum_{i=0}^{+\infty} a_i A^i.$$

Let us give some useful examples of matrix functions. The exponential function is analytic in $\mathbb{C}$. Accordingly, for all matrices A (without restriction on their spectral radii), we can define its exponential by the formula

$$e^A = \sum_{i=0}^{+\infty} \frac{A^i}{i!}.$$

Similarly, the function $1/(1 - z)$ is analytic in the unit disk and thus equal to $\sum_{i=0}^{+\infty} z^i$ for $z \in \mathbb{C}$ such that $|z| < 1$. We therefore deduce an expression for $(I - A)^{-1}$.

Proposition 3.3.1. *Let A be a matrix with spectral radius $\varrho(A) < 1$. The matrix $(I - A)$ is nonsingular and its inverse is given by*

$$(I - A)^{-1} = \sum_{i=0}^{+\infty} A^i.$$

Proof. We already know that the series $(A^i)_{i \geq 0}$ is convergent. We compute

$$(I - A) \sum_{i=0}^{p} A^i = \sum_{i=0}^{p} A^i (I - A) = I - A^{p+1}.$$

Since A^{p+1} converges to 0 as p tends to infinity, we deduce that the sum of the series $(A^i)_{i \geq 0}$ is equal to $(I - A)^{-1}$. $\square$

Remark 3.3.2. Proposition 3.3.1 shows in particular that the set of nonsingular matrices is an open set in $\mathcal{M}_n(\mathbb{C})$. Indeed, consider a subordinate norm $\|\cdot\|$. Given a nonsingular matrix M, any matrix N such that $\|M - N\| < \|M^{-1}\|^{-1}$ is nonsingular as the product of two nonsingular matrices:

$$N = M\left(I - M^{-1}(M - N)\right),$$

since $\|M^{-1}(M - N)\| < 1$. Hence for any nonsingular matrix M, there exists a neighborhood of M that also consists of nonsingular matrices. This proves that the set of nonsingular matrices is open.

3.4 Exercises

3.1. Let A be the matrix defined by `A=rand(n,n)`, where n is a fixed integer. Define $M = \max_{i,j} |A_{i,j}|$ (M may be computed by the `Matlab` function `max`). Compare M with the norms $\|A\|_1$, $\|A\|_2$, $\|A\|_\infty$, and $\|A\|_F$ (use the function `norm`). Justify.

3.2. Let T be a nonsingular triangular matrix defined by one of the instructions `T=LNonsingularMat(n)` and `T=UNonsingularMat(n)` (these functions have been defined in Exercise 2.2), where n is a fixed integer. Define $m = (\min_i |T_{i,i}|)^{-1}$ (m may be computed by the `Matlab` function `min`). Compare m with the norms $\|T^{-1}\|_1$, $\|T^{-1}\|_2$, $\|T^{-1}\|_\infty$, and $\|T^{-1}\|_F$.

3.3 (∗). Define a diagonal matrix A by
$$\text{u=rand(n,1); A=diag(u);.}$$
Compute the norm $\|A\|_p$ for $p = 1, 2, \infty$. Comment on the observed results.

3.4. For various values of n, define two vectors `u=rand(n,1)` and `v=rand(n,1)`. Compare the matrix norm $\|uv^t\|_2$ with the vector norms $\|u\|_2$ and $\|v\|_2$. Justify your observation. Same questions for the Frobenius norm as well as the norms $\|.\|_1$ and $\|.\|_\infty$.

3.5. For different values of n, define a nonsingular square matrix A by `A=NonsingularMat(n);` (see Exercise 2.2). Let P and v be the matrix and the vector defined by
$$\text{[P,D]=eig(A*A'); [d k]=min(abs(diag(D))); v=P(:,k);}$$
Compare $\|A^{-1}v\|_2$ and $\|A^{-1}\|_2$. Justify.

3.6. Let A be a real matrix of size $m \times n$.

1. What condition should A meet for the function $x \mapsto \|Ax\|$, where $\|.\|$ denotes some norm on $\mathbb{R}^n$, to be a norm on $\mathbb{R}^m$? Write a function `NormA` that computes $\|Ax\|_2$ for $x \in \mathbb{R}^n$.
2. Assume A is a square matrix. Write a function `NormAs` that computes $\sqrt{\langle Ax, x \rangle}$ for $x \in \mathbb{R}^n$. Does this function define a norm on $\mathbb{R}^n$?

3.7 (∗). Define a matrix A by `A=PdSMat(n)` (see Exercise 2.20) and denote by $\|.\|_A$ the norm defined on $\mathbb{R}^n$ by $\|x\|_A = \sqrt{\langle Ax, x \rangle}$ (see the previous exercise). Let S_A be the unit sphere of $\mathbb{R}^n$ for this norm:

$$S_A = \{x \in \mathbb{R}^n, \quad \|x\|_A = 1\}.$$

1. Prove (do not use `Matlab`) that S_A lies between two spheres (for the Euclidean norm) centered at the origin, and of respective radii $\frac{1}{\sqrt{\lambda_{\min}}}$ and $\frac{1}{\sqrt{\lambda_{\max}}}$, that is,

$$x \in S_A \implies \frac{1}{\sqrt{\lambda_{\max}}} \leq \|x\|_2 \leq \frac{1}{\sqrt{\lambda_{\min}}},$$

   where $\lambda_{\min}$ (respectively $\lambda_{\max}$) denotes the smallest (respectively largest) eigenvalue of A.
2. Plotting S_A for $n = 2$.
   (a) Let Γ_p be the line $x_2 = px_1$, $p \in \mathbb{R}$. Compute $\langle Ax, x \rangle$ for $x \in \Gamma_p$ (do not use `Matlab`). Compute the intersection of Γ_p and S_A.
   (b) Write a function `function [x,y]=UnitCircle(A,n)` whose input arguments are a 2×2 matrix A, and an integer n and that returns two vectors x and y, of size n, containing respectively the abscissas and the ordinates of n points of the curve S_A.
   (c) Plot on the same graph the unit circle for the Euclidean norm and S_A for `A= [7 5; 5 7]`. Prove rigorously that the curve obtained is an ellipse.
   (d) Plot on the same graph the unit circles for the norms defined by the matrices

$$A = \begin{pmatrix} 7 & 5 \\ 5 & 7 \end{pmatrix}, \quad B = \begin{pmatrix} 6.5 & 5.5 \\ 5.5 & 6.5 \end{pmatrix}, \quad C = \begin{pmatrix} 2 & 1 \\ 1 & 5 \end{pmatrix}. \tag{3.12}$$

   Comment on the results.

3.8 (∗). Define a matrix A by `A=rand(n,n)`. Compare the spectral radius of A and $\|A^k\|_2^{1/k}$ for $k = 10, 20, 30, \ldots, 100$. What do you notice? Does the result depend on the the chosen matrix norm? (Try the norms $\|.\|_1, \|.\|_\infty$, and $\|.\|_F$.)
Explanation.

1. Prove that $\varrho(A) \leq \|A^k\|^{1/k}$, $\forall k \in \mathbb{N}^*$.
2. For $\varepsilon > 0$, define $A_\varepsilon = \frac{1}{\varrho(A)+\varepsilon} A$. Prove that $\varrho(A_\varepsilon) < 1$. Deduce that there exists $k_0 \in \mathbb{N}$ such that $k \geq k_0$ implies $\varrho(A) \leq \|A^k\|^{1/k} \leq \varrho(A) + \varepsilon$.
3. Conclude.

3.9. Define a matrix `A=MatRank(m,n,r)` and let `[V S U] = svd(A)` be the SVD factorization of A (for example, take $m = 10$, $n = 7$, $r = 5$). For $k = 1, \ldots, r-1$, we compute the approximated SVD factorization of A by the instruction [v,s,u] = svds(A,k). We have seen in Proposition 3.2.1 that the best approximation (in $\|.\|_2$-norm) of matrix A by $m \times n$ matrices of rank k is the matrix $A_k = usv^t$, and that the approximation error is $\|A - A_k\|_2 = S_{k+1,k+1}$. We set out to compute the same error, but in the Frobenius norm.

1. For $k = r - 1 \searrow 1$, display $\|A - A_k\|_F^2$ and the square of the singular values of A. What relation do you observe between these two quantities?
2. Justify this relation rigorously.

3.10. We revisit the spring system in Section 1.3, assuming in addition that each mass is subjected to a damping, proportional to the velocity, with a given coefficient $c_i > 0$; the zero right-hand sides of (1.8) have to be replaced by $-c_i \dot{y}_i$ (damping or breaking term).

1. Show that the vector $y = (y_1, y_2, y_3)^t$ is a solution of

$$M\ddot{y} + C\dot{y} + Ky = 0, \qquad (3.13)$$

where C is a matrix to be specified.
2. Define a vector $z(t) = (y^t, \dot{y}^t)^t \in \mathbb{R}^6$. Prove that z is a solution of

$$\dot{z}(t) = Az(t), \qquad (3.14)$$

for some matrix A to be specified. For a given initial datum $z(0)$, i.e., for prescribed initial positions and speeds of the three masses, (3.14) admits a unique solution, which is precisely $z(t) = e^{At}z(0)$. (For the definition of the matrix exponential, see Section 3.3.)
3. Assume that the stiffness constants are equal ($k_1 = k_2 = k_3 = 1$), as well as the damping coefficients ($c_1 = c_2 = c_3 = 1/2$), and that the masses are $m_1 = m_3 = 1$ and $m_2 = 2$. The equilibrium positions of the masses are supposed to be $x_1 = -1$, $x_2 = 0$, and $x_3 = 1$. At the initial time, the masses m_1 and m_3 are moved away from their equilibrium position by $y_1(0) = -0.1$, $y_3(0) = 0.1$, with initial speeds $\dot{y}_1(0) = -1$, $\dot{y}_3(0) = 1$, while the other mass m_2 is at rest, $y_2(0) = 0$, $\dot{y}_2(0) = 0$. Plot on the same graph the time evolutions of the positions $x_i(t) = x_i + y_i(t)$ of the three masses. Vary the time t from 0 to 30, by a step of 1/10. Plot the speeds on another graph. Comment.
Hint: use the Matlab function expm to compute e^{At}.

4

Introduction to Algorithmics

This chapter is somewhat unusual in comparison to the other chapters of this course. Indeed, it contains no theorems, but rather notions that are at the crossroads of mathematics and computer science. However, the reader should note that this chapter is essential from the standpoint of applications, and for the understanding of the methods introduced in this course. For more details on the fundamental notions of algorithmics, the reader can consult [1].

4.1 Algorithms and pseudolanguage

In order to fully grasp the notion of a mathematical algorithm, we shall illustrate our purpose by the very simple, yet instructive, example of the multiplication of two matrices. We recall that the operation of matrix multiplication is defined by

$$\mathcal{M}_{n,p}(\mathbb{K}) \times \mathcal{M}_{p,q}(\mathbb{K}) \longrightarrow \mathcal{M}_{n,q}(\mathbb{K})$$

$$(A, B) \longmapsto C = AB,$$

where the matrix C is defined by its entries, which are given by the simple formula

$$c_{i,j} = \sum_{k=1}^{p} a_{i,k} b_{k,j}, \quad 1 \leq i \leq n, 1 \leq j \leq q. \tag{4.1}$$

Formula (4.1) can be interpreted in various ways as vector operations. The most "natural" way is to see (4.1) as the scalar product of the ith row of A with the jth column of B. Introducing $(a_i)_{1 \leq i \leq n}$, the rows of A (with $a_i \in \mathbb{R}^p$), and $(b_j)_{1 \leq j \leq q}$, the columns of B (with $b_j \in \mathbb{R}^p$), we successively compute the entries of C as

$$c_{i,j} = a_i \cdot b_j .$$

However, there is a "dual" way of computing the product C in which the prominent role of the rows of A and columns of B is inverted by focusing rather on the columns of A and the rows of B.

Let $(a^k)_{1 \leq k \leq p}$ be the columns of A (with $a^k \in \mathbb{R}^n$), and let $(b^k)_{1 \leq k \leq p}$ be the rows of B (with $b^k \in \mathbb{R}^q$). We note that C is also defined by the formula

$$C = \sum_{k=1}^{p} a^k \left(b^k\right)^t, \tag{4.2}$$

where we recall that the tensor product of a column vector by a row vector is defined by

$$xy^t = (x_i y_j)_{1 \leq i \leq n, \, 1 \leq j \leq q} \in \mathcal{M}_{n,q}(\mathbb{K}),$$

where the $(x_i)_{1 \leq i \leq n}$ are the entries of $x \in \mathbb{R}^n$, and $(y_j)_{1 \leq j \leq q}$ the entries of $y \in \mathbb{R}^q$. Formula (4.2) is no longer based the scalar product, but rather on the tensor product of vectors, which numerically amounts to multiplying each column of A by scalars that are the entries of each row of B.

Of course, in both computations of the product matrix C, the same multiplications of scalars are performed; what differentiates them is the order of the operations. In theory, this is irrelevant, however in computing, these two procedures are quite different! Depending on the way the matrices A, B, C are stored in the computer memory, access to their rows and columns can be more or less fast (this depends on a certain number of factors such as the memory, the cache, and the processor, none of which we shall consider here). In full generality, there are several ways of performing the same mathematical operation.

Definition 4.1.1. *We call the precise ordered sequence of elementary operations for carrying out a given mathematical operation an algorithm.*

This definition calls immediately for a number of comments.

✓ One has to distinguish between the mathematical operation, which is the goal or the task to be done, and the algorithm, which is the means to that end. In particular, there can be several different algorithms that perform the same operation.

✓ Two algorithms may carry out the same elementary operations and differ only in the order of the sequence (that is the case of the two algorithms above for matrix multiplication). All the same, two algorithms may also differ by the very nature of their elementary operations, while producing the same result (see below the Strassen algorithm).

✓ The notion of elementary operation is necessarily fuzzy. We can agree (as here and in the whole course) that it is an algebraic operation on a scalar. But after all, a computer knowns only binary numbers, and a product of real numbers or the extraction of a square root already requires algorithms! Nevertheless, we shall never go this far into details. By the same token, if a product of block matrices has to be computed, we can consider the multiplication of blocks as an elementary operation and not the scalar multiplication.

The script of an algorithm is an essential step, not only for writing a computer program out of a mathematical method, but also for the assessment of its performance and its efficiency, that is, for counting the number of elementary operations that are necessary to its realization (see Section 4.2). Of course, as soon as a rigorous measure of the efficiency of an algorithm is available, a key issue is to find the best possible algorithm for a given mathematical operation. This is a difficult problem, which we shall barely illustrate by the case of the Strassen algorithm for matrix multiplication (see Section 4.3).

Although the above definition stipulates that an algorithm is characterized by an "ordered" sequence of elementary operations, we have been up to now relatively vague in the description of the two algorithms proposed for matrix multiplication. More precisely, we need a notion of language, not so much for writing programs, but for arranging operations. We call it pseudolanguage. It allows one to accurately write the algorithm without going through purely computational details such as the syntax rules (which vary with languages), the declaration of arrays, and the passing of arguments. It is easy to transcribe (except for these details) into a computing language dedicated to numerical calculations (for instance, Fortran, Pascal, C, C++).

The reader will soon find out that the pseudolanguage is a description tool for algorithms that for the sake of convenience obeys no rigorous syntax. Even so, let us insist on the fact that although this pseudolanguage is not accurately defined, it complies with some basic rules:

1. The symbol $=$ is no longer the mathematical symbol of equality but the computer science symbol of allocation. When we write $a = b$ we allocate the value of b to the variable a by deleting the previous value of a. When we write $a = a + b$ we add to the value of a that of b, but by no means shall we infer that b is zero.

2. The elementary operations are performed on scalars. When vectors or matrices are handled, loops of elementary operations on their entries should be written. Note in passing that `Matlab` is able to perform elementary operations directly on matrices (this is actually much better, in terms of computing time, than writing loops on the matrix entries).

3. At the beginning of the algorithm, the data (entries) and the results (outputs) should be specified. In the course of an algorithm, intermediate computational variables may be used.

4. Redundant or useless operations should be avoided for the algorithm to be executed in a minimum number of operations.

As an example, we consider the two matrix multiplication algorithms that we have just presented.

Let us remark that these two algorithms differ only in the order of their loops (we have intentionally kept the same names for the indices in both algorithms). If this makes no difference from a mathematical viewpoint, it is not quite the same from the computer science viewpoint: the entries of A and

```
Data: A and B. Output: C = AB.
        For i = 1 ↗ n
            For j = 1 ↗ q
                C_{i,j} = 0
                For k = 1 ↗ p
                    C_{i,j} = C_{i,j} + A_{i,k}B_{k,j}
                End k
            End j
        End en i
```

Algorithm 4.1: Product of two matrices: "scalar product" algorithm.

```
Data: A and B. Output: C = AB.
        For i = 1 ↗ n
            For j = 1 ↗ q
                C_{i,j} = 0
            End j
        End i
        For k = 1 ↗ p
            For i = 1 ↗ n
                For j = 1 ↗ q
                    C_{i,j} = C_{i,j} + A_{i,k}B_{k,j}
                End j
            End i
        End k
```

Algorithm 4.2: Product of two matrices: "tensor product" algorithm.

B are accessed either along their rows or columns, which is not executed with the same speed depending on the way they are stored in the memory of the computer. In both algorithms, the order of the loops in i and j can be changed. Actually, we obtain as many algorithms as there are possible arrangements of the three loops in i, j, k (check it as an exercise).

Obviously, the matrix product is a too simple operation to convince the reader of the usefulness of writing an algorithm in pseudolanguage. Nevertheless, one should be reassured: more difficult operations will soon arrive! Let us emphasize again the essential contributions of the script in pseudolanguage: on the one hand, it provides a good understanding of the sequencing of the algorithm, and on the other hand, it enables one to accurately count the number of operations necessary to its execution.

4.2 Operation Count and Complexity

The performance or the cost of an algorithm is mainly appraised by the number of operations that are required to execute it. This cost also depends on other factors such as the necessary number of registers of memory and the number of memory accesses to look for new data, but we neglect them in the sequel. An algorithm will be the more efficient, the fewer operations it requiers.

Definition 4.2.1. *We call the number of multiplications and divisions required to execute an algorithm its complexity.*

We neglect all other operations such as additions (much quicker than multiplications on a computer) or square roots (much scarcer than multiplications in general), which makes simpler the counting of operations. If the algorithm is carried out on a problem of size n (for instance, the order of the matrix or the number of entries of a vector), we denote by $N_{op}(n)$ its complexity, or its number of operations. The exact computation of $N_{op}(n)$ is often complex or delicate (because of boundary effects in the writing of loops). We thus content ourselves in finding an equivalent of $N_{op}(n)$ when the dimension n of the problem is very large (we talk then about asymptotic complexity). In other words, we only look for the first term of the Taylor expansion of $N_{op}(n)$ as n tends to infinity.

Thanks to the transcription of the algorithm into pseudolanguage, it is easy to count its operations. At this stage we understand why a pseudolanguage script should avoid redundant or useless computations; otherwise, we may obtain a bad operation count that overestimates the actual number of operations. In both examples above (matrix product algorithms), the determination of $N_{op}(n)$ is easy: for each i, j, k, we execute a multiplication. Consequently, the number of operations is npq. If all matrices are square, $n = p = q$, we get the classical result $N_{op}(n) \approx n^3$.

4.3 The Strassen Algorithm

It was believed for a long time that the multiplication of matrices of order n could not be carried out in fewer than n^3 operations. So the discovery of a faster algorithm by Strassen in 1969 came as a surprise. Strassen devised a very clever algorithm for matrix multiplication that requires many fewer operations, on the order of

$$N_{op}(n) = \mathcal{O}(n^{\log_2 7}) \text{ with } \log_2 7 \approx 2.81.$$

It may seem pointless to seek the optimal algorithm for the multiplication of matrices. However, beyond the time that can be saved for large matrices (the Strassen algorithm has indeed been used on supercomputers), we shall see in the next section that the asymptotic complexity of matrix multiplication is equivalent to that of other operations clearly less trivial, such as matrix inversion. The Strassen algorithm relies on the following result.

Lemma 4.3.1. *The product of two matrices of order 2 may be done with 7 multiplications and 18 additions (instead of 8 multiplications and 4 additions by the usual rule).*

Proof.   A simple computation shows that

$$\begin{pmatrix} a & b \\ c & d \end{pmatrix} \begin{pmatrix} \alpha & \beta \\ \gamma & \delta \end{pmatrix} = \begin{pmatrix} m_1 + m_2 - m_4 + m_6 & m_4 + m_5 \\ m_6 + m_7 & m_2 - m_3 + m_5 - m_7 \end{pmatrix},$$

with

$$\begin{array}{ll} m_1 = (b - d)(\gamma + \delta), & m_5 = a(\beta - \delta), \\ m_2 = (a + d)(\alpha + \delta), & m_6 = d(\gamma - \alpha), \\ m_3 = (a - c)(\alpha + \beta), & m_7 = (c + d)\alpha, \\ m_4 = (a + b)\delta. & \end{array}$$

We count indeed 7 multiplications and 18 additions.   □

Remark 4.3.1. We note that the multiplication rule of Strassen in the above lemma is also valid if the entries of the matrices are not scalars but instead belong to a noncommutative ring. In particular, the rule holds for matrices defined by blocks.

Consider then a matrix of size $n = 2^k$. We split this matrix into 4 blocks of size 2^{k-1}, and we apply Strassen's rule. If we count not only multiplications but additions too, the number of operations $N_{op}(n)$ to determine the product of two matrices satisfies

$$N_{op}(2^k) = 7N_{op}(2^{k-1}) + 18(2^{k-1})^2,$$

since the addition of two matrices of size n requires n^2 additions. A simple induction yields

$$N_{op}(2^k) = 7^k N_{op}(1) + 18 \sum_{i=0}^{k-1} 7^i 4^{k-1-i} \leq 7^k \left(N_{op}(1) + 6 \right).$$

We easily deduce that the optimal number of operations $N_{op}(n)$ satisfies for all n,

$$N_{op}(n) \leq C n^{log_2 7} \quad \text{with} \quad \log_2 7 \approx 2.81.$$

Since Strassen's original idea, other increasingly complex algorithms have been devised, whose number of operations increases more slowly for n large. However, the best algorithm (in terms of complexity) has not yet been found. As of today, the best algorithm such that $N_{op}(n) \leq Cn^\alpha$ has an exponent α close to 2.37 (it is due to Coppersmith and Winograd). It has even been proved that if there exists an algorithm such that $P(n) \leq Cn^\alpha$, then there exists another algorithm such that $P(n) \leq C'n^{\alpha'}$ with $\alpha' < \alpha$. Unfortunately, these "fast" algorithms are tricky to program, numerically less stable, and the constant C in $N_{op}(n)$ is so large that no gains can be expected before a value of, say, $n = 100$.

4.4 Equivalence of Operations

We have just seen that for an operation as simple as the multiplication of matrices, there exist algorithms whose asymptotic complexities are quite different (n^3 for the standard algorithm, $n^{\log_2 7}$ for Strassen's algorithm). On the other hand, in most cases the best algorithm possible, in terms of complexity, is not known for mathematical operations such as matrix multiplication. Therefore, we cannot talk about the complexity of an operation in the sense of the complexity of its best algorithm. At most, we shall usually determine an upper bound for the number of operations (possibly improvable). Hence, we introduce the following definition.

Definition 4.4.1. *Let $N_{op}(n)$ be the (possibly unknown) complexity of the best algorithm performing a matrix operation. We call the following bound*

$$N_{op}(n) \le Cn^\alpha \quad \forall n \ge 0,$$

where C and α are positive constants independent of n, the asymptotic complexity of this operation, and we denote it by $\mathcal{O}(n^\alpha)$.

An amazing result is that many of matrix operations are equivalent in the sense that they have the same asymptotic complexity. For instance, although, at first glance, matrix inversion seems to be much more complex, it is equivalent to matrix multiplication.

Theorem 4.4.1. *The following operations have the same asymptotic complexity in the sense that if there exists an algorithm executing one of them with a complexity $\mathcal{O}(n^\alpha)$ where $\alpha \ge 2$, then it automatically yields an algorithm for every other operation with the same complexity $\mathcal{O}(n^\alpha)$:*

(i) product of two matrices: $(A, B) \longmapsto AB$,
(ii) inversion of a matrix: $A \longmapsto A^{-1}$,
(iii) computation of the determinant: $A \longmapsto \det A$,
(iv) solving a linear system: $(A, b) \longmapsto x = A^{-1}b$.

Proof. The difficulty is that Theorem 4.4.1 should be proved without knowing the algorithms, or the exact exponent α. We prove only the equivalence between (i) and (ii) (the other equivalences are much harder to prove). Let $I(n)$ be the number of operations required to compute A^{-1} by a given algorithm. We assume that there exist C and α such that $I(n) \le Cn^\alpha$. Let us show that there exists an algorithm that computes the product AB whose number of operations $P(n)$ satisfies $P(n) \le C'n^\alpha$ with the same exponent α and $C' > 0$. First of all, we note that

$$\begin{pmatrix} I & A & 0 \\ 0 & I & B \\ 0 & 0 & I \end{pmatrix}^{-1} = \begin{pmatrix} I & -A & AB \\ 0 & I & -B \\ 0 & 0 & I \end{pmatrix}.$$

Consequently, the product AB is obtained by inverting a matrix that is 3 times larger. Hence

$$P(n) \leq I(3n) \leq C3^\alpha n^\alpha.$$

Now let $P(n)$ be the number of operations needed to compute AB by a given algorithm. We assume that there exist C and α such that $P(n) \leq Cn^\alpha$. Let us show that there exists an algorithm that computes A^{-1} whose number of operations $I(n)$ satisfies $I(n) \leq C'n^\alpha$ with the same exponent α and $C' > 0$. In this case, we notice that

$$\begin{pmatrix} A & B \\ C & D \end{pmatrix}^{-1} = \begin{pmatrix} A^{-1} + A^{-1}B\Delta^{-1}CA^{-1} & -A^{-1}B\Delta^{-1} \\ -\Delta^{-1}CA^{-1} & \Delta^{-1} \end{pmatrix} \qquad (4.3)$$

with $\Delta = D - CA^{-1}B$ an invertible matrix (sometimes the called Schur complement). Therefore, to evaluate the inverse matrix on the left-hand side of (4.3), we can successively compute

- the inverse of A;
- the matrix $X_1 = A^{-1}B$;
- the Schur complement $\Delta = D - CX_1$;
- the inverse of Δ;
- the matrices $X_2 = X_1\Delta^{-1}$, $X_3 = CA^{-1}$, $X_4 = \Delta^{-1}X_3$, and $X_5 = X_2X_3$.

The left-hand side of (4.3) is then equal to

$$\begin{pmatrix} A^{-1} + X_5 & X_2 \\ X_4 & \Delta^{-1} \end{pmatrix}.$$

Since this method requires 2 inversions and 6 multiplications, we deduce that

$$I(2n) \leq 2I(n) + 6P(n),$$

if we neglect additions (for simplicity). By iterating this formula for $n = 2^k$, we get

$$I(2^k) \leq 2^k I(1) + 6 \sum_{i=0}^{k-1} 2^{k-i-1} P(2^i) \leq C \left(2^k + \sum_{i=0}^{k-1} 2^{k-i-1+\alpha i} \right).$$

Since $\alpha \geq 2$, we infer

$$I(2^k) \leq C'2^{\alpha k}.$$

If $n \neq 2^k$ for all k, then there exists k such that $2^k < n < 2^{k+1}$. We inscribe the matrix A in a larger matrix of size 2^{k+1}:

$$\tilde{A} = \begin{pmatrix} A & 0 \\ 0 & I \end{pmatrix} \quad \text{with} \quad \tilde{A}^{-1} = \begin{pmatrix} A^{-1} & 0 \\ 0 & I \end{pmatrix},$$

where I is the identity of order $2^{k+1} - n$. Applying the previous result to $\tilde{A}$ yields

$$I(n) \leq C'(2^{k+1})^\alpha \leq C'2^\alpha n^\alpha,$$

which is the desired result. □

4.5 Exercises

Warning: for the following exercises do not use Matlab except where explicitly requested.

4.1 (∗). Let u and v be two vectors of $\mathbb{R}^n$, and let A and B be two square matrices of $\mathcal{M}_n(\mathbb{R})$.

1. Find the numbers of operations required to compute the scalar product $\langle u, v \rangle$, the Euclidean norm $\|u\|_2$, and the rank-one matrix uv^t.
2. Find the numbers of operations required to compute the matrix-vector product Au, and the matrix product AB.
3. For $n = 100k$, with $k = 1, \ldots, 5$, estimate the running time of Matlab (use the functions tic and toc) for computing the product of two matrices A=rand(n,n) and B=rand(n,n). Plot this running time in terms of n.
4. Assume that this running time is a polynomial function of n, so that for n large enough, $T(n) \approx Cn^s$. In order to find a numerical approximation of the exponent s, plot the logarithm of T in terms of the logarithm of n. Deduce an approximate value of s.

4.2 (∗). In order to compute the product $C = AB$ of two real square matrices A and B, we use the usual algorithm

$$c_{i,j} = \sum_{k=1}^{n} a_{i,k} b_{k,j}, \quad 1 \le i, j \le n,$$

with the notation $A = (a_{i,j})_{1 \le i,j \le n}$, $B = (b_{i,j})_{1 \le i,j \le n}$, and $C = (c_{i,j})_{1 \le i,j \le n}$.

1. Prove that if A is lower triangular, then the computational complexity for the product $C = AB$ is equivalent to $n^3/2$ for n large (recall that only multiplications and divisions are counted).
2. Write, in pseudolanguage, an algorithm that makes it possible to compute the product $C = AB$ of a lower triangular matrix A with any matrix B that has the computational complexity $n^3/2$.
3. We assume henceforth that both matrices A and B are lower triangular. Taking into account their special structure, prove that the computational complexity for the product $C = AB$ is equivalent to $n^3/6$.
4. Write a function LowTriMatMult that performs the product of two lower triangular matrices, exploiting the sparse structure of these matrices. Compare the results obtained with those of Matlab.
5. Write a function MatMult that executes the product of two matrices (without any special structure). Compare the computational time of this function with that of LowTriMatMult for computing the product of two lower triangular matrices.
6. Fix $n = 300$. Define a=triu(rand(n,n)) and b=triu(rand(n,n)). Find the running time t_1 for computing the product a*b. In order to exploit the sparse structure of the matrices, we define sa=sparse(a),

sb=sparse(b). Find the running time t_2 for the command sa*sb. Compare t_1 and t_2.

4.3. Let A and B be square matrices of $\mathcal{M}_n(\mathbb{R})$, and u a vector of $\mathbb{R}^n$.

1. If A is a band matrix (see Definition 6.2.1), compute the computational complexity for computing Au (assuming n large) in terms of the half-bandwidth p and of n.
2. If A and B are two band matrices, of equal half-bandwidths p, prove that the product AB is a band matrix. Find the computational complexity for computing AB.

4.4. Write a function Strassen that computes the product of two matrices of size $n \times n$, with $n = 2^k$, by the Strassen algorithm. It is advised to use the recursiveness in Matlab, that is, the possibility of calling a function within its own definition. Check the algorithm by comparing its results with those provided by Matlab. Compare with the results obtained with the function MatMult.

4.5. Let A, B, C, and D be matrices of size $n \times n$. Define a matrix X of size $2n \times 2n$ by

$$X = \begin{pmatrix} A & B \\ C & D \end{pmatrix}.$$

We assume that A is nonsingular as well as the matrix $\Delta = D - CA^{-1}B$ (Δ is called the Schur complement of X). Under these two assumptions, check that the matrix X is nonsingular, and that its inverse is

$$X^{-1} = \begin{pmatrix} A^{-1} + A^{-1}B\Delta^{-1}CA^{-1} & -A^{-1}B\Delta^{-1} \\ -\Delta^{-1}CA^{-1} & \Delta^{-1} \end{pmatrix}. \tag{4.4}$$

Compute the inverse of a $2n \times 2n$ matrix by implementing (4.4) in Matlab. Use the command inv to compute the inverses of the blocks of size $n \times n$, then try to minimize the number of block products. Compare your results with the standard command inv(X).

5

Linear Systems

We call the problem that consists in finding the (possibly multiple) solution $x \in \mathbb{K}^p$, if any, of the following algebraic equation

$$Ax = b \tag{5.1}$$

a linear system. The matrix $A \in \mathcal{M}_{n,p}(\mathbb{K})$, called the "system matrix," and the vector $b \in \mathbb{K}^n$, called the "right-hand side," are the data of the problem; the vector $x \in \mathbb{K}^p$ is the unknown. As usual, $\mathbb{K}$ denotes the field $\mathbb{R}$ or $\mathbb{C}$. The matrix A has n rows and p columns: n is the number of equations (the dimension of b) and p is the number of unknowns (the dimension of x).

In this chapter, we study existence and uniqueness of solutions for the linear system (5.1) and we discuss some issues concerning stability and precision for any practical method to be used on a computer for solving it.

5.1 Square Linear Systems

In this section, we consider only linear systems with the same number of equations and unknowns: $n = p$. Such a linear system is said to be square (like the matrix A). This particular case, $n = p$, is very important, because it is the most frequent in numerical practice. Furthermore, the invertibility of A provides an easy criterion for the existence and uniqueness of the solution. Note that it is only in the case $n = p$ that the inverse of a matrix can be defined.

Theorem 5.1.1. *If the matrix A is nonsingular, then there exists a unique solution of the linear system $Ax = b$. If A is singular, then one of the following alternatives holds: either the right-hand side b belongs to the range of A and there exists an infinity of solutions that differ one from the other by addition of an element of the kernel of A, or the right-hand side b does not belong to the range of A and there are no solutions.*

The proof of this theorem is obvious (see [10] if necessary), but it does not supply a formula to compute the solution when it exists. The next proposition gives such a formula, the so-called Cramer formulas.

Proposition 5.1.1 (Cramer formulas). *Let* $A = (a_1| \ldots |a_n)$ *be a nonsingular matrix with columns* $a_i \in \mathbb{R}^n$. *The solution of the linear system* $Ax = b$ *is given by its entries:*

$$x_i = \frac{\det (a_1| \ldots |a_{i-1}| b |a_{i+1}| \ldots |a_n)}{\det A}, \quad 1 \leq i \leq n.$$

Proof. Since the determinant is an alternate multilinear form, we have for all $j \neq i$,

$$\det (a_1| \ldots |a_{i-1}|\lambda a_i + \mu a_j|a_{i+1}| \ldots |a_n) = \lambda \det A$$

for all λ and μ in $\mathbb{K}$. The equality $Ax = b$ is equivalent to $b = \sum_{i=1}^n x_i a_i$, that is, the x_i are the entries of b in the basis formed by the columns a_i of the matrix A. We deduce that

$$\det (a_1| \ldots |a_{i-1}| b |a_{i+1}| \ldots |a_n) = x_i \det A,$$

which is the aforementioned formula.   □

Let us claim right away that the Cramer formulas are not of much help in computing the solution of a linear system. Indeed, they are very expensive in terms of CPU time on a computer. To give an idea of the prohibitive cost of the Cramer formulas, we give a lower bound c_n for the number of multiplications required to compute the determinant (by the classical row (or column) development method) of a square matrix of order n. We have $c_n = n(1 + c_{n-1}) \geq nc_{n-1}$ and thus $c_n \geq n! = n(n-1)(n-2) \cdots 1$. Therefore, more than $(n + 1)!$ multiplications are needed to compute the solution of problem (5.1) by the Cramer method, which is huge. For $n = 50$, and if the computations are carried out on a computer working at 1 gigaflop (i.e., one billion operations per second), the determination of the solution of (6.1) by the Cramer method requires at least

$$\frac{51!}{(365 \cdot 24 \cdot 60 \cdot 60) \cdot (10^9)} \approx 4.8 \times 10^{49} \text{ years!!!}$$

Even if we use a clever way of computing determinants, requiring say the order of n^α operations, the Cramer formulas would yield a total cost of order $n^{\alpha+1}$, which is still prohibitive, since Theorem 4.4.1 claims that computing a determinant or solving a linear system should have the same asymptotic complexity. We shall study in Chapter 6 methods that require a number of operations on the order of n^3, which is much less than $n!$; the same computer would take 10^{-4} seconds to execute the n^3 operations for $n = 50$. Let us check that `Matlab` actually solves linear systems of size n in $\mathcal{O}(n^3)$ operations. In Figure 5.1 we display the computational time required by the command `A\b`, where the entries of A, an $n \times n$ matrix, are chosen randomly. The results are displayed with a log-log scale (i.e., $\ln(time)$ in terms of $\ln(n)$) in Figure 5.1 for 10 values of n in the range $(100, 1000)$. If the dependence is of the form

$$time(n) = an^p + \mathcal{O}(n^{p-1}), \quad a > 0,$$

taking the logarithms yields

$$\ln(time(n)) = p\ln(n) + \ln(a) + \mathcal{O}(1/n).$$

The plotted curve is indeed close to a line of slope 3; hence the approximation $p \approx 3$. In practice, the slope p is larger than 3 (especially for large values of n) because in addition to the execution time of operations, one should take into account the time needed to access the data in the computer memory.

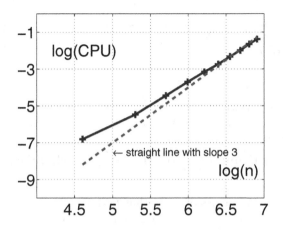

Fig. 5.1. Running time for solving $Ax = b$ (command A\b of Matlab) in terms of n.

To conclude, the Cramer formulas are never used, because they are too expensive in terms of computational time. It is an example, among many others, that an elegant concept from a theoretical viewpoint is not necessarily practical and efficient from a numerical viewpoint.

The next chapters are devoted to various methods for solving linear systems like (5.1). Before studying the general case, corresponding to any non-singular matrix, let us review some simple particular cases:

✗ If A is a diagonal matrix, it is clear that the computation of the solution x of (5.1) is performed in just n operations. Recall that we take into account only multiplications and divisions and not additions and subtractions.

✗ If A is a unitary matrix, the solution is given by $x = A^{-1}b = A^*b$. Since transposition and complex conjugation require no operations, such a computation boils down to a matrix multiplied by a vector, which is performed in n^2 operations.

✗ If A is a triangular matrix (lower for instance), then (5.1) reads

$$
\begin{pmatrix}
a_{1,1} & 0 & \cdots & & \cdots & 0 \\
a_{2,1} & a_{2,2} & \ddots & & \cdots & \vdots \\
\vdots & \vdots & \ddots & \ddots & & \vdots \\
a_{n-1,1} & a_{n-1,2} & & a_{n-1,n-1} & 0 \\
a_{n,1} & a_{n,2} & \cdots & a_{n,n-1} & a_{n,n}
\end{pmatrix}
\begin{pmatrix}
x_1 \\ x_2 \\ \vdots \\ x_{n-1} \\ x_n
\end{pmatrix}
=
\begin{pmatrix}
b_1 \\ b_2 \\ \vdots \\ b_{n-1} \\ b_n
\end{pmatrix}.
$$

The solution can be computed by the so-called forward substitution algorithm (Algorithm 5.1). We first compute $x_1 = b_1/a_{1,1}$; then, using this value of x_1, we obtain $x_2 = (b_2 - a_{2,1}x_1)/a_{2,2}$; and so on up to x_n. For an upper triangular matrix, we use a "back substitution algorithm," i.e., we compute the entries of x in reverse order, starting from x_n up to x_1.

Data: A, b. Output: $x = A^{-1}b$.
    For $i = 1 \nearrow n$
        $s = 0$
        For $j = 1 \nearrow i - 1$
            $s = s + A_{i,j}x_j$
        End j
        $x_i = (b_i - s)/A_{i,i}$
    End i

Algorithm 5.1: Forward substitution algorithm.

To compute the n entries of x, we thus perform
- $1 + 2 + \cdots + n - 1 = n(n-1)/2$ multiplications
- n divisions

This is a total number of order $n^2/2$ operations. Note that this algorithm gives the solution x without having to compute the inverse matrix A^{-1}: this will always be the case for any efficient algorithm for solving linear systems.

Since it is very easy to solve a linear system whose matrix is triangular, many solution methods consist in reducing the problem to solving a triangular system. Examples of such methods are given in the next chapters. We shall remember at this stage that solving a triangular linear system requires (on the order of) $n^2/2$ operations.

Remark 5.1.1. To solve a linear system, it is not necessary to compute A^{-1}. We have just seen this with the Cramer formulas and the case of a triangular matrix. In general, solving a linear system does not require that one compute the inverse matrix A^{-1} because it is too expensive.

Conversely, if we have at our disposal an algorithm to solve the linear system $Ax = b$, then we easily infer a method for determining the inverse matrix A^{-1}. Indeed, if we denote by $(e_i)_{1 \le i \le n}$ the canonical basis of $\mathbb{K}^n$ and x_i the solution of the system with e_i as right-hand side (i.e., $Ax_i = e_i$), then the matrix with columns $(x_i)_{1 \le i \le n}$ is nothing but A^{-1}.

Remark 5.1.2. When solving a linear system, we use neither the diagonalization nor the triangularization properties of the matrix A. Obviously, if we explicitly knew such a factorization $A = PTP^{-1}$, solving the linear system would be easy. However, the trouble is that computing the diagonal or triangular form of a matrix comes down to determining its eigenvalues, which is much harder and much more costly than solving a linear system by classical methods.

Remark 5.1.3. If one wants to minimize the storage size of a triangular (or symmetric) matrix, it is not a good idea to represent such matrices $A \in \mathcal{M}_n(\mathbb{R})$ by a square array of dimension $n \times n$. Actually, half the memory space would suffice to store the $n(n+1)/2$ nonzero entries of A. It is thus enough to declare a vector array STOREA of dimension $n(n+1)/2$ and to manage the correspondence between indices (i, j) and index k such that $A(i, j) =$ STOREA(k). We easily check that $k(i, j) = j + i(i-1)/2$ if the upper triangular matrix A is stored row by row; see Exercise 5.3.

5.2 Over- and Underdetermined Linear Systems

In this section we consider linear systems of the form (5.1) with different numbers of equations and unknowns, $n \neq p$. When $n < p$, we say that the system is underdetermined: there are more unknowns than equations (which allows more "freedom" for the existence of solutions). When $n > p$, we say that the system is overdetermined: there are fewer unknowns than equations (which restricts the possibility of existence of solutions). In both cases, let us recall a very simple but fundamental result [10].

Theorem 5.2.1. *There exists at least one solution of linear system (5.1) if and only if the right-hand side b belongs to the range of A. The solution is unique if and only if the kernel of A is reduced to the zero vector. Two solutions differ by an element of the kernel of A.*

When $n \neq p$, there is no simpler criterion for the existence of solutions to a linear system. We can only indicate, in a heuristic fashion, that it is more likely for these to be solutions to an underdetermined system than to an overdetermined one. Let us recall in any case the following obvious consequence of the rank theorem.

Lemma 5.2.1. *If $n < p$, then $\dim \operatorname{Ker} A \ge p - n \ge 1$, and if there exists a solution of the linear system (5.1), there exists an infinity of them.*

To avoid this delicate existence issue of the solution to a nonsquare linear system, there is another way of looking at it by considering it as a "least squares problem," in other words, a "generalized" or approximate solution, in the sense of least squares fitting, is called for. Chapter 7 is dedicated to this topic.

5.3 Numerical Solution

We now focus on the practical aspects for solving numerically a linear system on a computer. We have already seen that solution algorithms have to be very efficient, that is, fast (by minimizing the number of performed operations), and sparing memory storage. Furthermore, there is another practical requisite for numerical algorithms: their accuracy. Indeed, in scientific computing there are no exact computations! As we shall see below in Section 5.3.1, a computer's accuracy is limited due to the number of bits used to represent real numbers: usually 32 or 64 bits (which makes about 8 or 16 significant digits). Therefore, utmost attention has to be paid to the inevitable rounding errors and to their propagation during the course of a computation. The example below is a particularly striking illustration of this issue.

Example 5.3.1. Consider the following linear system:

$$\begin{pmatrix} 8 & 6 & 4 & 1 \\ 1 & 4 & 5 & 1 \\ 8 & 4 & 1 & 1 \\ 1 & 4 & 3 & 6 \end{pmatrix} x = \begin{pmatrix} 19 \\ 11 \\ 14 \\ 14 \end{pmatrix} \Rightarrow x = \begin{pmatrix} 1 \\ 1 \\ 1 \\ 1 \end{pmatrix}.$$

If we slightly modify the right-hand side, we obtain a very different solution:

$$\begin{pmatrix} 8 & 6 & 4 & 1 \\ 1 & 4 & 5 & 1 \\ 8 & 4 & 1 & 1 \\ 1 & 4 & 3 & 6 \end{pmatrix} x = \begin{pmatrix} 19.01 \\ 11.05 \\ 14.07 \\ 14.05 \end{pmatrix} \Rightarrow x = \begin{pmatrix} -2.34 \\ 9.745 \\ -4.85 \\ -1.34 \end{pmatrix}.$$

This example shows that small errors in the data or in intermediate results may lead to unacceptable errors in the solution. Actually, the relative error in the solution, computed in the $\|.\|_\infty$ norm, is about 2373 times larger than the relative error on the right-hand side of the equation. This amplification of errors depends on the considered matrix (for instance for the identity matrix there are no amplifications of errors). One has therefore to make sure that numerical algorithms do not favor such an amplification. Such a property is called stability.

Remark 5.3.1. Numerical methods (or algorithms) for solving linear systems have to be at the same time efficient and stable. This is really a crucial issue, especially for so-called iterative methods (see Chapter 8). Their name is in opposition to direct methods (see Chapter 6), which would compute the exact solution if there were no rounding errors (perfect or exact arithmetic). On the other hand, iterative methods compute a sequence of approximate solutions that converges to the exact solution: in such a case, stability is a necessary condition.

5.3.1 Floating-Point System

We briefly discuss the representation of numbers in digital computers and its associated arithmetic, which is not exact and is of limited accuracy, as we anticipated. Since digital computers have a finite memory, integers and real numbers can be (approximately) represented by only a finite number of bits. Let us describe a first naive representation system.

Fixed-Point Representation. Suppose that p bits are available to code an integer. Here is a simple way to do it. The first bit is used to indicate the sign of the integer (0 for a positive integer and 1 for a negative one), the $p - 1$ other bits contain the base-2 representation of the integer. For example, for $p = 8$, the positive integers $11 = 1 \times 2^3 + 1 \times 2^1 + 1 \times 2^0$ and $43 = 1 \times 2^5 + 1 \times 2^3 + 1 \times 2^1 + 1 \times 2^0$ are encoded as

$$\boxed{0|0|0|0|1|0|1|1} \quad \text{and} \quad \boxed{0|0|1|0|1|0|1|1}$$

For negatives integers, the complement representation can be used: it consists in reversing the bits (0 becomes 1 and 1 becomes 0). For example, -11 and -43 are encoded as

$$\boxed{1|1|1|1|0|1|0|0} \quad \text{and} \quad \boxed{1|1|0|1|0|1|0|0}$$

Integers outside the interval $[-2^{p-1}, 2^{p-1} - 1]$ cannot be encoded in the fixed-point representation, which is a severe limitation! The same difficulty arises for real numbers too. Therefore, there is a need for another representation. We now describe a more elaborate representation system, which is used by all computers dedicated to scientific computing.

Floating-Point Representation. For given integers b, p, n_{min}, and n_{max}, we define the floating-point numbers as real numbers of the form

$$\pm (0.d_1 \ldots d_p) \times b^n,$$

with $d_1 \neq 0$, $0 \leq d_i \leq b - 1$, and $-n_{min} \leq n \leq n_{max}$. We denote by $\mathcal{F}$ the (finite) set of all floating-point numbers. In this notation,

1. b is the base. The most common bases are $b = 2$ (binary base), $b = 10$ (decimal base), and $b = 16$ (hexadecimal base).
2. $n \in [-n_{min}, n_{max}]$ is the exponent that defines the order of magnitude of the numbers to be encoded.

3. The integers $d_i \in [0, b-1]$ are called the digits and p is the number of significant digits. The mantissa or significand is the integer $m = d_1 \ldots d_p$. Note that

$$m \times b^{n-p} = (0.d_1 \ldots d_p) \times b^n = b^n \sum_{k=1}^{p} d_k b^{-k}.$$

The following bounds hold for floating-point numbers

$$a_{\min} \leq |a| \leq a_{\max}, \qquad \forall a \in \mathcal{F},$$

where $a_{\min} = b^{-(n_{\min}+1)}$ corresponds to the case $d_1 = 1$, $d_2 = \cdots = d_p = 0$, and $n = -n_{\min}$, $a_{\max} = b^{n_{\max}}(1 - b^{-p})$ corresponds to the case $n = n_{\max}$ and $d_1 = \ldots d_p = (b-1)$. In other words, $a_{\min}$ is the smallest positive real number, and $a_{\max}$ the largest one, that can be represented in the set of floating-point numbers. Smaller numbers produce an *underflow* and larger ones an *overflow*. Computers usually support simple precision (i.e., representation with 32 bits) and double precision (i.e., representation with 64 bits). In the single-precision representation, 1 bit is used to code the sign, 8 bits for the exponent, and 23 bits for the mantissa (for a total of 32 bits). In the double-precision representation, 1 bit is used to code the sign, 11 bits for the exponent, and 52 bits for the mantissa (for a total of 64 bits). The precise encoding is system dependent. For example, in a single precision representation, 356.728 is encoded

| 0 | 0 | 0 | 0 | 0 | 0 | 0 | 3 | 3 | 5 | 6 | 7 | 2 | 8 | 0 | 0 | 0 | 0 | 0 | 0 | 0 | 0 | 0 | 0 | 0 | 0 | 0 | 0 | 0 | 0 | 0 | 0 |

in the decimal base and $81.625 = 2^6 + 2^4 + 2^0 + 2^{-1} + 2^{-3} = (2^{-1} + 2^{-3} + 2^{-7} + 2^{-8} + 2^{-10})2^7$ is encoded

| 0 | 0 | 0 | 0 | 0 | 0 | 1 | 1 | 1 | 1 | 1 | 0 | 1 | 0 | 0 | 0 | 1 | 1 | 0 | 1 | 0 | 0 | 0 | 0 | 0 | 0 | 0 | 0 | 0 | 0 | 0 | 0 |

in the binary base. In practice, the floating-point representation is more elaborate than these simple examples. For example, in the binary base, the first digit of the mantissa is always 1, hence it is not necessary to store it. By the same token, the exponent is encoded as an unsigned number by adding to it a fixed "bias" (127 is the usual bias in single precision). Let us consider again the real number 81.625 written this time as $81.625 = (2^0 + 2^{-2} + 2^{-6} + 2^{-7} + 2^{-9})2^6$. The biased exponent to be stored is $127 + 6 = 2^7 + 2^2 + 2^0$, and the complete encoding is (compare with the previous encoding)

| 0 | 1 | 0 | 0 | 0 | 0 | 1 | 0 | 1 | 0 | 1 | 0 | 0 | 0 | 1 | 1 | 0 | 1 | 0 | 0 | 0 | 0 | 0 | 0 | 0 | 0 | 0 | 0 | 0 | 0 | 0 | 0 |

The mapping from real number to floating-point numbers is called the floating-point representation or the rounding. Let $fl(x)$ be the floating-point number associated to the real number x. The following equality holds for all real numbers $x \in [a_{\min}, a_{\max}] \cup \{0\}$:

$$fl(x) = x(1 + \varepsilon),$$

with $|\varepsilon| \leq \varepsilon_{machine}$, and $\varepsilon_{machine} = \frac{1}{2}b^{1-p}$, a number called the machine precision. Typically, for a computer with binary 32-bit single precision ($b = 2$, $p = 23$), the machine precision is $\varepsilon_{machine} = 2^{-23} \approx 11.9 \times 10^{-8}$, while for a binary 64-bit double precision computer $\varepsilon_{machine} = 2^{-52} \approx 2.2 \times 10^{-16}$ (the exponents in $\varepsilon_{machine}$ explain the 8 or 16 significant digits for single or double-precision arithmetic). Concerning other properties of the floating-point numbers (for example their distribution and the effect of rounding), we refer the reader to the sections in [6], [13], and [15] devoted to the floating-point system.

Floating-Point Arithmetic. A key issue is to quantify the precision of the computer realization of an elementary arithmetic operation. Let us consider the case of the operation $+$. The same holds true for the other operations $-$, $\times$, and $\div$. Of course, the sum of two floating-point numbers is usually not a floating-point number. We denotes by $\widetilde{+}$ the computer realization of the addition: for real numbers x and y,

$$x \widetilde{+} y = fl\left(fl(x) + fl(y)\right).$$

Unlike the operation $+$, this operation is not associative: $(x \widetilde{+} y) \widetilde{+} z \neq x \widetilde{+} (y \widetilde{+} z)$. Overflow occurs if the addition produces a too-large number, $|x \widetilde{+} y| > a_{max}$, and underflow occurs if it produces a too-small number, $|x \widetilde{+} y| < a_{min}$. Most computer implementations of addition (including the widely used IEEE arithmetic) satisfy the property that the relative error is less than the machine precision:

$$\left| \frac{(x \widetilde{+} y) - (x + y)}{x + y} \right| \leq \varepsilon_{machine},$$

assuming $x + y \neq 0$ and $a_{min} < |x|, |y| < a_{max}$. Hence the relative error on one single operation is very small, but this is not always the case when a sequence of many operations is performed.

We will not study the precise roundoff, or error propagation, of vectorial operations (such as scalar product and matrix-vector product) or numerical algorithms presented in this book, and we refer the reader to [7] for more details.

Let us conclude this section by saying that underflow and overflow are not the only warning or error messages produced by a floating-point representation. Forbidden operations (like dividing by zero) or unresolved operations (like $0 \div 0$) produce, as an output, NaN (which means not a number) or Inf (infinity). In practice, obtaining a NaN or Inf is a clear indication that something is going wrong in the algorithm!

5.3.2 Matrix Conditioning

To quantify the rounding error phenomenon, we introduce the notion of matrix conditioning. It helps to measure the sensitivity of the solution x of the linear system $Ax = b$ to perturbations of the data A and b (we assume that A is a nonsingular square matrix and that b is nonzero). Let $\varepsilon \geq 0$ be a small parameter of data perturbation. We define A_ε and b_ε, perturbations of A and b, by

$$A_\varepsilon = A + \varepsilon B, \; B \in \mathcal{M}_n(\mathbb{K}), \qquad b_\varepsilon = b + \varepsilon\gamma, \; \gamma \in \mathbb{K}^n. \qquad (5.2)$$

Since A is nonsingular, A_ε is also nonsingular for ε small enough (see Remark 3.3.2), and we denote by x_ε the solution of the system

$$A_\varepsilon x_\varepsilon = b_\varepsilon. \qquad (5.3)$$

We remark that $A_\varepsilon^{-1} = (I + \varepsilon A^{-1}B)^{-1}A^{-1}$, and using Proposition 3.3.1 for ε small, we have $(I + \varepsilon A^{-1}B)^{-1} = I - \varepsilon A^{-1}B + \mathcal{O}(\varepsilon^2)$. Consequently, we can write an asymptotic expansion of x_ε in terms of ε:

$$\begin{aligned} x_\varepsilon &= (I + \varepsilon A^{-1}B)^{-1}A^{-1}(b + \varepsilon\gamma) \\ &= \left[I - \varepsilon A^{-1}B + \mathcal{O}(\varepsilon^2)\right](x + \varepsilon A^{-1}\gamma) \\ &= x + \varepsilon A^{-1}(\gamma - Bx) + \mathcal{O}(\varepsilon^2), \end{aligned}$$

where $\mathcal{O}(\varepsilon^2)$ denotes a vector $y \in \mathbb{K}^n$ such that $\|y\| = \mathcal{O}(\varepsilon^2)$ in a given vector norm. Noting that $\|b\| \leq \|A\|\,\|x\|$, we have the following upper bounds for a vector norm and its corresponding matrix norm:

$$\|x_\varepsilon - x\| \leq \varepsilon\|x\|\,\|A^{-1}\|\,\|A\|\left\{\frac{\|\gamma\|}{\|b\|} + \frac{\|B\|}{\|A\|}\right\} + \mathcal{O}(\varepsilon^2). \qquad (5.4)$$

Definition 5.3.1. *The condition number of a matrix $A \in \mathcal{M}_n(\mathbb{K})$, relative to a subordinate matrix norm $\|.\|$, is the quantity defined by*

$$\mathrm{cond}(A) = \|A\|\,\|A^{-1}\|.$$

Note that we always have $\mathrm{cond}(A) \geq 1$, since $1 = \|I\| = \|AA^{-1}\| \leq \|A\|\,\|A^{-1}\|$. Inequality (5.4) reads then as

$$\frac{\|x_\varepsilon - x\|}{\|x\|} \leq \mathrm{cond}(A)\left\{\frac{\|A_\varepsilon - A\|}{\|A\|} + \frac{\|b_\varepsilon - b\|}{\|b\|}\right\} + \mathcal{O}(\varepsilon^2). \qquad (5.5)$$

This upper bound shows that the relative error (to first-order in ε) in x is bounded from above by $\mathrm{cond}(A)$ times the relative error in A and b. The condition number $\mathrm{cond}(A)$ thus measures the conditioning or the sensitivity of the problem $Ax = b$ to perturbations in the data A or b. Even if the relative error in data A and b is small, the relative error in the solution x may be large if the quantity $\mathrm{cond}(A)$ is large. In other words, the condition number measures the amplification of errors in the data (right-hand side b or matrix A). We can establish a more accurate upper bound than (5.5) if we perturb only one datum, b or A.

Proposition 5.3.1. *Let A be a nonsingular matrix and $b \neq 0$ a vector.*

1. *If x and $x + \delta x$ are respectively the solutions of the systems*

$$Ax = b \text{ and } A(x + \delta x) = b + \delta b,$$

we have

$$\frac{\|\delta x\|}{\|x\|} \leq \text{cond}(A) \frac{\|\delta b\|}{\|b\|}. \tag{5.6}$$

2. *If x and $x + \delta x$ are respectively the solutions of the systems*

$$Ax = b \text{ and } (A + \delta A)(x + \delta x) = b, \tag{5.7}$$

we have

$$\frac{\|\delta x\|}{\|x + \delta x\|} \leq \text{cond}(A) \frac{\|\delta A\|}{\|A\|}.$$

Furthermore, these inequalities are optimal.

Proof. To prove the first result, we observe that $A\delta x = \delta b$ implies that $\|\delta x\| \leq \|A^{-1}\| \cdot \|\delta b\|$. However, we also have $\|b\| \leq \|A\| \|x\|$, which yields

$$\frac{\|\delta x\|}{\|x\|} \leq \text{cond}(A) \frac{\|\delta b\|}{\|b\|}.$$

This inequality is optimal in the following sense: for every matrix A, there exists δb and x (which depend on A) such that

$$\frac{\|\delta x\|}{\|x\|} = \text{cond}(A) \frac{\|\delta b\|}{\|b\|}. \tag{5.8}$$

In fact, according to a property of subordinate matrix norms (cf. Proposition 3.1.1 in Chapter 3) there exist $x_0 \neq 0$ such that $\|Ax_0\| = \|A\| \|x_0\|$ and $x_1 \neq 0$ such that $\|A^{-1}x_1\| = \|A^{-1}\| \|x_1\|$. For $b = Ax_0$ and $\delta b = x_1$, we have $x = x_0$ and $\delta x = A^{-1}x_1$, and equality (5.8) holds.

To obtain the second result, we observe that

$$A\delta x + \delta A(x + \delta x) = 0 \Rightarrow \|\delta x\| \leq \|A^{-1}\| + \|\delta A\| \|x + \delta x\|,$$

from which we deduce

$$\frac{\|\delta x\|}{\|x + \delta x\|} \leq \text{cond}(A) \frac{\|\delta A\|}{\|A\|}.$$

To prove the optimality, we show that for any matrix A, there exist a perturbation δA and a right-hand side b that satisfy the equality. Thanks to Proposition 3.1.1 there exists $y \neq 0$ such that $\|A^{-1}y\| = \|A^{-1}\| \|y\|$. Let ε be a nonzero scalar. We set $\delta A = \varepsilon I$ and $b = (A + \delta A)y$. We then check that $y = y + \delta x$ and $\delta x = -\varepsilon A^{-1}y$, and since $\|\delta A\| = |\varepsilon|$, we infer the desired equality. $\qquad\square$

In practice, the most frequently used conditionings are

$$\text{cond}_p(A) = \|A\|_p \|A^{-1}\|_p \text{ for } p = 1, 2, +\infty,$$

where the matrix norms are subordinate to the vector norms $\|.\|_p$. For instance, for the matrix in Example 5.3.1, we have $\text{cond}_\infty(A) \approx 5367$, which accounts for the strong amplification of small perturbations of the right-hand side on the solution. Let us note at once that the upper bound (5.6), while optimal, is in general very pessimistic; see Remark 5.3.3.

We now establish some properties of the condition number.

Proposition 5.3.2. *Consider a matrix* $A \in \mathcal{M}_n(\mathbb{C})$.

1. $\text{cond}(A) = \text{cond}(A^{-1})$, $\text{cond}(\alpha A) = \text{cond}(A) \ \forall \alpha \neq 0$.
2. *For any matrix* A,

$$\text{cond}_2(A) = \frac{\mu_1(A)}{\mu_n(A)}, \tag{5.9}$$

   where $\mu_n(A)$ *and* $\mu_1(A)$ *are respectively the smallest and the largest singular values of* A.
3. *For a normal matrix* A,

$$\text{cond}_2(A) = \frac{|\lambda_{\max}(A)|}{|\lambda_{\min}(A)|} = \varrho(A)\varrho(A^{-1}), \tag{5.10}$$

   where $|\lambda_{\min}(A)|$ *and* $|\lambda_{\max}(A)|$ *are respectively the modulus of the smallest and largest eigenvalues of* A.
4. *For any unitary matrix* U, $\text{cond}_2(U) = 1$.
5. *For any unitary matrix* U, $\text{cond}_2(AU) = \text{cond}_2(UA) = \text{cond}_2(A)$.

The proof of this proposition follows directly from the properties of the subordinate norm $\| \cdot \|_2$.

Remark 5.3.2. Equality (5.10) is optimal, in the sense that for any matrix norm, we have

$$\text{cond}(A) = \|A\| \|A^{-1}\| \geq \varrho(A)\varrho(A^{-1}).$$

In particular, for a normal matrix A, we always have $\text{cond}(A) \geq \text{cond}_2(A)$.

A matrix A is said to be "well conditioned" if for a given norm, $\text{cond}(A) \approx 1$; it is said to be "ill conditioned" if $\text{cond}(A) \gg 1$. Unitary matrices are very well conditioned, which explains why one has to manipulate, whenever possible, these matrices rather than others.

Since all norms are equivalent in a vector space of finite dimension, the condition numbers of matrices in $\mathcal{M}_n(\mathbb{K})$ are equivalent in the following sense.

Proposition 5.3.3. *Conditionings* cond_1, cond_2, *and* cond_∞ *are equivalent:*

$$\begin{aligned}
n^{-1} \text{cond}_2(A) &\leq \text{cond}_1(A) \leq n \, \text{cond}_2(A), \\
n^{-1} \text{cond}_\infty(A) &\leq \text{cond}_2(A) \leq n \, \text{cond}_\infty(A), \\
n^{-2} \text{cond}_1(A) &\leq \text{cond}_\infty(A) \leq n^2 \, \text{cond}_1(A).
\end{aligned}$$

Proof. The inequalities follow from the equivalences between the matrix norms $\|\cdot\|_1$, $\|\cdot\|_2$, and $\|\cdot\|_\infty$, which in turn follow from the equivalences between the corresponding vector norms. □

Remark 5.3.3. The upper bound (5.6), while optimal, is in general very pessimistic, as is shown by the following argument, based on the SVD decomposition. Let $A = V\Sigma U^*$ be the SVD decomposition of the nonsingular square matrix A with $\Sigma = \text{diag}(\mu_i)$ and $\mu_1 \geq \cdots \geq \mu_n > 0$. We expand b (respectively x) in the basis of columns v_i of V (respectively u_i of U)

$$b = \sum_{i=1}^n b_i v_i, \quad x = \sum_{i=1}^n x_i u_i.$$

We consider a perturbation δb that we write in the form $\delta b = \varepsilon \|b\|_2 \sum_{i=1}^n \delta_i v_i$, with $\varepsilon > 0$ and $\sum_{i=1}^n |\delta_i|^2 = 1$, so that $\|\delta b\|_2 = \varepsilon \|b\|_2$. Let us show that equality in (5.6) may occur only exceptionally if we use the Euclidean norm. Observing that $Au_i = \mu_i v_i$, we have

$$x_i = \frac{b_i}{\mu_i} \quad \text{and} \quad \delta x_i = \varepsilon \|b\|_2 \frac{\delta_i}{\mu_i},$$

and since the columns of U and V are orthonormal, the equality in (5.6) occurs if and only if

$$\varepsilon^2 \|b\|_2^2 \frac{\sum_{i=1}^n |\frac{\delta_i}{\mu_i}|^2}{\|x\|_2^2} = \varepsilon^2 \frac{\mu_1^2}{\mu_n^2}.$$

Setting $c_i = b_i/\|b\|_2$, this equality becomes

$$\mu_n^2 \sum_{i=1}^n \left|\frac{\delta_i}{\mu_i}\right|^2 = \mu_1^2 \sum_{i=1}^n \left|\frac{c_i}{\mu_i}\right|^2,$$

that is,

$$|\delta_n|^2 + \sum_{i=1}^{n-1} \frac{\mu_n^2}{\mu_i^2} |\delta_i|^2 = |c_1|^2 + \sum_{i=2}^n \frac{\mu_1^2}{\mu_i^2} |c_i|^2.$$

And since $\sum_{i=1}^n |\delta_i|^2 = \sum_{i=1}^n |c_i|^2 = 1$, we have

$$\sum_{i=1}^{n-1} \left(\frac{\mu_n^2}{\mu_i^2} - 1\right) |\delta_i|^2 = \sum_{i=2}^n \left(\frac{\mu_1^2}{\mu_i^2} - 1\right) |c_i|^2.$$

Since the left sum is nonpositive and the right sum is nonnegative, both sums are zero. Now, all the terms of these two sums have the same sign; hence every term of these sums is zero:

$$\left(\frac{\mu_n^2}{\mu_i^2} - 1\right) |\delta_i|^2 = 0 \quad \text{and} \quad \left(\frac{\mu_1^2}{\mu_i^2} - 1\right) |c_i|^2 = 0 \text{ for } 1 \leq i \leq n.$$

We deduce that if $\mu_i \neq \mu_n$ then $\delta_i = 0$, and if $\mu_i \neq \mu_1$ then $c_i = 0$. In other words, the equality in (5.6) may occur only if the right-hand side b belongs to the first eigenspace (corresponding to μ_1) of A^*A and if the perturbation δb belongs to the last eigenspace (corresponding to μ_n) of A^*A. This coincidence seldom takes place in practice, which accounts for the fact that in general, $\frac{\|\delta x\|}{\|x\|}$ is much smaller than its upper bound $\operatorname{cond}(A)\frac{\|\delta b\|}{\|b\|}$. See on this topic an example in Section 5.3.3 below.

Geometric interpretation of $\operatorname{cond}_2(A)$. We have seen (see Figure 2.2) that the range, by the matrix A, of the unit sphere of $\mathbb{R}^n$ is an ellipsoid, whose semiaxes are the singular values μ_i. Proposition 5.3.2 shows that the (2-norm) conditioning of a matrix measures the flattening of the ellipsoid. A matrix is therefore well conditioned when this ellipsoid is close to a sphere.

Another interpretation of $\operatorname{cond}_2(A)$. The condition number $\operatorname{cond}_2(A)$ of a nonsingular matrix A turns out to be equal to the inverse of the relative distance from A to the subset of singular matrices in $\mathcal{M}_n(\mathbb{C})$. Put differently, the more a matrix is ill conditioned, the closer it is to being singular (and thus difficult to invert numerically).

Lemma 5.3.1. *The condition number* $\operatorname{cond}_2(A)$ *of a nonsingular matrix* A *is equivalently defined as*

$$\frac{1}{\operatorname{cond}_2(A)} = \inf_{B \in \mathcal{S}_n} \left\{ \frac{\|A - B\|_2}{\|A\|_2} \right\},$$

where $\mathcal{S}_n$ *is the set of singular matrices of* $\mathcal{M}_n(\mathbb{C})$.

Proof. Multiplying the above equality by $\|A\|_2$ we have to prove that

$$\frac{1}{\|A^{-1}\|_2} = \inf_{B \in \mathcal{S}_n} \{\|A - B\|_2\}. \tag{5.11}$$

If there were $B \in \mathcal{S}_n$ such that $\|A - B\|_2 < 1/\|A^{-1}\|_2$, we would have

$$\|A^{-1}(A - B)\|_2 \leq \|A^{-1}\|_2 \|A - B\|_2 < 1,$$

and by Proposition 3.3.1 (and Lemma 3.3.1) the matrix $I - A^{-1}(A - B) = A^{-1}B$ would be nonsingular, whereas by assumption, $B \in \mathcal{S}_n$. Hence, we have

$$\inf_{B \in \mathcal{S}} \{\|A - B\|_2\} \geq \frac{1}{\|A^{-1}\|_2}.$$

Now we show that the infimum in (5.11) is attained by a matrix $B_0 \in \mathcal{S}_n$ satisfying

$$\|A - B_0\|_2 = \frac{1}{\|A^{-1}\|_2}.$$

By virtue of Proposition 3.1.1 there exists a unit vector $u \in \mathbb{C}^n$, $\|u\|_2 = 1$, such that $\|A^{-1}\|_2 = \|A^{-1}u\|_2$. Let us check that the matrix

$$B_0 = A - \frac{u(A^{-1}u)^*}{\|A^{-1}\|_2^2}$$

satisfies the condition. We have

$$\|A - B_0\|_2 = \frac{1}{\|A^{-1}\|_2^2}\|u(A^{-1}u)^*\|_2 = \frac{1}{\|A^{-1}\|_2^2}\max_{x\neq 0}\frac{\|u(A^{-1}u)^*x\|_2}{\|x\|_2}.$$

Since $u(A^{-1}u)^*x = \langle x, A^{-1}u\rangle u$ and $\|A^{-1}u\|_2 = \|A^{-1}\|_2$, we deduce

$$\|A - B_0\|_2 = \frac{1}{\|A^{-1}\|_2^2}\max_{x\neq 0}\frac{|\langle x, A^{-1}u\rangle|}{\|x\|_2} = \frac{\|A^{-1}u\|_2}{\|A^{-1}\|_2^2} = \frac{1}{\|A^{-1}\|_2}.$$

The matrix B_0 indeed belongs to $\mathcal{S}_n$, i.e., is singular because $A^{-1}u \neq 0$ and

$$B_0A^{-1}u = u - \frac{u(A^{-1}u)^*A^{-1}u}{\|A^{-1}\|_2^2} = u - \frac{\langle A^{-1}u, A^{-1}u\rangle u}{\|A^{-1}\|_2^2} = 0.$$

$\square$

Remark 5.3.4 (Generalization of the conditioning). If a nonzero matrix A is not square or singular, we define its condition number relative to a given norm by

$$\mathrm{cond}(A) = \|A\|\,\|A^\dagger\|,$$

where $A^\dagger$ is the pseudoinverse of the matrix A. By Definition 2.7.2 we have

$$\|A^\dagger\|_2 = \|\tilde{\Sigma}^\dagger\|_2 = \frac{1}{\mu_p},$$

where μ_p is the smallest nonzero singular value of the matrix A. Denoting by $\mu_1(A)$ the largest singular value of A, we obtain the following generalization of (5.9):

$$\mathrm{cond}_2(A) = \frac{\mu_1(A)}{\mu_p(A)}.$$

5.3.3 Conditioning of a Finite Difference Matrix

We return to the differential equation (1.1) and its discretization by finite differences (see Section 1.1). When the coefficient $c(x)$ is identically zero on the interval $[0, 1]$, (1.1) is called the "Laplacian." In this case, the matrix A_n, resulting from the discretization by finite difference of the Laplacian, and defined by (1.2), reads

$$A_n = n^2 \begin{pmatrix} 2 & -1 & 0 & \cdots & 0 \\ -1 & 2 & \ddots & \ddots & \vdots \\ 0 & \ddots & \ddots & -1 & 0 \\ \vdots & \ddots & -1 & 2 & -1 \\ 0 & \cdots & 0 & -1 & 2 \end{pmatrix}. \tag{5.12}$$

Lemma 5.3.2. *For any $n \geq 2$, the linear system $A_n u^{(n)} = b^{(n)}$ has a unique solution.*

Proof. It suffices to show that A_n is nonsingular. An easy calculation shows that

$$\langle A_n v, v \rangle = \sum_{i=1}^{n-1} c_i v_i^2 + n^2 \left(v_1^2 + v_{n-1}^2 + \sum_{i=2}^{n-1} (v_i - v_{i-1})^2 \right),$$

which proves that the matrix A_n is positive definite, since $c_i \geq 0$. Hence this matrix is nonsingular. □

Its particularly simple form allows us to explicitly determine its 2-norm conditioning. Recall that the reliability of the linear system solution associated with A_n, and thus of the approximation of the solution of the Laplacian, is linked to the condition number of A_n. Being symmetric and positive definite (see the proof of Lemma 5.3.2), its 2-norm conditioning is given by (5.10), so it is equal to the quotient of its extreme eigenvalues.

We thus compute the eigenvalues and secondarily the eigenvectors of A_n. Let $p = (p_1, \ldots, p_{n-1})^t$ be an eigenvector of A_n corresponding to an eigenvalue λ. Equation $A_n p = \lambda p$ reads, setting $h = 1/n$,

$$-p_{k-1} + (2 - \lambda h^2) p_k - p_{k+1} = 0, \qquad 1 \leq k \leq n-1, \qquad (5.13)$$

with $p_0 = p_n = 0$. We look for special solutions of (5.13) in the form $p_k = \sin(k\alpha)$, $0 \leq k \leq n$, where α is a real number to be determined. Relation (5.13) implies that

$$\{2 - 2\cos\alpha - \lambda h^2\} \sin(k\alpha) = 0.$$

In particular, for $k = 1$,

$$\{2 - 2\cos\alpha - \lambda h^2\} \sin(\alpha) = 0.$$

Since α is not a multiple of π, i.e., $\sin(\alpha) \neq 0$ (otherwise p would be zero), we infer that

$$\lambda = \frac{2(1 - \cos\alpha)}{h^2} = \frac{4\sin^2(\alpha/2)}{h^2}.$$

Moreover, the boundary conditions, $p_0 = 0$ and $p_n = 0$ imply that $\sin(n\alpha) = 0$. Therefore we find $(n-1)$ different possible values of α, which we denote by $\alpha_\ell = \ell\pi/n$ for $1 \leq \ell \leq n-1$, and which yield $(n-1)$ distinct eigenvalues of A_n (i.e., all the eigenvalues of A_n, since it is of order $n-1$):

$$\lambda_\ell = \frac{4}{h^2} \sin^2\left(\ell\frac{\pi}{2n}\right). \qquad (5.14)$$

Each eigenvalue λ_ℓ is associated with an eigenvector p_ℓ whose entries are $\left(\sin[\ell k\pi/n]\right)_{k=1}^{n-1}$. The $n-1$ eigenvalues of A_n are positive (and distinct),

$0 < \lambda_1 < \cdots < \lambda_{n-1}$, which is consistent with the fact that the matrix A_n is positive definite. The condition number of the symmetric matrix A_n is thus

$$\mathrm{cond}_2(A_n) = \frac{\sin^2\left(\frac{\pi}{2}\frac{n-1}{n}\right)}{\sin^2\left(\frac{\pi}{2n}\right)}.$$

When n tends to $+\infty$ (or h tends to 0), we have

$$\mathrm{cond}_2(A_n) \approx \frac{4n^2}{\pi^2} = \frac{4}{\pi^2 h^2}.$$

We deduce that $\lim_{n\to+\infty} \mathrm{cond}_2(A_n) = +\infty$, so A_n is ill conditioned for n large. We come to a dilemma: when n is large, the vector u_n (discrete solution of the linear system) is close to the exact solution of the Laplacian (see Theorem 1.1.1). However, the larger n is, the harder it is to accurately determine u_n: because of rounding errors, the computer provides an approximate solution $\tilde{u}_n$ that may be very different from $u^{(n)}$ if we are to believe Proposition 5.6, which states that the relative error on the solution is bounded from above as follows:

$$\frac{\|\tilde{u}^{(n)} - u^{(n)}\|_2}{\|u^{(n)}\|_2} \leq \mathrm{cond}(A_n)\frac{\|\Delta b^{(n)}\|_2}{\|b^{(n)}\|_2} \approx \frac{4}{\pi^2}n^2\frac{\|\Delta b^{(n)}\|_2}{\|b^{(n)}\|_2},$$

where $\Delta b^{(n)}$ measures the variation of the right-hand side due to rounding errors. This upper bound is however (and fortunately) very pessimistic! Actually, we shall show for this precise problem, when the boundary conditions are $u(0) = u(1) = 0$, that we have

$$\frac{\|\Delta u^{(n)}\|_2}{\|u^{(n)}\|_2} \leq C\frac{\|\Delta b^{(n)}\|_2}{\|b^{(n)}\|_2}, \tag{5.15}$$

with a constant C independent of n. This outstanding improvement of the bound on the relative error is due to the particular form of the right-hand side b of the linear system $A_n u^{(n)} = b^{(n)}$. Let us recall that b is obtained by discretization of the right-hand side $f(x)$ of (1.1), that is, $b_i^{(n)} = f(x_i)$. We have $h\|b^{(n)}\|_2^2 = h\sum_{i=1}^{n-1} f^2(ih)$, and we recognize here a Riemann sum discretizing the integral $\int_0^1 f^2(x)dx$. Since the function f is continuous, we know that

$$\lim_{n\to+\infty} h\|b^{(n)}\|_2^2 = \int_0^1 f^2(x)dx.$$

Similarly,

$$\lim_{n\to+\infty} h\|u^{(n)}\|_2^2 = \int_0^1 u^2(x)dx.$$

Recall that $\Delta u^{(n)} = A_n^{-1}\Delta b^{(n)}$, and hence $\|\Delta u^{(n)}\|_2 \leq \|A_n^{-1}\|_2\|\Delta b^{(n)}\|_2$. Thus we deduce the following upper bound for the relative error on the solution

$$\frac{\|\Delta u^{(n)}\|_2}{\|u^{(n)}\|_2} \leq \|A_n^{-1}\|_2 \frac{\|\Delta b^{(n)}\|_2}{\|u^{(n)}\|_2} = \frac{1}{\lambda_1} \frac{\|b^{(n)}\|_2}{\|u^{(n)}\|_2} \frac{\|\Delta b^{(n)}\|_2}{\|b^{(n)}\|_2},$$

since $\|A_n^{-1}\|_2 = 1/\lambda_1 \approx \pi^2$, according to (5.14). Using the above convergence of Riemann sums we claim that

$$\frac{1}{\lambda_1} \frac{\|b^{(n)}\|_2}{\|u^{(n)}\|_2} \approx \pi^2 \left(\frac{\int_0^1 f^2(x)dx}{\int_0^1 u^2(x)dx} \right)^{1/2},$$

which yields the announced upper bound (5.15) with a constant C independent of n.

5.3.4 Approximation of the Condition Number

In general, it is too difficult to compute the condition number of a matrix A exactly. For the conditioning in the 1- or ∞-norm, we have simple formulas at our disposal for the norm of A (see Proposition 3.1.2), but it requires the explicit computation of A^{-1} to find its norm, which is far too expensive for large matrices. Computing the 2-norm conditioning, given by formula (5.9) or (5.10), is also very costly because it requires the extreme singular values or eigenvalues of A, which is neither easy nor cheap, as we shall see in Chapter 10.

Fortunately, in most cases we do not need an exact value of a matrix condition number, but only an approximation that will allow us to predict beforehand the quality of the expected results (what really matters is the order of magnitude of the conditioning more than its precise value). We look therefore for an approximation of $\mathrm{cond}_p(A) = \|A^{-1}\|_p \|A\|_p$. The case $p = 1$ is handled, in Exercise 5.13, as a maximization of a convex function on a convex set. The case $p = \infty$ is an easy consequence of the case $p = 1$ since $\mathrm{cond}_\infty(A) = \mathrm{cond}_1(A^t)$. In both cases ($p = 1$ and $p = \infty$) the main difficulty is the computation of $\|A^{-1}\|_p$.

We now focus on the case $p = 2$. Consider the SVD decomposition of the nonsingular matrix $A \in \mathcal{M}_n(\mathbb{R})$, $A = V\Sigma U^t$, where $\Sigma = \mathrm{diag}\,(\mu_i)$ with $\mu_1 \geq \cdots \geq \mu_n > 0$, the singular values of A. It furnishes two orthonormal bases of $\mathbb{R}^n$: one made up of the columns u_i of the orthogonal matrix U, the other of the columns v_i of the orthogonal matrix V. We expand a vector $x \in \mathbb{R}^n$ in the v_i basis, $x = \sum_i x_i v_i$. Since $Au_i = \mu_i v_i$, we have the following relations:

$$\|A\|_2 = \max_{x \neq 0} \frac{\|Ax\|}{\|x\|} = \mu_1 = \|Au_1\| \tag{5.16}$$

and

$$\|A^{-1}\|_2 = \max_{x \neq 0} \frac{\|A^{-1}x\|}{\|x\|} = \frac{1}{\mu_n} = \|A^{-1}v_n\|. \tag{5.17}$$

We can thus easily determine the norm of A by computing the product Au_1, and the norm of A^{-1} by solving a linear system $Ax = v_n$. The trouble is that in practice, computing u_1 and v_n is a very costly problem, similar to finding the SVD decomposition of A.

We propose a heuristic evaluation of $\|A\|_2$ and $\|A^{-1}\|_2$ by restricting the maximum in (5.16) and (5.17) to the subset $\{x \in \mathbb{R}^n, x_i = \pm 1\}$. The scheme is the following: we compute an approximation α of $\|A\|_2$, an approximation β of $\|A^{-1}\|_2$, and deduce an approximation $\alpha\beta$ of $\mathrm{cond}_2(A)$. Note that these approximations are actually lower bound for we make a restriction on the maximization set. Hence, we always get a lower bound for the conditioning

$$\mathrm{cond}_2(A) = \|A\|_2 \, \|A^{-1}\|_2 \geq \alpha\beta.$$

In practice, we can restrict our attention to triangular matrices. Indeed, we shall see in the next chapter that most (if not all) efficient algorithms for solving linear systems rely on the factorization of a matrix A as a product of two simple matrices (triangular or orthogonal) $A = BC$. The point in such a manipulation is that solving a linear system $Ax = b$ is reduced to two easy solutions of triangular or orthogonal systems $By = b$ and $Cx = y$ (see Section 5.1). The following upper bound on conditioning holds:

$$\mathrm{cond}(A) \leq \mathrm{cond}(B) \, \mathrm{cond}(C).$$

The condition number cond_2 of an orthogonal matrix being equal to 1, if we content ourselves with an upper bound, it is enough to compute condition numbers for triangular matrices only. We shall therefore assume in the sequel of this section that the matrix A is (for instance, lower) triangular.

Data: A. Output: $r \approx \|A\|_2$
$\quad\quad\quad x_1 = 1; y_1 = a_{1,1}$
$\quad\quad\quad$ **For** $i = 2 \nearrow n$
$\quad\quad\quad\quad\quad s = 0$
$\quad\quad\quad\quad\quad$ **For** $j = 1 \nearrow i - 1$
$\quad\quad\quad\quad\quad\quad\quad s = s + a_{i,j} x_j$
$\quad\quad\quad\quad\quad$ **End** j
$\quad\quad\quad\quad\quad$ **If** $|a_{i,i} + s| > |a_{i,i} - s|$
$\quad\quad\quad\quad\quad\quad\quad$ **then**
$\quad\quad\quad\quad\quad\quad\quad\quad\quad x_i = 1$
$\quad\quad\quad\quad\quad\quad\quad$ **otherwise**
$\quad\quad\quad\quad\quad\quad\quad\quad\quad x_i = -1$
$\quad\quad\quad\quad\quad$ **End If**
$\quad\quad\quad\quad\quad y_i = a_{i,i} x_i + s$
$\quad\quad\quad$ **End** i
$\quad\quad\quad r = \|y\|_2/\sqrt{n}$

Algorithm 5.2: Approximation of $\|A\|_2$.

Approximation of $\|A\|_2$. When A is lower triangular, the entries of $y = Ax$ are deduced (from $i = 1$ to n) from those of x by the formulas

$$y_i = a_{i,i}x_i + \sum_{j=1}^{i-1} a_{i,j}x_j.$$

A popular computing heuristic runs as follows. We fix $x_1 = 1$ and accordingly $y_1 = a_{1,1}$. At each next step $i \geq 2$, we choose x_i equal to either 1 or -1, in order to maximize the modulus of y_i; see Algorithm 5.2. Observe that we do not maximize among all vectors x whose entries are equal to ± 1, since at each step i we do not change the previous choices x_k with $k < i$ (this is typical of so-called greedy algorithms). Since the norm of x is equal to $\sqrt{n}$, we obtain the approximation $\|A\|_2 \approx \|y\|_2/\sqrt{n}$.

Approximation of $\|A^{-1}\|_2$. Similarly, we seek a vector x whose entries x_i are all equal to ± 1 and that heuristically maximizes the norm of $y = A^{-1}x$. Since A is lower triangular, y will be computed by the forward substitution algorithm previously studied (see Algorithm 5.1). The proposed computing heuristic consists in fixing $x_1 = 1$ and choosing each entry x_i for $i \geq 2$ equal to 1 or -1 so that the modulus of the corresponding entry y_i is maximal. It yields the approximation $\|A^{-1}\|_2 \approx \|y\|_2/\sqrt{n}$; see Algorithm 5.3.

Data: A. Output: $r \approx \|A^{-1}\|_2$.
$\quad y_1 = 1/a_{1,1}$
$\quad$ **For** $i = 2 \nearrow n$
$\quad\quad s = 0$
$\quad\quad$ **For** $j = 1 \nearrow i - 1$
$\quad\quad\quad s = s + a_{i,j}y_j$
$\quad\quad$ **End** j
$\quad\quad y_i = -(\text{sign}(s) + s)/a_{i,i}$
$\quad$ **End** i
$\quad r = \|y\|_2/\sqrt{n}$

Algorithm 5.3: Computation of $\|A^{-1}\|_2$.

Finally, Algorithm 5.4 computes an approximation, at low cost, of the 2-norm condition number of a matrix. We can arguably criticize Algorithm 5.4 (see the numerical tests in Exercise 5.12) for being based on a local criterion: each entry y_i is maximized without taking into account the other ones. There exist less local variants of Algorithm 5.4 in the sense that they *simultaneously* take into account several entries.

Data: A. Output: $c \approx \text{cond}_2(A)$.
$\quad\bullet$ compute r_1 by Algorithm 5.2
$\quad\bullet$ compute r_2 by Algorithm 5.3
$\quad\bullet$ set $c = r_1 r_2$.

Algorithm 5.4: Approximation of $\text{cond}_2(A)$.

5.3.5 Preconditioning

Instead of solving a linear system $Ax = b$ with an ill-conditioned matrix A, it may be more efficient to solve the equivalent linear system $C^{-1}Ax = C^{-1}b$ with a nonsingular matrix C that is easily invertible and such that $C^{-1}A$ is better conditioned than A. All the trouble is to find such a matrix C, called a *preconditioner*. The best choice would be such that $C^{-1}A$ is close to the identity (whose conditioning is minimal, equal to 1), that is, C is close to A, but computing A^{-1} is at least as difficult as solving the linear system!

We already know that conditioning is important for the stability and sensitivity to rounding errors in solving linear systems. We shall see later, in Chapter 9, that conditioning is also crucial for the convergence of iterative methods for solving linear systems (especially the conjugate gradient method). Thus it is very important to find good preconditioners. However it is a difficult problem for which there is no universal solution. Here are some examples.

✗ **Diagonal preconditioning.** The simplest example of a preconditioner is given by the diagonal matrix whose diagonal entries are the inverses of the diagonal entries of A. For example, we numerically compare the conditionings of matrices

$$A = \begin{pmatrix} 8 & -2 \\ -2 & 50 \end{pmatrix} \quad \text{and } B = D^{-1}A, \text{ where } D = \text{diag}\,(8, 50),$$

for which Matlab gives the approximate values

$$\text{cond}_2(A) = 6.3371498 \quad \text{and} \quad \text{cond}_2(B) = 1.3370144.$$

Thus, the diagonal preconditioning allows us to reduce the condition number, at least for some problems, but certainly not for all of them (think about matrices having a constant diagonal entry)!

✗ **Polynomial preconditioning.** The idea is to define $C^{-1} = p(A)$, where p is a polynomial such that $\text{cond}(C^{-1}A) \ll \text{cond}(A)$. A good choice of $C^{-1} = p(A)$ is to truncate the expansion in power series of A^{-1},

$$A^{-1} = \left(I - (I - A)\right)^{-1} = I + \sum_{k \geq 1}(I - A)^k,$$

which converges if $\|I - A\| < 1$. In other words, we choose the polynomial $p(x) = 1 + \sum_{k=1}^{d}(1 - x)^k$. We suggest that the reader program this preconditioner in Exercise 5.15.

✗ **Right preconditioning.** Replacing the system $Ax = b$ by $C^{-1}Ax = C^{-1}b$ is called left preconditioning since we multiply the system on its left by C^{-1}. A symmetric idea is to replace $Ax = b$ by the so-called right preconditioned system $AD^{-1}y = b$ with $x = D^{-1}y$ and D (easily) invertible. Of course, we can mix these two kinds of preconditioning, and solve $C^{-1}AD^{-1}y = C^{-1}b$, then compute x by $D^{-1}y = x$.

This is interesting if A is symmetric and if we choose $C = D^t$ because $C^{-1}AD^{-1}$ is still symmetric. Of course, this preconditioning is efficient when $\text{cond}(C^{-1}AD^{-1}) \ll \text{cond}(A)$ and solving $Dx = y$ is easy.

We shall return to preconditioning in Chapter 9, which features more efficient preconditioners.

5.4 Exercises

5.1. Floating-point representation, floating-point arithmetic.
Run the following instructions and comment the results.

1. Floating-point accuracy (machine precision).

   ```
   a=eps;b=0.5*eps;X=[2, 1;2, 1];
   A=[2, 1;2, 1+a];norm(A-X)
   B=[2, 1;2, 1+b];norm(X-B)
   ```

2. Floating-point numbers bounds.

   ```
   rM=realmax, 1.0001*rM, rm=realmin, .0001*rm
   ```

3. Infinity and "Not a number."

   ```
   A=[1 2 0 3]; B=1./A, isinf(B),   C=A.*B
   ```

4. Singular or not?

   ```
   A=[1 1; 1 1+eps];inv(A), rank(A)
   B=[1 1; 1 1+.5*eps];inv(B), rank(B)
   ```

5.2 (∗). How to solve a triangular system.

1. Write a function whose heading is `function x=ForwSub(A,b)` computing by forward substitution (Algorithm 5.1) the solution, if it exists, of the system $Ax = b$, where A is a lower triangular square matrix.
2. Write similarly a function `BackSub(A,b)` computing the solution of a system whose matrix is upper triangular.

5.3 (∗). How to store a lower triangular matrix.

1. Write a program `StoreL` for storing a lower triangular square matrix.
2. Write a program `StoreLpv` for computing the product of a lower triangular square matrix and a vector. The matrix is given in the form `StoreL`.
3. Write a forward substitution program `ForwSubL` for computing the solution of a lower triangular system with matrix given by `StoreL`.

5.4. How to store an upper triangular matrix. In the spirit of the previous exercise, write programs `StoreU`, `StoreUpv`, and `ForwSubU` for an upper triangular matrix.

5.5. Write a program `StoreLpU` computing the product of two matrices, the first one being lower triangular and the second upper triangular. The matrices are given by the programs `StoreL` and `StoreU`.

5.6. We define a matrix `A=[1:5;5:9;10:14]`.

1. Compute a matrix Q whose columns form a basis of the null space of A^t.
2. (a) Consider `b=[5; 9; 4]` and the vector $x \in \mathbb{R}^5$ defined by the instruction `x=A\b`. Compute x, $Ax - b$, and $Q^t b$.
   (b) Same question for `b=[1; 1; 1]`. Compare both cases.
   (c) Justification. Let A be a real matrix of size $m \times n$. Let $b \in \mathbb{R}^m$. Prove the equivalence
   $$b \in \text{Im}(A) \iff Q^t b = 0.$$

   (d) Write a function `InTheImage(A,b)` whose input arguments are a matrix A and a vector b and whose output argument is "yes" if $b \in \text{Im } A$ and "no" otherwise. Application:
   `A=[1 2 3; 4 5 6; 7 8 9]`, `b=[1;1;1]`, then `b=[1 ;2;1]`.

5.7. The goal of this exercise is to show that using the Cramer formulas is a bad idea for solving the linear system $Ax = b$, where A is a nonsingular $n \times n$ matrix and $b \in \mathbb{R}^n$. Denoting by $a_1, \ldots, a_n$ the columns of A, Cramer's formula for the entry x_i of the solution x is $x_i = \det(a_1|\ldots|a_{i-1}|b|a_{i+1}|\ldots|a_n)/\det A$ (see Proposition 5.1.1). Write a function `Cramer` computing the solution x by means of Cramer's formulas and compare the resulting solution with that obtained by the instruction `A\b`.
Hint. Use the `Matlab` function `det` for computing the determinant of a matrix. Application: for $n = 20, 40, 60, 80, \ldots$ consider the matrix A and vector b defined by the instructions

```
b=ones(n,1);c=1:n; A=c'*ones(size(c));A=A+A';
s=norm(A,'inf'); for i=1:n, A(i,i)=s;end;
```

Conclude about the efficiency of this method.

5.8. Let A and B be two matrices defined by the instructions

```
n=10;B=rand(n,n);A=[eye(size(B)) B; zeros(size(B)) eye(size(B))];
```

Compute the Frobenius norm of B as well as the condition number of A (in the Frobenius norm). Compare the two quantities for various values of n. Justify the observations.

5.9. The goal of this exercise is to empirically determine the asymptotic behavior of $\text{cond}_2(H_n)$ as n goes to ∞, where $H_n \in \mathcal{M}_n(\mathbb{R})$ is the Hilbert matrix of order n, defined by its entries $(H_n)_{i,j} = 1/(i+j-1)$. Compute $\text{cond}_2(H_5)$, $\text{cond}_2(H_{10})$. What do you notice? For n varying from 2 to 10, plot the curve $n \mapsto \ln(\text{cond}_2(H_n))$. Draw conclusions about the experimental asymptotic behavior.

5.10. Write a function `Lnorm` that computes the approximate 2-norm of a lower triangular matrix by Algorithm 5.2. Compare its result with the norm computed by `Matlab`.

5.11. Write a function `LnormAm1` that computes the approximate 2-norm of the inverse of a lower triangular matrix by Algorithm 5.3. Compare its result with the norm computed by `Matlab`.

5.12. Write a function `Lcond` that computes an approximate 2-norm conditioning of a lower triangular matrix by Algorithm 5.4. Compare its result with the conditioning computed by `Matlab`.

5.13 (∗). The goal of this exercise is to implement Hager's algorithm for computing an approximate value of $\mathrm{cond}_1(A)$. We denote by $S = \{x \in \mathbb{R}^n, \quad \|x\|_1 = 1\}$ the unit sphere of $\mathbb{R}^n$ for the 1-norm, and for $x \in \mathbb{R}^n$, we set $f(x) = \|A^{-1}x\|_1$ with $A \in \mathcal{M}_n(\mathbb{R})$ a nonsingular square matrix. The 1-norm conditioning is thus given by

$$\mathrm{cond}_1(A) = \|A\|_1 \max_{x \in S} f(x).$$

1. Explain how to determine $\|A\|_1$.
2. Prove that f attains its maximum value at one of the vectors e_j of the canonical basis of $\mathbb{R}^n$.
3. From now on, for a given $x \in \mathbb{R}^n$, we denote by $\tilde{x}$ the solution of $A\tilde{x} = x$ and by $\bar{x}$ the solution of $A^t\bar{x} = s$, where s is the "sign" vector of $\tilde{x}$, defined by $s_i = -1$ if $\tilde{x}_i < 0$, $s_i = 0$ if $\tilde{x}_i = 0$, and $s_i = 1$ if $\tilde{x}_i > 0$. Prove that $f(x) = \langle \tilde{x}, s \rangle$.
4. Prove that for any $a \in \mathbb{R}^n$, we have $f(x) + \bar{x}^t(a - x) \le f(a)$.
5. Show that if $\bar{x}_j > \langle x, \bar{x} \rangle$ for some index j, then $f(e_j) > f(x)$.
6. Assume that $\tilde{x}_j \ne 0$ for all j.
   (a) Show that for y close enough to x, we have $f(y) = f(x) + s^t A^{-1}(y-x)$.
   (b) Show that if $\|\bar{x}\|_\infty \le \langle x, \bar{x} \rangle$, then x is a local maximum of f on the unit sphere S.
7. Deduce from the previous questions an algorithm for computing the 1-norm conditioning of a matrix.
8. Program this algorithm (function `Cond1`). Compare its result with the conditioning computed by `Matlab`.

5.14. We define $n \times n$ matrices C, D, and E by
```
C=NonsingularMat(n);D=rand(m,n);E=D*inv(C)*D';
```
We also define $(n + m) \times (n + m)$ block matrices A and M
```
A=[C D';D zeros(m,m)];M=[C zeros(n,m);zeros(m,n) E];
```

1. For different values of n, compute the spectrum of $M^{-1}A$. What do you notice?
2. What is the point in replacing system $Ax = b$ by the equivalent system $M^{-1}Ax = M^{-1}b$?

3. We now want to give a rigorous explanation of the numerical results of the first question. We assume that $A \in \mathcal{M}_{n+m}(\mathbb{R})$ is a nonsingular matrix that admits the block structure $A = \begin{pmatrix} C & D^t \\ D & 0 \end{pmatrix}$, where $C \in \mathcal{M}_n(\mathbb{R})$ and $D \in \mathcal{M}_{m,n}(\mathbb{R})$ are such that C and $DC^{-1}D^t$ are nonsingular too.

   (a) Show that the assumption "A is nonsingular" implies $m \leq n$.

   (b) Show that for $m = n$, the matrix D is invertible.

4. From now on, we assume $m < n$. Let $x = (x_1, x_2)^t$ be the solution of the system $Ax = b = (b_1, b_2)^t$. The matrix D is not assumed to be invertible, so that we cannot first compute x_1 by relation $Dx_1 = b_2$, then x_2 by $Cx_1 + D^t x_2 = b_1$. Therefore, the relation $Dx_1 = b_2$ has to be considered as a constraint to be satisfied by the solutions x_1, x_2 of the system $Cx_1 + D^t x_2 = b_1$. We study the preconditioning of the system $Ax = b$ by the matrix M^{-1} with $M = \begin{pmatrix} C & 0 \\ 0 & DC^{-1}D^t \end{pmatrix}$.

   (a) Let λ be an eigenvalue of $M^{-1}A$ and $(u, v)^t \in \mathbb{R}^{n+m}$ a corresponding eigenvector. Prove that $(\lambda^2 - \lambda - 1)Du = 0$.

   (b) Deduce the spectrum of the matrix $M^{-1}A$.

   (c) Compute the 2-norm conditioning of $M^{-1}A$, assuming that it is a symmetric matrix.

5.15 ($*$). Program the polynomial preconditioning algorithm presented in Section 5.3.5 on page 91 (function `PrecondP`).

5.16 ($*$). The goal of this exercise is to study the numerical solution of the linear system that stems from the finite difference approximation of the Laplace equation. According to Section 1.1, the Laplace equation is the following second-order differential equation:

$$\begin{cases} -u''(x) + c(x)u(x) = f(x), \\ u(0) = 0, \quad u(1) = 0, \end{cases} \tag{5.18}$$

where $u : [0, 1] \to \mathbb{R}$ denotes the solution (which is assumed to exist and be unique) and f and c are given functions. We first study the case $c \equiv 0$. We recall that a finite difference discretization at points $x_k = k/n$, $k = 1, \ldots, n-1$, leads to the linear system

$$A_n u^{(n)} = b^{(n)}, \tag{5.19}$$

where $A_n \in \mathcal{M}_{n-1}(\mathbb{R})$ is the matrix defined by (5.12), $b^{(n)} \in \mathbb{R}^{n-1}$ is the right-hand side with entries $(f(x_i))_{1 \leq i \leq n-1}$, and $u^{(n)} \in \mathbb{R}^{n-1}$ is the discrete solution approximating the exact solution at the points x_k, i.e., $u^{(n)} = (u_1, \ldots, u_{n-1})^t$ with $u_k \approx u(x_k)$. We also recall that the $n-1$ eigenvalues of A_n are given by

$$\lambda_k = 4n^2 \sin^2 \left(k \frac{\pi}{2n} \right), \quad k = 1, \ldots, n-1. \tag{5.20}$$

1. Computation of the matrix and right-hand side of (5.19).

   (a) Write a function `Laplacian1dD(n)` with input argument n and output argument A_n.

(b) Write a function `InitRHS(n)` with input argument n and output argument $b^{(n)}$.

2. Validation.

(a) Give the exact solution $\tilde{u}^e(x)$ of problem (5.18) when the function f is constant, equal to 1. Write a function constructing the vector $u_n^e = (\tilde{u}^e(x_1), \ldots, \tilde{u}^e(x_{n-1}))^t$. Solve system (5.19) by `Matlab`. Compare the vectors $u^{(n)}$ and u_n^e. Explain.

(b) Convergence of the method. We choose

$$\tilde{u}^e(x) = (x-1)\sin(10x) \text{ and } f(x) = -20\cos(10\,x) + 100(x-1)\sin(10\,x).$$

Plot the norm of the error $u^{(n)} - u_n^e$ in terms of n. What do you notice?

3. Eigenvalues and eigenvectors of the matrix A_n.

(a) Compare the eigenvalues of A_n with those of the operator $u \mapsto -u''$ endowed with the boundary conditions defined by (5.18). These are real numbers λ for which we can find nonzero functions φ satisfying the boundary conditions and such that $-\varphi'' = \lambda\varphi$.

(b) Numerically compute the eigenvalues of A_n with `Matlab` and check that the results are close to the values given by formula (5.20).

(c) Plot the 2-norm conditioning of A_n in terms of n. Comment.

4. We now assume that the function c is constant but nonzero.

(a) Give a formula for the new finite difference matrix $\tilde{A}_n$. How do its eigenvalues depend on the constant c?

(b) From now on we fix $n = 100$, and c is chosen equal to the negative of the first eigenvalue of A_n. Solve the linear system associated to the matrix $\tilde{A}_n$ and a right-hand side with constant entries equal to 1. Check the result. Explain.

5.17. Reproduce Figures 1.2 and 1.3 of Chapter 1. Recall that these figures display the approximation in the least squares sense of the values specified in Table 1.1 by a first-degree polynomial and a fourth-degree one respectively.

5.18. Define $f(x) = \sin(x) - \sin(2x)$ and let X be an array of 100 entries $(x_i)_{i=1}^n$ chosen randomly between 0 and 4 by the function `rand`. Sort this array in increasing order using the function `sort`.

1. Find an approximation of f in the least squares sense by a second-degree polynomial p. Compute the discrete error $\sqrt{\sum_{i=1}^n |f(x_i) - p(x_i)|^2}$.

2. Find another approximation of f in the least squares sense by a trigonometric function $q(x) = a + b\cos(x) + c\sin(x)$. Compare q and p.

6

Direct Methods for Linear Systems

This chapter is devoted to the solution of systems of linear equations of the form

$$Ax = b, \qquad (6.1)$$

where A is a nonsingular square matrix with real entries, b is a vector called the "right-hand side," and x is the unknown vector. For simplicity, we invariably assume that $A \in \mathcal{M}_n(\mathbb{R})$ and $b \in \mathbb{R}^n$. We call a method that allows for computing the solution x within a finite number of operations (in exact arithmetic) a direct method for solving the linear system $Ax = b$. In this chapter, we shall study some direct methods that are much more efficient than the Cramer formulas in Chapter 5. The first method is the celebrated Gaussian elimination method, which reduces any linear system to a triangular one. The other methods rely on the factorization of the matrix A as a product of two matrices $A = BC$. The solution of the system $Ax = b$ is then replaced by the solution of two easily invertible systems (the matrices B and C are triangular or orthogonal) $By = b$, and $Cx = y$.

6.1 Gaussian Elimination Method

The main idea behind this method is to reduce the solution of a general linear system to one whose matrix is triangular. As a matter of fact, we have seen in Chapter 5 that in the case of an upper triangular system (respectively, lower), the solution is straightforward by mere back substitution (respectively, forward substitution) in the equations.

Let us recall the Gaussian elimination method through an example.

Example 6.1.1. Consider the following 4×4 system to be solved:

$$\begin{cases} 2x & +4y & -4z & +t & = & 0, \\ 3x & +6y & +z & -2t & = & -7, \\ -x & +y & +2z & +3t & = & 4, \\ x & +y & -4z & +t & = & 2, \end{cases} \qquad (6.2)$$

which can also be written in matrix form as

$$
\begin{pmatrix} 2 & 4 & -4 & 1 \\ 3 & 6 & 1 & -2 \\ -1 & 1 & 2 & 3 \\ 1 & 1 & -4 & 1 \end{pmatrix} \begin{pmatrix} x \\ y \\ z \\ t \end{pmatrix} = \begin{pmatrix} 0 \\ -7 \\ 4 \\ 2 \end{pmatrix}.
$$

The Gaussian elimination method consists in first, removing x from the second, third, and fourth equations, then y from the third, and fourth equations, and finally, z from the fourth equation. Hence, we compute t with the fourth equation, then z with the third equation, y with the second equation, and lastly, x with the first equation.

Step 1. We denote by $p = 2$ the entry $1, 1$ of the system matrix, we shall call it the pivot (of the first step). Substituting

- the second equation by itself "minus" the first equation multiplied by $\frac{3}{p}$,
- the third equation by itself "minus" the first equation multiplied by $\frac{-1}{p}$,
- the third equation by itself "minus" the first equation multiplied by $\frac{1}{p}$,

we get the following system:

$$
\left\{
\begin{array}{rrrrcr}
2x & +4y & -4z & +t & = & 0, \\
 & & 7z & -7t/2 & = & -7, \\
 & 3y & & +7t/2 & = & 4, \\
 & -y & -2z & +t/2 & = & 2.
\end{array}
\right.
$$

Step 2. This time around, the pivot (entry $2, 2$ of the new matrix) is zero. We swap the second row and the third one in order to get a nonzero pivot:

$$
\left\{
\begin{array}{rrrrcr}
2x & +4y & -4z & +t & = & 0, \\
 & 3y & & +7t/2 & = & 4, \\
 & & 7z & -7t/2 & = & -7, \\
 & -y & -2z & +t/2 & = & 2.
\end{array}
\right.
$$

The new pivot is $p = 3$:

- the third equation is unchanged,
- substituting the fourth equation by itself "minus" the second equation multiplied by $\frac{-1}{p}$, we obtain the system

$$
\left\{
\begin{array}{rrrrcr}
2x & +4y & -4z & +t & = & 0, \\
 & 3y & & +7t/2 & = & 4, \\
 & & 7z & -7t/2 & = & -7, \\
 & & -2z & +5t/3 & = & \frac{10}{3}.
\end{array}
\right.
$$

Step 3. Entry $3, 3$ of the matrix is nonzero, we set $p = 7$, and we substitute the fourth equation by itself "minus" the third equation multiplied by $\frac{-2}{p}$:

$$\begin{cases} 2x & +4y & -4z & +t & = & 0, \\ & 3y & & +7t/2 & = & 4, \\ & & 7z & -7t/2 & = & -7, \\ & & & 2t/3 & = & \frac{4}{3}. \end{cases} \tag{6.3}$$

The last system is triangular. It is easily solved through back substitution; we obtain $t = 2$, $z = 0$, $y = -1$, and $x = 1$.

We now present a matrix formalism allowing to convert any matrix (or system such as (6.2)) into a triangular matrix (or system such as (6.3)). The idea is to find a nonsingular matrix M such that the product MA is upper triangular, then to solve through back substitution the triangular system $MAx = Mb$. To implement this idea, the Gaussian elimination method is broken into three steps:

- elimination: computation of a nonsingular matrix M such that $MA = T$ is upper triangular;
- right-hand-side update: simultaneous computation of Mb;
- substitution: solving the triangular system $Tx = Mb$ by mere back substitution.

The existence of such a matrix M is ensured by the following result to which we shall give a constructive proof that is nothing but the Gaussian elimination method itself.

Theorem 6.1.1 (Gaussian elimination theorem). *Let A be a square matrix (invertible or not). There exists at least one nonsingular matrix M such that the matrix $T = MA$ is upper triangular.*

Proof. The outline of the method is as follows: we build a sequence of matrices A^k, for $1 \leq k \leq n$, in such a way that we go from $A^1 = A$ to $A^n = T$, by successive alterations. The entries of the matrix A^k are denoted by

$$A^k = \left(a_{i,j}^k \right)_{1 \leq i,j \leq n},$$

and the entry $a_{k,k}^k$ is called the pivot of A^k. To pass from A^k to A^{k+1}, we shall first make sure that the pivot $a_{k,k}^k$ is nonzero. If it is not so, we permute the kth row with another row in order to bring a nonzero element into the pivot position. The corresponding permutation matrix is denoted by P^k (see Section 2.2.4 on row permutations). Then, we proceed to the elimination of all entries of the kth column below the kth row by linear combinations of the current row with the kth row. Namely, we perform the following steps.
Step 1: We start off with $A^1 = A$. We build a matrix $\tilde{A}^1$ of the form

$$\tilde{A}^1 = P^1 A^1,$$

where P^1 is a permutation matrix such that the new pivot $\tilde{a}_{1,1}^1$ is nonzero. If the pivot $a_{1,1}^1$ is nonzero, we do not permute, i.e., we take $P^1 = I$. If $a_{1,1}^1 = 0$

and if there exists an entry in the first column $a_{i,1}^1 \neq 0$ (with $2 \leq i \leq n$), then we swap the first row with the ith and P^1 is equal to the elementary permutation matrix $P(1, i)$ (we recall that the elementary permutation matrix $P(i, j)$ is equal to the identity matrix whose rows i and j have been swapped). Next, we multiply $\tilde{A}^1$ by the matrix E^1 defined by

$$
E^1 = \begin{pmatrix} 1 & & & 0 \\ -\frac{\tilde{a}_{2,1}^1}{\tilde{a}_{1,1}^1} & \ddots & & \\ \vdots & & \ddots & \\ -\frac{\tilde{a}_{n,1}^1}{\tilde{a}_{1,1}^1} & & & 1 \end{pmatrix} ;
$$

this removes all the entries of the first column but the first. We set

$$
A^2 = E^1 \tilde{A}^1 = \begin{pmatrix} \tilde{a}_{1,1}^1 & \cdots & \tilde{a}_{1,n}^1 \\ 0 & & \\ \vdots & \left(a_{i,j}^2 \right) \\ 0 & \end{pmatrix}
$$

with $a_{i,j}^2 = \tilde{a}_{i,j}^1 - \frac{\tilde{a}_{i,1}^1}{\tilde{a}_{1,1}^1} \tilde{a}_{1j}^1$ for $2 \leq i, j \leq n$. The matrix A^2 has therefore a first column with only zeros below its diagonal.

During the permutation step, it may happen that all the elements of the first column $a_{i,1}^1$ vanish in which case it is not possible to find a nonzero pivot. This is not a problem, since this first column has already the desired properties of having zeros below its diagonal! We merely carry on with the next step by setting $A^2 = A^1$ and $E^1 = P^1 = I$. Such an instance occurs only if A is singular; otherwise, its first column is inevitably nonzero.

Step K: We assume that A^k has its $(k - 1)$ first columns with zeros below its diagonal. We multiply A^k by a permutation matrix P^k to obtain

$$
\tilde{A}^k = P^k A^k
$$

such that its pivot $\tilde{a}_{k,k}^k$ is nonzero. If $a_{k,k}^k \neq 0$, then we take $P^k = I$. Otherwise, there exists $a_{i,k}^k \neq 0$ with $i \geq k + 1$, so we swap the kth row with the ith by taking $P^k = P(i, k)$. Next, we multiply $\tilde{A}^k$ by a matrix E^k defined by

$$E^k = \begin{pmatrix} 1 & & & & & & \\ 0 & \ddots & & & & & \\ \vdots & & 1 & & & & \\ \vdots & & -\frac{\tilde{a}^k_{k+1,k}}{\tilde{a}^k_{k,k}} & 1 & & & \\ \vdots & & \vdots & & \ddots & & \\ \vdots & & \vdots & & & \ddots & \\ 0 & & -\frac{\tilde{a}^k_{n,k}}{\tilde{a}^k_{k,k}} & & & & 1 \end{pmatrix},$$

which removes all the entries of the kth column below the diagonal. We set

$$A^{k+1} = E^k \tilde{A}^k = \begin{pmatrix} \tilde{a}^1_{1,1} \cdots \cdots & & \cdots & & \cdots & \tilde{a}^1_{1,n} \\ 0 & \ddots & & & & \vdots \\ \vdots & \ddots & \tilde{a}^k_{k,k} & \tilde{a}^k_{k,k+1} & \cdots & \tilde{a}^k_{k,n} \\ \vdots & & 0 & a^{k+1}_{k+1,k+1} & \cdots & a^{k+1}_{k+1,n} \\ \vdots & & \vdots & \vdots & & \vdots \\ 0 & \cdots & 0 & a^{k+1}_{n,k+1} & \cdots & a^{k+1}_{n,n} \end{pmatrix}$$

with $a^{k+1}_{i,j} = \tilde{a}^k_{i,j} - \frac{\tilde{a}^k_{i,k}}{\tilde{a}^k_{k,k}}\tilde{a}^k_{k,j}$ for $k+1 \leq i,j \leq n$. The matrix A^{k+1} has its first k columns with only zeros below the diagonal. During the permutation step, it may happen that all elements of the kth column below the diagonal, $a^k_{i,k}$ with $i \geq k$, are zeros. Then, this kth column has already the desired form and there is nothing to be done! We carry on with the next step by setting $A^{k+1} = A^k$ and $E^k = P^k = I$. Such an instance occurs only if A is singular. Indeed, if A is nonsingular, then so is, A^k, and its kth column cannot have zeros from the kth line to the last one, since its determinant would then be zero.

After $(n-1)$ steps, the matrix A^n is upper triangular:

$$A^n = (E^{n-1}P^{n-1} \cdots E^1 P^1)A.$$

We set $M = E^{n-1}P^{n-1} \cdots E^1 P^1$. It is indeed a nonsingular matrix, since

$$\det M = \prod_{i=1}^{n-1} \det E^i \det P^i,$$

with $\det P^i = \pm 1$ and $\det E^i = 1$.

We can refresh the right-hand side (that is, compute Mb) sequentially while computing the matrices P^k and E^k. We build a sequence of right-hand sides $(b^k)_{1 \leq k \leq n}$ defined by

$$b^1 = b, \quad b^{k+1} = E^k P^k b^k, \quad \text{for } 1 \le k \le n - 1,$$

which satisfies $b^n = Mb$ in the end.

To solve the linear system $Ax = b$, it suffices now to solve the system $A^n x = Mb$, where A^n is an upper triangular matrix. If A is singular, we can still perform the elimination step, that is, compute A^n. However, there is no guarantee that we can solve the system $A^n x = Mb$, since one of the diagonal entries of A^n is zero.

Remark 6.1.1. The proof of Theorem 6.1.1 is indeed exactly the Gaussian elimination method that is used in practice. It is therefore important to emphasize some practical details.

1. We never compute M! We need not multiply matrices E^i and P^i to determine Mb and A^n.
2. If A is singular, one of the diagonal entries of $A^n = T$ is zero. As a result, we cannot always solve $Tx = Mb$. Even so, elimination is still possible.
3. At step k, we only modify rows from $k+1$ to n between columns $k+1$ to n.
4. A byproduct of Gaussian elimination is the easy computation of the determinant of A. Actually, we have $\det A = \pm \det T$ depending on the number of performed permutations.
5. In order to obtain better numerical stability in computer calculations, we may choose the pivot $\tilde{a}_{k,k}^k$ in a clever way. To avoid the spreading of rounding errors, the largest possible pivot (in absolute value) is preferred. The same selection can be done when the usual pivot $a_{k,k}^k$ is nonzero, i.e., we swap rows and/or columns to substitute it with a larger pivot $\tilde{a}_{k,k}^k$. We call the process of choosing the largest possible pivot in the kth column under the diagonal (as we did in the above proof) partial pivoting. We call the process of choosing the largest possible pivot in the lower diagonal submatrix of size $(n - k + 1) \times (n - k + 1)$ formed by the intersection of the last $(n - k + 1)$ rows and the last $(n - k + 1)$ columns (in such a case, we swap rows and columns) complete pivoting. These two variants of Gaussian elimination are studied in Exercise 6.3.

Let us go back to Example 6.1.1 and describe the different steps with the matrix formalism. The initial system reads as $Ax = b$ with

$$A = \begin{pmatrix} 2 & 4 & -4 & 1 \\ 3 & 6 & 1 & -2 \\ -1 & 1 & 2 & 3 \\ 1 & 1 & -4 & 1 \end{pmatrix}, \quad b = \begin{pmatrix} 0 \\ -7 \\ 4 \\ 2 \end{pmatrix}.$$

Set $A_1 = A$ and $b_1 = b$. In the first step, the pivot is nonzero so we take $P_1 = I$,

$$E_1 = \begin{pmatrix} 1 & 0 & 0 & 0 \\ -\frac{3}{2} & 1 & 0 & 0 \\ \frac{1}{2} & 0 & 1 & 0 \\ -\frac{1}{2} & 0 & 0 & 1 \end{pmatrix}, \quad A_2 = E_1 P_1 A_1 = \begin{pmatrix} 2 & 4 & -4 & 1 \\ 0 & 0 & 7 & -\frac{7}{2} \\ 0 & 3 & 0 & \frac{7}{2} \\ 0 & -1 & -2 & \frac{1}{2} \end{pmatrix},$$

and

$$b_2 = E_1 P_1 b_1 = \begin{pmatrix} 0 \\ -7 \\ 4 \\ 2 \end{pmatrix}.$$

In the second step, the pivot is zero, so we take

$$P_2 = \begin{pmatrix} 1 & 0 & 0 & 0 \\ 0 & 0 & 1 & 0 \\ 0 & 1 & 0 & 0 \\ 0 & 0 & 0 & 1 \end{pmatrix}, \quad E_2 = \begin{pmatrix} 1 & 0 & 0 & 0 \\ 0 & 1 & 0 & 0 \\ 0 & 0 & 1 & 0 \\ 0 & \frac{1}{3} & 0 & 1 \end{pmatrix},$$

and

$$A_3 = E_2 P_2 A_2 = \begin{pmatrix} 2 & 4 & -4 & 1 \\ 0 & 3 & 0 & \frac{7}{2} \\ 0 & 0 & 7 & -\frac{7}{2} \\ 0 & 0 & -2 & \frac{5}{3} \end{pmatrix}, \quad b_3 = E_2 P_2 b_2 = \begin{pmatrix} 0 \\ 4 \\ -7 \\ \frac{10}{3} \end{pmatrix}.$$

In the third step, the pivot is nonzero, so we take $P_3 = I$,

$$E_3 = \begin{pmatrix} 1 & 0 & 0 & 0 \\ 0 & 1 & 0 & 0 \\ 0 & 0 & 1 & 0 \\ 0 & 0 & \frac{2}{7} & 1 \end{pmatrix}, \quad A_4 = E_3 P_3 A_3 = \begin{pmatrix} 2 & 4 & -4 & 1 \\ 0 & 3 & 0 & \frac{7}{2} \\ 0 & 0 & 7 & -\frac{7}{2} \\ 0 & 0 & 0 & \frac{2}{3} \end{pmatrix},$$

and

$$b_4 = E_3 P_3 b_3 = \begin{pmatrix} 0 \\ 4 \\ -7 \\ \frac{4}{3} \end{pmatrix}.$$

The solution x is computed by solving the upper triangular system $A_4 x = b_4$, and we obtain $x = (1, -1, 0, 2)^t$.

6.2 LU Decomposition Method

The LU decomposition method consists in factorizing A into a product of two triangular matrices

$$A = LU,$$

where L is lower triangular and U is upper triangular. This decomposition allows us to reduce the solution of the system $Ax = b$ to solving two triangular systems $Ly = b$ and $Ux = y$. It turns out to be nothing else than Gaussian elimination in the case without pivoting.

The matrices defined by

$$\Delta^k = \begin{pmatrix} a_{1,1} & \cdots & a_{1,k} \\ \vdots & \ddots & \vdots \\ a_{k,1} & \cdots & a_{k,k} \end{pmatrix}$$

are called the diagonal submatrices of order k of $A \in \mathcal{M}_n$. The next result gives a sufficient condition on the matrix A to have no permutation during Gaussian elimination.

Theorem 6.2.1 (LU factorization). *Let $A = (a_{i,j})_{1 \leq i,j \leq n}$ be a matrix of order n all of whose diagonal submatrices of order k are nonsingular. There exists a unique pair of matrices (L, U), with U upper triangular and L lower triangular with a unit diagonal (i.e., $l_{i,i} = 1$), such that $A = LU$.*

Remark 6.2.1. The condition stipulated by the theorem is often satisfied in practice. For example, it holds true if A is positive definite, i.e.,

$$x^t A x > 0, \quad \forall x \neq 0.$$

Indeed, if Δ^k were singular, then there would exist a vector

$$\begin{pmatrix} x_1 \\ \vdots \\ x_k \end{pmatrix} \neq 0 \quad \text{such that} \quad \Delta^k \begin{pmatrix} x_1 \\ \vdots \\ x_k \end{pmatrix} = 0.$$

Now let x_0 be the vector whose k first entries are $(x_1, \ldots, x_k)$, and whose last $n - k$ entries are zero. We have

$$x_0^t A x_0 = 0 \quad \text{and} \quad x_0 \neq 0,$$

which contradicts the assumption that A is positive definite. Consequently, Δ^k is nonsingular. Note that the converse is not true: namely, a matrix A, such that all its diagonal submatrices are nonsingular is not necessarily positive definite, as in the following instance:

$$A = \begin{pmatrix} 1 & 0 \\ 0 & -1 \end{pmatrix}.$$

Hence the assumption of Theorem 6.2.1 is more general than positive definiteness.

Proof of Theorem 6.2.1. Assume that during Gaussian elimination, there is no need to permute in order to change the pivot, that is, all natural pivots $a_{k,k}^k$ are nonzero. Then we have

$$A^n = E^{n-1} \cdots E^1 A,$$

with

$$E^k = \begin{pmatrix} 1 & & & & & & \\ 0 & \ddots & & & & & \\ \vdots & & 1 & & & & \\ \vdots & & -l_{k+1,k} & 1 & & & \\ \vdots & & \vdots & & \ddots & & \\ \vdots & & \vdots & & & \ddots & \\ 0 & & -l_{n,k} & & & & 1 \end{pmatrix},$$

and for $k + 1 \leq i \leq n$,

$$l_{i,k} = \frac{a_{i,k}^k}{a_{k,k}^k}.$$

Set $U = A^n$ and $L = (E^1)^{-1} \cdots (E^{n-1})^{-1}$, so that we have $A = LU$. We need to check that L is indeed lower triangular. A simple computation shows that $(E^k)^{-1}$ is easily deduced from E^k by changing the sign of the entries below the diagonal:

$$(E^k)^{-1} = \begin{pmatrix} 1 & & & & & & \\ 0 & \ddots & & & & & \\ \vdots & & 1 & & & & \\ \vdots & & +l_{k+1,k} & 1 & & & \\ \vdots & & \vdots & & \ddots & & \\ \vdots & & \vdots & & & \ddots & \\ 0 & & +l_{n,k} & & & & 1 \end{pmatrix}.$$

Another computation shows that L is lower triangular and that its kth column is the kth column of $(E^k)^{-1}$:

$$L = \begin{pmatrix} 1 & 0 & \cdots & 0 \\ l_{2,1} & \ddots & \ddots & \vdots \\ \vdots & \ddots & \ddots & 0 \\ l_{n,1} & \cdots & l_{n,n-1} & 1 \end{pmatrix}.$$

It remains to prove that the pivots do not vanish under the assumption made on the submatrices Δ^k. We do so by induction. The first pivot $a_{1,1}$ is nonzero, since it is equal to Δ^1, which is nonsingular. We assume the first $k - 1$ pivots to be nonzero. We have to show that the next pivot $a_{k,k}^k$ is nonzero too. Since the first $k - 1$ pivots are nonzero, we have computed without permutation the matrix A^k, which is given by $(E^1)^{-1} \cdots (E^{k-1})^{-1} A^k = A$. We write this equality with block matrices:

$$\begin{pmatrix} L_{1,1}^k & 0 \\ L_{2,1}^k & I \end{pmatrix} \begin{pmatrix} U_{1,1}^k & A_{1,2}^k \\ A_{2,1}^k & A_{2,2}^k \end{pmatrix} = \begin{pmatrix} \Delta^k & A_{1,2} \\ A_{2,1} & A_{2,2} \end{pmatrix},$$

where $U_{1,1}^k$, $L_{1,1}^k$, and Δ^k are square blocks of size k, and $A_{2,2}^k$, I, and $A_{2,2}$ square blocks of size $n - k$. Applying the block matrix product rule yields

$$L_{1,1}^k U_{1,1}^k = \Delta^k,$$

where $U_{1,1}^k$ is an upper triangular matrix, and $L_{1,1}^k$ is a lower triangular matrix with 1 on the diagonal. We deduce that $U_{1,1}^k = (L_{1,1}^k)^{-1} \Delta^k$ is nonsingular as a product of two nonsingular matrices. Its determinant is therefore nonzero,

$$\det U_{1,1}^k = \prod_{i=1}^k a_{i,i}^k \neq 0,$$

which implies that the pivot $a_{k,k}^k$ at the kth step is nonzero.

Finally, let us check the uniqueness of the decomposition. Let there be two LU factorizations of A:

$$A = L_1 U_1 = L_2 U_2.$$

We infer that

$$L_2^{-1} L_1 = U_2 U_1^{-1},$$

where $L_2^{-1} L_1$ is lower triangular, and $U_2 U_1^{-1}$ is upper triangular. By virtue of Lemma 2.2.5, the inverse and product of two upper (respectively, lower) triangular matrices are upper (respectively, lower) triangular too. Hence, both matrices are diagonal, and since the diagonal of $L_2^{-1} L_1$ consists of 1's, we have

$$L_2^{-1} L_1 = U_2 U_1^{-1} = I,$$

which proves the uniqueness.   □

Determinant of a matrix. As for Gaussian elimination, a byproduct of the LU factorization is the computation of the determinant of the matrix A, since $\det A = \det U$. As we shall check in Section 6.2.3, it is a much more efficient method than the usual determinant formula (cf. Definition 2.2.8).

Conditioning of a matrix. Knowing the LU decomposition of A yields an easy upper bound on its conditioning, $\mathrm{cond}(A) \leq \mathrm{cond}(L)\, \mathrm{cond}(U)$, where the conditionings of triangular matrices are computed by Algorithm 5.4.

Incomplete LU preconditioning. Since the LU factorization provides a way of computing A^{-1}, an approximate LU factorization yields an approximation of A^{-1} that can be used as a preconditioner. A common approximation is the so-called incomplete LU factorization, which computes triangular matrices $\tilde{L}$ and $\tilde{U}$ such that $\tilde{L}\tilde{U}$ is an approximation of A that is cheap to compute. To have a fast algorithm for obtaining $\tilde{L}$ and $\tilde{U}$ we modify the standard LU algorithm as follows: the entries $\tilde{L}_{i,j}$ and $\tilde{U}_{i,j}$ are computed only if the element $A_{i,j}$ is nonzero (or larger than a certain threshold). The sparse structure of A is thus conserved; see Exercise 6.10.

6.2.1 Practical Computation of the LU Factorization

A practical way of computing the LU factorization (if it exists) of a matrix A is to set

$$
L = \begin{pmatrix} 1 & 0 & \cdots & 0 \\ l_{2,1} & \ddots & \ddots & \vdots \\ \vdots & \ddots & \ddots & 0 \\ l_{n,1} & \cdots & l_{n,n-1} & 1 \end{pmatrix}, \quad U = \begin{pmatrix} u_{1,1} & \cdots & \cdots & u_{1,n} \\ 0 & u_{2,2} & & \vdots \\ \vdots & \ddots & \ddots & \vdots \\ 0 & \cdots & 0 & u_{n,n} \end{pmatrix},
$$

and then identify the product LU with A. Since L is lower triangular and U upper triangular, for $1 \leq i, j \leq n$, it entails

$$
a_{i,j} = \sum_{k=1}^{n} l_{i,k} u_{k,j} = \sum_{k=1}^{\min(i,j)} l_{i,k} u_{k,j}.
$$

A simple algorithm is thus to read in increasing order the columns of A and to deduce the entries of the columns of L and U.

Column 1. We fix $j = 1$ and vary i:

$$
a_{1,1} = l_{1,1} u_{1,1} \Rightarrow u_{1,1} = a_{1,1};
$$
$$
a_{2,1} = l_{2,1} u_{1,1} \Rightarrow l_{2,1} = \frac{a_{2,1}}{a_{1,1}};
$$
$$
\vdots \qquad\qquad \vdots
$$
$$
a_{n,1} = l_{n,1} u_{1,1} \Rightarrow l_{n,1} = \frac{a_{n,1}}{a_{1,1}}.
$$

We have thereby computed all the entries of the first column of L and of the first column of U.

Column j. We assume that we have computed the first $(j-1)$ columns of L and U. Then we read the jth column of A:

$$
\begin{aligned}
a_{1,j} &= l_{1,1} u_{1,j} & &\Rightarrow u_{1,j} = a_{1,j}; \\
a_{2j} &= l_{2,1} u_{1,j} + l_{2,2} u_{2,j} & &\Rightarrow u_{2,j} = a_{2,j} - l_{2,1} a_{1,j}; \\
&\vdots & &\vdots \\
a_{j,j} &= l_{j,1} u_{1,j} + \cdots + l_{j,j} u_{j,j} & &\Rightarrow u_{j,j} = a_{j,j} - \sum_{k=1}^{j-1} l_{j,k} u_{k,j}; \\
a_{j+1,j} &= l_{j+1,1} u_{1,j} + \cdots + l_{j+1,j} u_{j,j} & &\Rightarrow l_{j+1,j} = \frac{a_{j+1,j} - \sum_{k=1}^{j-1} l_{j+1,k} u_{k,j}}{u_{j,j}}; \\
&\vdots & &\vdots \\
a_{n,j} &= l_{n,1} u_{1,j} + \cdots + l_{n,j} u_{j,j} & &\Rightarrow l_{n,j} = \frac{a_{n,j} - \sum_{k=1}^{j-1} l_{n,k} u_{k,j}}{u_{j,j}}.
\end{aligned}
$$

We compute in this way the first j entries of the jth column of U and the last $n - j$ entries of the jth column of L in terms of the first $(j - 1)$ columns. Note the division by the pivot $u_{j,j}$, which should not be zero.

Example 6.2.1. The LU factorization of the matrix in Example 6.1.1, *by swapping its second and third rows,* is

$$\begin{pmatrix} 1 & 0 & 0 & 0 \\ -\frac{1}{2} & 1 & 0 & 0 \\ \frac{3}{2} & 0 & 1 & 0 \\ \frac{1}{2} & -\frac{1}{3} & -\frac{2}{7} & 1 \end{pmatrix} \begin{pmatrix} 2 & 4 & -4 & 1 \\ 0 & 3 & 0 & \frac{7}{2} \\ 0 & 0 & 7 & -\frac{7}{2} \\ 0 & 0 & 0 & \frac{2}{3} \end{pmatrix} = \begin{pmatrix} 2 & 4 & -4 & 1 \\ -1 & 1 & 2 & 3 \\ 3 & 6 & 1 & -2 \\ 1 & 1 & -4 & 1 \end{pmatrix} = P(2,3)A.$$

6.2.2 Numerical Algorithm

We now write in pseudolanguage the algorithm corresponding to the LU factorization. We have just seen that the matrix A is scanned column by column. At the kth step we change its kth column so that the entries below the diagonal vanish by performing linear combinations of the kth row with every row from the $(k + 1)$th to the nth. At the kth step, the first k rows and the first $k - 1$ columns of the matrix are no longer modified. We exploit this property in order to store in the same array, which initially contains the matrix A, the matrices A^k and $L^k = (E^1)^{-1} \cdots (E^{k-1})^{-1}$. More precisely, the zeros of A^k in its first $(k - 1)$ columns below the diagonal are replaced by the corresponding nontrivial entries of L^k, which all lie below the diagonal in the first $(k - 1)$ columns. At the end of the process, the array will contain the two triangular matrices L and U (i.e., the lower part of L without the diagonal of 1's and the upper part of U). We implement this idea in Algorithm 6.1, where the columns of L are also precomputed before the linear combinations of rows, which saves some operations.

Data: A. Output: A containing U and L (but its diagonal)
    For $k = 1 \nearrow n - 1$ *step* k
        For $i = k + 1 \nearrow n$ *row* i
            $a_{i,k} = \dfrac{a_{i,k}}{a_{k,k}}$ *new column of* L
            For $j = k + 1 \nearrow n$
                $a_{i,j} = a_{i,j} - a_{i,k}a_{kj}$ *combination of rows* i *and* k
            End j
        End i
    End k

Algorithm 6.1: LU factorization algorithm.

6.2.3 Operation Count

To assess the efficiency of the LU factorization algorithm, we count the number of operations $N_{op}(n)$ its execution requires (which will be proportional to the running time on a computer). We do not accurately determine the number of operations, and we content ourselves with the first-order term of its asymptotic expansion when the dimension n is large. Moreover, for simplicity, we count only multiplications and divisions, but not additions, as usual (see Section 4.2).

✗ LU factorization: the number of operations is

$$N_{op}(n) = \sum_{k=1}^{n-1} \sum_{i=k+1}^{n} \left(1 + \sum_{j=k+1}^{n} 1 \right),$$

which, to first order yields

$$N_{op}(n) \approx \frac{n^3}{3}.$$

✗ Back substitution (on the triangular system): the number of operations is

$$N_{op}(n) = \sum_{j=1}^{n} j \approx \frac{n^2}{2}.$$

✗ Solution of a linear system $Ax = b$: an LU factorization of A is followed by two substitutions, $Ly = b$, then $Ux = y$. Since n^2 is negligible compared to n^3 when n is large, the number of operations is

$$N_{op}(n) \approx \frac{n^3}{3}.$$

The LU method (or Gaussian elimination) is therefore much more efficient than Cramer's formulas for solving a linear system. Once the LU factorization of the matrix is performed, we can easily compute its determinant, as well as its inverse matrix.

✗ Computing $\det A$: we compute the determinant of U (that of L is equal to 1), which requires only the product of the diagonal entries of U ($n - 1$ multiplications). As a consequence, the number of operations is again

$$N_{op}(n) \approx \frac{n^3}{3}.$$

✗ Computing A^{-1}: the columns of A^{-1}, denoted by x_i, are the solutions of the n systems $Ax_i = e_i$, where $(e_i)_{1 \le i \le n}$ is the canonical basis of $\mathbb{R}^n$. A naive count of operations for computing A^{-1} is $4n^3/3$, which is the sum of $n^3/3$ for a single LU factorization and n^3 for solving $2n$ triangular systems. We can improve this number by taking into account the fact that the basis vectors e_i have many zero entries, which decreases the cost of the forward substitution step with L, because the solution of $Ly_i = e_i$ has its first $(i - 1)$ entries equal to zero. The number of operations becomes

$$N_{op}(n) \approx \frac{n^3}{3} + \sum_{j=1}^{n} \frac{j^2}{2} + n \left(\frac{n^2}{2} \right) \approx n^3.$$

6.2.4 The Case of Band Matrices

Band matrices appear in many applications (such as the discretization of partial differential equations, see Section 1.1) and their special structure allows us to spare memory and computational cost during the LU factorization.

Definition 6.2.1. *A matrix $A \in M_n(\mathbb{R})$ satisfying $a_{i,j} = 0$ for $|i - j| > p$ with $p \in \mathbb{N}$ is said to be a band matrix of bandwidth $2p + 1$.*

For instance, a tridiagonal matrix has half-bandwidth $p = 1$.

Proposition 6.2.1. *The LU factorization preserves the band structure of matrices.*

Proof. Let A be a band matrix of bandwidth $2p + 1$ and let $A = LU$ be its LU decomposition. We want to prove that L and U are band matrices of bandwidth $2p + 1$ too. By definition we have

$$a_{i,j} = \sum_{k=1}^{\min(i,j)} l_{i,k} u_{k,j}.$$

We proceed by induction on $i = 1, \ldots, n$. For $i = 1$, we have

- $a_{1,j} = l_{1,1} u_{1,j} = u_{1,j}$; we infer that $u_{1,j} = 0$ for all $j > p + 1$;
- $a_{j,1} = l_{j,1} u_{1,1}$; in particular, $a_{1,1} = l_{1,1} u_{1,1}$, which entails that $u_{1,1} = a_{1,1} \neq 0$, so $l_{j,1} = a_{j,1}/a_{1,1}$, which shows that $l_{j,1} = 0$ for all $j > p + 1$.

Assume now that for all $i = 1, \ldots, I - 1$, we have

$$j > i + p \Longrightarrow u_{i,j} = l_{j,i} = 0,$$

and let us prove that this property holds for $i = I$.

- For $j > I + p$, we have

$$a_{I,j} = \sum_{k=1}^{I} l_{I,k} u_{k,j} = l_{I,I} u_{I,j} + \sum_{k=1}^{I-1} l_{I,k} u_{k,j}.$$

Observing that for all $k = 1, \ldots, I-1$, we have $j > I+p \geq k+1+p > k+p$, we deduce, by application of the induction hypothesis, that $u_{k,j} = 0$, $j > k + p$. This implies that $a_{I,j} = u_{I,j}$, hence proving that $u_{I,j} = 0$ for all $j > I + p$.
- Similarly, for $j > I + p$, we have

$$a_{j,I} = l_{j,I} u_{I,I} + \sum_{k=1}^{I-1} l_{j,k} u_{k,I} = l_{j,I} u_{I,I}.$$

The proof of Theorem 6.2.1 has shown that $u_{I,I} \neq 0$. Thus $l_{j,I} = a_{j,I}/u_{I,I}$, which implies that $l_{j,I} = 0$ for $j > I + p$.

So the matrices L and U have the same band structure as the matrix A.   □

Remark 6.2.2. Proposition 6.2.1 does not state that the matrices A, on the one hand, and L and U, on the other, have the same "sparse" structure. It may happen that the band of A is sparse (contains a lot of zeros), whereas the bands of L and U are full. We say that the LU decomposition has filled the matrix's band; see Exercise 6.8. Reducing the bandwidth of a matrix is a good way of minimizing the computational and the memory requirements of its LU factorization; see Exercise 6.9.

Example 6.2.2. To compute the LU factorization of the matrix

$$A = \begin{pmatrix} 1 & 2 & 0 & 0 & 0 \\ 2 & 6 & 1 & 0 & 0 \\ 0 & 2 & -2 & -1 & 0 \\ 0 & 0 & -9 & 1 & 2 \\ 0 & 0 & 0 & -4 & 3 \end{pmatrix},$$

we look for the matrices L and U:

$$L = \begin{pmatrix} 1 & 0 & 0 & 0 & 0 \\ a & 1 & 0 & 0 & 0 \\ 0 & b & 1 & 0 & 0 \\ 0 & 0 & c & 1 & 0 \\ 0 & 0 & 0 & d & 1 \end{pmatrix}, \quad U = \begin{pmatrix} e & f & 0 & 0 & 0 \\ 0 & g & h & 0 & 0 \\ 0 & 0 & i & j & 0 \\ 0 & 0 & 0 & k & l \\ 0 & 0 & 0 & 0 & m \end{pmatrix}.$$

By identification of the entries of the product $A = LU$, we obtain

$$A = \begin{pmatrix} 1 & 0 & 0 & 0 & 0 \\ 2 & 1 & 0 & 0 & 0 \\ 0 & 1 & 1 & 0 & 0 \\ 0 & 0 & 3 & 1 & 0 \\ 0 & 0 & 0 & -1 & 1 \end{pmatrix} \begin{pmatrix} 1 & 2 & 0 & 0 & 0 \\ 0 & 2 & 1 & 0 & 0 \\ 0 & 0 & -3 & -1 & 0 \\ 0 & 0 & 0 & 4 & 2 \\ 0 & 0 & 0 & 0 & 5 \end{pmatrix}.$$

Storage of a band matrix. To store a band matrix A of half-bandwidth p, we use a vector array STOREA of dimension $(2p+1)n$. The matrix A is stored row by row, starting with the first. Let k be the index such that $A(i,j) =$ STOREA (k). To determine k, we notice that the first entry of row i of A (i.e., $a_{i,i-p}$) has to be placed in position $(i-1)(2p+1)+1$, from which we deduce that the element $a_{i,j}$ has to be stored in STOREA in position $k(i,j) = (i-1)(2p+1)+j-i+p+1 = (2i-1)p+j$. Be aware that some entries of the vector STOREA are not allocated; however, their number is equal to $p(p+1)$, which is negligible; see Exercise 6.5.

6.3 Cholesky Method

The Cholesky method applies only to positive definite real symmetric matrices. Recall that a real symmetric matrix A is positive definite if all its eigenvalues are positive. The Cholesky method amounts to factorizing $A = BB^t$ with B a lower triangular matrix, so that solving the linear system $Ax = b$ boils down to two triangular systems $By = b$ and $B^t x = y$.

Theorem 6.3.1 (Cholesky factorization). *Let A be a real symmetric positive definite matrix. There exists a unique real lower triangular matrix B, having positive diagonal entries, such that*

$$A = BB^*.$$

Proof. By the LU factorization theorem, there exists a unique pair of matrices (L, U) satisfying $A = LU$ with

$$L = \begin{pmatrix} 1 & & & \\ \times & \ddots & & \\ \vdots & & \ddots & \\ \times & \cdots & \times & 1 \end{pmatrix} \quad \text{and} \quad U = \begin{pmatrix} u_{1,1} & \times & \cdots & \times \\ & \ddots & & \vdots \\ & & \ddots & \times \\ & & & u_{nn} \end{pmatrix}.$$

We introduce the diagonal matrix $D = \mathrm{diag}\,(\sqrt{u_{i,i}})$. The square root of the diagonal entries of U are well defined, since $u_{i,i}$ is positive, as we now show. By the same argument concerning the product of block matrices as in the proof of Theorem 6.2.1, we have

$$\prod_{i=1}^{k} u_{i,i} = \det \Delta^k > 0,$$

with Δ^k the diagonal submatrix of order k of A. Therefore, by induction, each $u_{i,i}$ is positive. Next, we set $B = LD$ and $C = D^{-1}U$, so that $A = BC$. Since $A = A^*$, we deduce

$$C(B^*)^{-1} = B^{-1}(C^*).$$

By virtue of Lemma 2.2.5, $C(B^*)^{-1}$ is upper triangular, while $B^{-1}C^*$ is lower triangular. Both of them are thus diagonal. Furthermore, the diagonal entries of B and C are the same. Therefore, all diagonal entries of $B^{-1}C^*$ are equal to 1. We infer $C(B^*)^{-1} = B^{-1}C^* = I$, which implies that $C = B^*$.

To prove the uniqueness of the Cholesky factorization, we assume that there exist two such factorizations:

$$A = B_1 B_1^* = B_2 B_2^*.$$

Then

$$B_2^{-1} B_1 = B_2^* (B_1^*)^{-1},$$

and Lemma 2.2.5 implies again that there exists a diagonal matrix $D = \mathrm{diag}\,(d_1, \ldots, d_n)$ such that $B_2^{-1}B_1 = D$. We deduce that $B_1 = B_2 D$ and

$$A = B_2 B_2^* = B_2(DD^*)B_2^* = B_2 D^2 B_2^*.$$

Since B_2 is nonsingular, it yields $D^2 = I$, so $d_i = \pm 1$. However, all the diagonal entries of a Cholesky factorization are positive by definition. Therefore $d_i = 1$, and $B_1 = B_2$. $\qquad\square$

6.3.1 Practical Computation of the Cholesky Factorization

We now give a practical algorithm for computing the Cholesky factor B for a positive definite symmetric matrix A. This algorithm is different from that of LU factorization. Take $A = (a_{i,j})_{1\le i,j\le n}$ and $B = (b_{i,j})_{1\le i,j\le n}$ with $b_{i,j} = 0$ if $i < j$. We identify the entries on both sides of the equality $A = BB^*$. For $1 \le i, j \le n$, we get

$$a_{i,j} = \sum_{k=1}^{n} b_{i,k}b_{j,k} = \sum_{k=1}^{\min(i,j)} b_{i,k}b_{j,k}.$$

By reading, in increasing order, the columns of A (or equivalently its rows, since A is symmetric) we derive the entries of the columns of B.
Column 1. Fix $j = 1$ and vary i:

$$a_{1,1} = (b_{1,1})^2 \Rightarrow b_{1,1} = \sqrt{a_{1,1}},$$

$$a_{2,1} = b_{1,1}b_{2,1} \Rightarrow b_{2,1} = \tfrac{a_{2,1}}{b_{1,1}},$$

$$\vdots \qquad\qquad \vdots$$

$$a_{n1} = b_{1,1}b_{n1} \Rightarrow b_{n,1} = \tfrac{a_{n,1}}{b_{1,1}}.$$

We have thus determined the entries of the first column of B.
Column j. We assume that we have computed the first $(j-1)$ columns of B. Then, we read the jth column of A below the diagonal:

$$a_{j,j} = (b_{j,1})^2 + (b_{j,2})^2 + \cdots + (b_{j,j})^2 \qquad\qquad \Rightarrow b_{j,j} = \sqrt{a_{j,j} - \textstyle\sum_{k=1}^{j-1}(b_{j,k})^2};$$

$$a_{j+1,j} = b_{j,1}b_{j+1,1} + b_{j,2}b_{j+1,2} + \cdots + b_{j,j}b_{j+1,j} \Rightarrow b_{j+1,j} = \frac{a_{j+1,j} - \sum_{k=1}^{j-1} b_{j,k}b_{j+1,k}}{b_{j,j}};$$

$$\vdots \qquad\qquad\qquad\qquad\qquad \vdots$$

$$a_{n,j} = b_{j,1}b_{n,1} + b_{j,2}b_{n,2} + \cdots + b_{j,j}b_{n,j} \qquad \Rightarrow b_{n,j} = \frac{a_{n,j} - \sum_{k=1}^{j-1} b_{j,k}b_{n,k}}{b_{j,j}}.$$

We have thus obtained the jth column of B in terms of its first $(j-1)$ columns. Theorem 6.3.1 ensures that when A is symmetric positive definite, the terms underneath the square roots are positive and the algorithm does not break down.

Remark 6.3.1. In practice, we don't need to check whether A is positive definite before starting the algorithm (we merely verify that it is symmetric). Actually, if at step j of the Cholesky algorithm we cannot compute a square root because we find that $b_{j,j}^2 = a_{j,j} - \sum_{k=1}^{j-1}(b_{j,k})^2 < 0$, this proves that A is not nonnegative. If we find that $b_{j,j}^2 = 0$, which prevents us from computing the entries $b_{i,j}$ for $i > j$, then A is not positive definite. However, if the Cholesky algorithm terminates "without trouble," we deduce that A is positive definite.

Determinant of a matrix. The Cholesky factorization is also used for computing the determinant of a matrix A, since $\det A = (\det B)^2$.

Conditioning of a matrix. Knowing the Cholesky decomposition of A allows us to easily compute the 2-norm conditioning of A, which is $\operatorname{cond}(A)_2 = \operatorname{cond}(BB^*) = \operatorname{cond}_2(B)^2$, since for any square matrix, $\|XX^*\|_2 = \|X^*X\|_2 = \|X\|_2^2$. The conditioning of the triangular matrix B is computed by Algorithm 5.4.

Incomplete Cholesky preconditioning. By extending the idea of the incomplete LU factorization, we obtain an incomplete Cholesky factorization that can be used as a preconditioner.

6.3.2 Numerical Algorithm

The Cholesky algorithm is written in a compact form using the array that initially contained A and is progressively filled by B. At each step j, only the jth column of this array is modified: it contains initially the jth column, which is overridden by the jth column of B. Note that it suffices to store the lower half of A, since A is symmetric.

Data: A. Output: A containing B in its lower triangular part
    For $j = 1 \nearrow n$
        For $k = 1 \nearrow j - 1$
            $a_{j,j} = a_{j,j} - (a_{j,k})^2$
        End k
        $a_{j,j} = \sqrt{a_{j,j}}$
        For $i = j + 1 \nearrow n$
            For $k = 1 \nearrow j - 1$
                $a_{i,j} = a_{i,j} - a_{j,k} a_{i,k}$
            End k
            $a_{i,j} = \frac{a_{i,j}}{a_{j,j}}$
        End i
    End j

Algorithm 6.2: Cholesky Algorithm.

6.3.3 Operation Count

To assess the efficiency of the Cholesky method, we count the number of operations its execution requires (which will be proportional to its running time on a computer). Once again, we content ourselves with the asymptotic first-order term when the dimension n is large. We take into account only multiplications and divisions. Although taking a square root is a more expensive operation than a multiplication, we neglect them because their number is n, which is negligible in comparison to n^3.

- Cholesky factorization: The number of operations is

$$N_{op}(n) = \sum_{j=1}^{n} \left((j-1) + \sum_{i=j+1}^{n} j \right) \approx \frac{n^3}{6}.$$

- Substitution: a forward and a back substitution are performed on the triangular systems associated with B and B^*. The number of operations is $N_{op}(n) \approx n^2$, which is thus negligible compared to the $n^3/6$ of the factorization.

The Cholesky method is thus approximately twice as fast as the Gauss method for a positive definite symmetric matrix.

Example 6.3.1. Let us compute the Cholesky factorization of

$$A = \begin{pmatrix} 1 & 2 & 1 & 2 \\ 2 & 13 & 2 & 4 \\ 1 & 2 & 2 & 3 \\ 2 & 4 & 3 & 9 \end{pmatrix}.$$

We look for a lower triangular matrix B of the form

$$B = \begin{pmatrix} b_{1,1} & 0 & 0 & 0 \\ b_{2,1} & b_{2,2} & 0 & 0 \\ b_{3,1} & b_{3,2} & b_{3,3} & 0 \\ b_{4,1} & b_{4,2} & b_{4,3} & b_{4,4} \end{pmatrix}.$$

The algorithm of Section 6.3.1 yields

✗ *computing the first column of B:*
- $b_{1,1}^2 = 1 \Longrightarrow b_{1,1} = 1,$
- $b_{1,1}b_{2,1} = 2 \Longrightarrow b_{2,1} = 2,$
- $b_{1,1}b_{3,1} = 1 \Longrightarrow b_{3,1} = 1,$
- $b_{1,1}b_{4,1} = 2 \Longrightarrow b_{4,1} = 2;$

✗ *computing the second column of B:*
- $b_{2,1}^2 + b_{2,2}^2 = 13 \Longrightarrow b_{2,2} = 3,$
- $b_{3,1}b_{2,1} + b_{3,2}b_{2,2} = 2 \Longrightarrow b_{3,2} = 0,$
- $b_{4,1}b_{2,1} + b_{4,2}b_{2,2} = 4 \Longrightarrow b_{4,2} = 0;$

✗ *computing the third column of B:*
- $b_{3,1}^2 + b_{3,2}^2 + b_{3,3}^2 = 2 \Longrightarrow b_{3,3} = 1,$
- $b_{4,1}b_{3,1} + b_{4,2}b_{3,2} + b_{4,3}b_{3,3} = 3 \Longrightarrow b_{4,3} = 1,$

✗ *computing the fourth column of B:*
- $b_{4,1}^2 + b_{4,2}^2 + b_{4,3}^2 + b_{4,4}^2 = 9 \Longrightarrow b_{4,4} = 2.$

Eventually we obtain

$$B = \begin{pmatrix} 1 & 0 & 0 & 0 \\ 2 & 3 & 0 & 0 \\ 1 & 0 & 1 & 0 \\ 2 & 0 & 1 & 2 \end{pmatrix}.$$

We solve the linear system $Ax = b$ with $b = (4, 8, 5, 17)^t$ by the Cholesky method. We first determine the solution y of $By = b$, next the solution x of $B^t x = y$. We obtain $y = (4, 0, 1, 4)^t$ and $x = (1, 0, -1, 2)^t$.

By a simple adaptation of the proof of Proposition 6.2.1 we can prove the following result.

Proposition 6.3.1. *The Cholesky factorization preserves the band structure of matrices.*

Remark 6.3.2. Computing the inverse A^{-1} of a symmetric matrix A by the Cholesky method costs $n^3/2$ operations (which improves by a factor of 2 the previous result in Section 6.2.3). We first pay $n^3/6$ for the Cholesky factorization, then compute the columns x_i of A^{-1} by solving $Ax_i = e_i$. This is done in two steps: first solve $By_i = e_i$, then solve $B^* x_i = y_i$. Because of the zeros in e_i, solving $By_i = e_i$ costs $(n - i)^2/2$, while because of the symmetry of A^{-1}, we need to compute only the $(n - i + 1)$ last components of x_i, which costs again of the order of $(n - i)^2/2$. The total cost of solving the triangular linear systems is thus of order $n^3/3$. The addition of $n^3/6$ and $n^3/3$ yields the result $N_{op}(n) \approx n^3/2$.

6.4 QR Factorization Method

The main idea of the QR factorization is again to reduce a linear system to a triangular one. However, the matrix is not factorized as the product of two triangular matrices (as previously), but as the product of an upper triangular matrix R and an orthogonal (unitary) matrix Q, which, by definition, is easy to invert, since $Q^{-1} = Q^*$.

In order to solve the linear system $Ax = b$ we proceed in three steps.

(i) Factorization: finding an orthogonal matrix Q such that $Q^* A = R$ is upper triangular.
(ii) Updating the right-hand side: computing $Q^* b$.
(iii) Back substitution: solving the triangular system $Rx = Q^* b$.

If A is nonsingular, the existence of such an orthogonal matrix Q is guaranteed by the following result, for which we give a constructive proof by the Gram–Schmidt orthonormalization process.

Theorem 6.4.1 (QR factorization). *Let A be a real nonsingular matrix. There exists a unique pair (Q, R), where Q is an orthogonal matrix and R is an upper triangular matrix, whose diagonal entries are positive, satisfying*

$$A = QR.$$

Remark 6.4.1. This factorization will be generalized to rectangular and singular square matrices in Section 7.3.3.

Proof of Theorem 6.4.1.   Let $a_1, \ldots, a_n$ be the column vectors of A. Since they form a basis of $\mathbb{R}^n$ (because A is nonsingular), we apply to them the Gram–Schmidt orthonormalization process, which produces an orthonormal basis $q_1, \ldots, q_n$ defined by

$$q_i = \frac{a_i - \sum_{k=1}^{i-1} \langle q_k, a_i \rangle q_k}{\| a_i - \sum_{k=1}^{i-1} \langle q_k, a_i \rangle q_k \|}, \quad 1 \le i \le n.$$

We deduce

$$a_i = \sum_{k=1}^{i} r_{ki} q_k, \quad \text{with } r_{ki} = \langle q_k, a_i \rangle, \text{ for } 1 \le k \le i-1, \qquad (6.4)$$

and

$$r_{i,i} = \left\| a_i - \sum_{k=1}^{i-1} \langle q_k, a_i \rangle q_k \right\| > 0.$$

We set $r_{ki} = 0$ if $k > i$, and we denote by R the upper triangular matrix with entries (r_{ki}). We denote by Q the matrix with columns $q_1, \ldots, q_n$, which is precisely an orthogonal matrix. With this notation, (6.4) is equivalent to

$$A = QR.$$

To prove the uniqueness of this factorization, we assume that there exist two factorizations

$$A = Q_1 R_1 = Q_2 R_2.$$

Then $Q_2^* Q_1 = R_2 R_1^{-1}$ is upper triangular with positive diagonal entries as a product of two upper triangular matrices (see Lemma 2.2.5). Let $T = R_2 R_1^{-1}$. We have

$$TT^* = (Q_2^* Q_1)(Q_2^* Q_1)^* = I.$$

Hence T is a Cholesky factorization of the identity, and since it is unique, we necessarily have $T = I$. $\qquad \square$

Remark 6.4.2. The above proof of uniqueness of the QR factorization relies crucially on the positivity assumption of the diagonal entries of R. Let us investigate the case that the diagonal entries of R have no specific sign (this turns out to be useful in the proof of Theorem 10.6.1). Consider two QR factorizations of the same nonsingular matrix

$$A = Q_1 R_1 = Q_2 R_2.$$

The upper triangular matrix $R_2 R_1^{-1}$ is thus equal to the orthogonal one $Q_2^t Q_1$, hence it is diagonal (see the proof of Theorem 2.5.1): there exists a diagonal matrix D such that $R_2 R_1^{-1} = Q_2^t Q_1 = D$. That is to say, $R_2 = D R_1$ and $Q_1 = Q_2 D$. The last equality implies $|D_{i,i}| = 1$. In other words, the QR factorization of a real nonsingular matrix is always unique up to the multiplication of each column k of Q and each row k of R by the factor $r_k = \pm 1$. In the complex case, the multiplication factor is a complex number of unit modulus, e^{is}, where s is a real number.

Example 6.4.1. The QR factorization of the matrix

$$A = \begin{pmatrix} 1 & -1 & 2 \\ -1 & 1 & 0 \\ 0 & -2 & 1 \end{pmatrix}$$

is

$$Q = \begin{pmatrix} 1/\sqrt{2} & 0 & 1/\sqrt{2} \\ -1/\sqrt{2} & 0 & 1/\sqrt{2} \\ 0 & -1 & 0 \end{pmatrix}, \quad R = \begin{pmatrix} \sqrt{2} & -\sqrt{2} & \sqrt{2} \\ 0 & 2 & -1 \\ 0 & 0 & \sqrt{2} \end{pmatrix}.$$

To determine the solution of $Ax = (-3, 1, 5)^t$, we first compute $y = Q^t b = \frac{1}{\sqrt{2}}(-4, -5\sqrt{2}, -2)^t$, then solve $Rx = y$ to obtain $x = (-4, -3, -1)^t$.

6.4.1 Operation Count

We assess the efficiency of the Gram–Schmidt algorithm for the QR method by counting the number of multiplications that are necessary to its execution. The number of square roots is n, which is negligible in this operation count.

- Gram–Schmidt factorization: the number of operations is

$$N_{op}(n) = \sum_{i=1}^{n} ((i-1)(2n) + (n+1)) \approx n^3.$$

- Updating the right-hand side: to compute the matrix-vector product $Q^* b$ requires $N_{op}(n) \approx n^2$.
- Back substitution: to solve the triangular system associated with R requires $N_{op} \approx n^2/2$.

The Gram–Schmidt algorithm for the QR method is thus three times slower than Gaussian elimination. It is therefore not used in practice to solve linear systems. Nonetheless, the QR method may be generalized, and it is useful for solving least squares fitting problems (see Chapter 7).

Remark 6.4.3. In numerical practice the Gram–Schmidt procedure is not used to find the QR factorization of a matrix because it is an unstable algorithm (rounding errors prevent the matrix Q from being exactly orthogonal). We shall see in the next chapter a better algorithm, known as the Householder algorithm, to compute the QR factorization of a matrix.

Conditioning of a matrix. If one knows the QR factorization of a matrix A, its 2-norm conditioning is easy to compute, since $\mathrm{cond}_2(A) = \mathrm{cond}_2(QR) = \mathrm{cond}_2(R)$ because Q is unitary.

6.5 Exercises

6.1. We define a matrix A=[1 2 3; 4 5 6; 7 8 9]. Compute its determinant using the Matlab function det. Explain why the result is not an integer.

6.2. The goal of this exercise is to compare the performances of the LU and Cholesky methods.

1. Write a function LUfacto returning the matrices L and U determined via Algorithm 6.1. If the algorithm cannot be executed (division by 0), return an error message.
2. Write a function Cholesky returning the matrix B computed by Algorithm 6.2. If the algorithm cannot be executed (nonsymmetric matrix, division by 0, negative square root), return an error message. Compare with the Matlab function chol.
3. For $n = 10, 20, \ldots, 100$, we define a matrix A=MatSdp(n) (see Exercise 2.20) and a vector b=ones(n,1). Compare:
   - On the one hand, the running time for computing the matrices L and U given by the function LUFacto, then the solution x of the system $Ax = b$. Use the functions BackSub and ForwSub defined in Exercise 5.2.
   - On the other hand, the running time for computing the matrix B given by the function Cholesky, then the solution x of the system $Ax = b$. Use the functions BackSub and ForwSub.

   Plot on the same graph the curves representing the running times in terms of n. Comment.

6.3 (∗). The goal of this exercise is to program the following variants of the Gauss algorithm:

- the Gauss algorithm with partial pivoting (by row), which consists, at each step k of the Gauss elimination, in determining an index i_0 ($k \leq i_0 \leq n$) such that

$$|a_{i_0,k}| = \max_{k \leq i \leq n} |a_{i,k}|, \tag{6.5}$$

then swapping rows k and i_0,
- the Gauss algorithm with complete pivoting, which consists in determining indices i_0 and j_0 ($k \leq i_0, j_0 \leq n$), such that

$$|a_{i_0,j_0}| = \max_{k \leq i,j \leq n} |a_{i,j}|, \tag{6.6}$$

then swapping rows k and i_0, and columns k and j_0.

Let A_k be the matrix obtained at the end of step k of the Gauss elimination. In the first $k - 1$ columns of A_k, all the entries below the diagonal are zero.

1. Write a function x=Gauss(A,b) solving the linear system $Ax = b$ by the Gauss method outlined in Section 6.1. Recall that if the pivot $A_{k,k}^{(k)}$ is zero, this method permutes row k with the next row i ($i \geq k$) such that $A_{i,i}^{(k)}$ is nonzero.
2. Write a function x=GaussWithoutPivot(A,b) solving the system $Ax = b$ by the Gauss method without any pivoting strategy. If the algorithm cannot proceed (because of a too-small pivot $A_{k,k}^{(k)}$), return an error message.
3. Write a function x=GaussPartialPivot(A,b) solving the linear system $Ax = b$ by the Gauss method with partial pivoting by row.
4. Write a function x=GaussCompletePivot(A,b) solving the linear system $Ax = b$ by the Gauss method with complete pivoting.
5. Comparison of the algorithms.
   (a) Check on the following example that it is sometimes necessary to use a pivoting strategy. Define the matrix A, and the vectors b and x by

$$A = \begin{pmatrix} \varepsilon & 1 & 1 \\ 1 & 1 & -1 \\ 1 & 1 & 2 \end{pmatrix}, \quad x = \begin{pmatrix} 1 \\ -1 \\ 1 \end{pmatrix}, \quad \text{and} \quad b = Ax.$$

   For $\varepsilon = 10^{-15}$, compare the solutions obtained by Gauss(A,b) and GaussPartialPivot(A,b). Comment.
   (b) In order to compare the Gauss pivoting algorithms, we define the following ratio ϱ, which we shall call growth rate:

$$\varrho = \frac{\max_{i,j} |A_{i,j}^{(n-1)}|}{\max_{i,j} |A_{i,j}|},$$

   where $A^{(n-1)}$ denotes the upper triangular matrix generated by Gaussian elimination. The growth rate measures the amplification of the matrix entries during Gauss elimination. For numerical stability reasons, the ratio ϱ should not be too large.

i. Modify the programs GaussWithoutPivot, GaussPartialPivot, GaussCompletePivot, and Gauss to compute respectively the rates ϱ_{GWP}, ϱ_{GCP}, ϱ_{GPP} and ϱ_{G}.

ii. For different values of n, compute the growth rates for the matrices A, B, and C defined by
A=DiagDomMat(n); B=SpdMat(n); C=rand(n,n);
Conclude.

iii. Comparison of ϱ_{GPP} and ϱ_{GCP}.

A. For each $n = 10\,k$ ($1 \leq k \leq 10$), compute ϱ_{GPP} and ϱ_{GCP} for three (or more) matrices randomly generated A=rand(n,n). Plot these values on the same graph in terms of the matrix dimension n. What do you notice?

B. For each $n = 2\,k$ ($1 \leq k \leq 5$), compute ϱ_{GPP} and ϱ_{GCP} for the matrix defined by
A=-tril(ones(n,n))+2*diag(ones(n,1));
A=A+[zeros(n,n-1) [ones(n-1,1);0]];
What do you notice?

6.4. The goal of this exercise is to evaluate the influence of row permutation in Gaussian elimination. Let A and b be defined by
e=1.E-15;A=[e 1 1;1 -1 1; 1 0 1];b=[2 0 1]';

1. Compute the matrices L and U given by the function LUFacto of Exercise 6.2.
2. We define two matrices l and u by [l u]=LUFacto(p*A), where p is the permutation matrix defined by the instruction [w z p]=lu(A). Display the matrices l and u. What do you observe?
3. Determine the solution of the system $Ax = b$ computed by the instruction BackSub (U,ForwdSub(L,b)), then the solution computed by the instruction BackSub (u,ForwSub(l,p*b)). Compare with the exact solution $x = (0, 1, 1)^t$. Conclude.

6.5 (∗).

1. Write a program StoreB to store a band matrix.
2. Write a program StoreBpv to compute the product of a band matrix with a vector. The matrix is given in the form StoreB.

6.6 (∗). Write a program LUBand that computes the LU factorization of a band matrix given in the form StoreB. The resulting matrices L and U have to be returned in the form StoreB.

6.7. The goal of this exercise is to study the resolution of the finite difference discretization of the 2D Laplace equation. For given smooth functions f and g we seek a solution $u(x,y)$ of the following partial differential equation:

$$-\Delta u(x, y) = f(x, y), \quad \text{for } (x, y) \in \Omega =]0, 1[\times]0, 1[, \tag{6.7}$$

together with the boundary condition

$$u(x,y) = g(x,y), \qquad \text{for } (x,y) \in \partial\Omega = \text{boundary of } \Omega, \qquad (6.8)$$

where $\Delta u = \partial^2 u/\partial x^2 + \partial^2 u/\partial y^2$ is the Laplacian of u. As for the one-dimensional problem, we discretize the domain Ω: given the space step $h = 1/(N+1)$ (respectively, $k = 1/(M+1)$) in the direction of x (respectively, y), we define the points

$$x_i = ih, \; i = 0, \ldots, N+1, \quad y_j = jk, \; j = 0, \ldots, M+1.$$

The goal is to compute an approximation $(u_{i,j})$ of u at the points in Ω, (x_i, y_j), $1 \le i \le N$, and $1 \le j \le M$.

1. Finite difference approximation of the Laplacian.
   (a) Combining the Taylor expansions of $u(x_i - h, y_j)$ and $u(x_i + h, y_j)$, show that

   $$\frac{\partial^2 u}{\partial x^2}(x_i, y_j) = \frac{u(x_{i-1}, y_j) - 2u(x_i, y_j) + u(x_{i+1}, y_j)}{h^2} + \mathcal{O}(h^2).$$

   We say that $\big(u(x_{i-1}, y_j) - 2u(x_i, y_j) + u(x_{i+1}, y_j)\big)/h^2$ is a second-order approximation of $\partial^2 u/\partial x^2$ at point (x_i, y_j).
   (b) Same question for

   $$\frac{\partial^2 u}{\partial y^2}(x_i, y_j) = \frac{u(x_i, y_{j-1}) - 2u(x_i, y_j) + u(x_i, y_{j+1})}{k^2} + \mathcal{O}(k^2).$$

   (c) Justify the finite difference method for solving the Laplace equation (6.7):

   $$\frac{-u_{i-1,j} + 2u_{i,j} - u_{i+1,j}}{h^2} + \frac{-u_{i,j-1} + 2u_{i,j} - u_{i,j+1}}{k^2} = f_{i,j}, \qquad (6.9)$$

   where $u_{i,j}$ denotes the approximation of $u(x_i, y_j)$, and $f_{i,j} = f(x_i, y_j)$. Formula (6.9) is called the 5-point discretization of the Laplacian, because it couples 5 values of u at 5 neighboring points.
2. Taking into account the boundary condition (6.8), formula (6.9) has to be modified for the points (x_i, y_j) close to the boundary $\partial\Omega$, that is, for $i = 1$ and N, or $j = 1$ and M. For instance, for $j = 1$, the term $u_{i,j-1}$ appearing in (6.9) is known and equal to $g_{i,0}$, according to (6.8). Therefore, this term moves to the right-hand side of the equality:

$$\frac{-u_{i-1,1} + 2u_{i,1} - u_{i+1,1}}{h^2} + \frac{2u_{i,1} - u_{i,2}}{k^2} = f_{i,1} + \frac{g_{i,0}}{k^2}. \qquad (6.10)$$

For $i = 1$ or $i = N$, there is yet another term of (6.10) that is known:

$$\frac{2u_{1,1} - u_{2,1}}{h^2} + \frac{2u_{1,1} - u_{1,2}}{k^2} = f_{1,1} + \frac{g_{1,0}}{k^2} + \frac{g_{0,1}}{h^2};$$

$$\frac{-u_{N-1,1} + 2u_{N,1}}{h^2} + \frac{2u_{N,1} - u_{N,2}}{k^2} = f_{N,1} + \frac{g_{N,0}}{k^2} + \frac{g_{N+1,1}}{h^2}.$$

Write the corresponding equations for the other points close to the boundary.

3. We now solve the linear system corresponding to (6.9) and (6.10) and assume, for simplicity, that $h = k$. Let $\bar{u}_j$ be the vector whose entries are the n unknowns located on row j, $\bar{u}_j = (u_{1,j}, u_{2,j}, \dots, u_{N,j})^t$, and $\bar{f}_j = (f_{1,j}, f_{2,j}, \dots, f_{N,j})^t$. Determine the matrix B such that the vectors $\bar{u}_j$ for $j = 1, \dots, M$ satisfy the equations

$$\frac{-\bar{u}_{j-1} + B\bar{u}_j - \bar{u}_{j+1}}{h^2} = \bar{f}_j.$$

For $j = 1$ or $j = M$, $\bar{f}_j$ must be modified in order to take into account the boundary values $u_{i,0}$ and $u_{i,M+1}$, which are known. For simplicity again, we assume $g = 0$. Prove that the complete system reads

$$A\bar{u} = \bar{f}, \tag{6.11}$$

where the unknown is $\bar{u} = (\bar{u}_1, \dots, \bar{u}_M)^t$, the right-hand side is $\bar{f} = (\bar{f}_1, \dots, \bar{f}_M)^t$, and the matrix A is to be determined. Exhibit the band structure of this matrix.

(a) Write a function `Laplacian2dD(n)` returning the matrix A (of order n^2, where $n = N = M$). Use the `Matlab` function `spy` to visualize the matrix A. *Hint:* we do not request at this stage of the problem to use the `Matlab` instruction `sparse`.

(b) Write a function `Laplacian2dDRHS(n,f)` returning the right-hand side $\bar{f}$ of equation (6.11), given n and the function f defined as a `Matlab` function.

4. Validation. Set $f(x, y) = 2x(1 - x) + 2y(1 - y)$, so that the solution of (6.7) is $u(x, y) = x(1 - x)y(1 - y)$. For $N = 10$, compute the approximate solution and compare it with the exact solution, plotting them on the same graph using the function `plot3`.

5. Convergence. We now choose f such that the solution u is not a polynomial.

(a) How should one choose f so that the solution is
$$u(x, y) = (x - 1)(y - 1)\sin(\pi x)\sin(\pi y)?$$

(b) What is the maximal value N_0 for which `Matlab` can carry out the computations (before a memory size problem occurs)?

(c) Taking into account the sparse nature of the matrix A, we define a function `Laplacian2dDSparse`. The command `sparse` should be used to define and store the matrix A in sparse form: larger problems (i.e., with N larger than N_0) can be solved accordingly. Let N_e be the total number of nonzero entries of A. Define three vectors of size N_e:
   • a vector `ii` of integers containing the indices of the rows of the nonzero entries of A;

- a vector jj of integers containing the indices of the columns of the nonzero entries of A;
- a vector u containing the nonzero entries of A.

For any $k = 1, \ldots, N_e$, they satisfy $u(k) = A_{ii(k),jj(k)}$. Next, define a matrix spA=sparse(ii,jj,u). For every value $N = 5, 10, 15, \ldots, 50$, compute the error between the numerical solution and the exact solution. Plot the error in terms of N on a log-log scale. The error is computed in the ∞-norm, i.e., is equal to the maximum of the error between the exact and approximate solutions at the $N \times N$ mesh points. Comment on the results.

6. Spectrum of A. We fix $N = 20$.
   (a) Compute (using eig) the eigenvalues and the eigenvectors of A.
   (b) Use the instruction sort to find the four smallest eigenvalues. Plot the corresponding eigenvectors (using surfc). *Hint:* The eigenvectors computed by Matlab are vectors of size $N \times N$, which have to be represented as a function of (x, y) given on an $N \times N$ regular grid.
   (c) The eigenvalue λ and eigenfunction φ of the Laplacian on the unit square with homogeneous Dirichlet boundary conditionsare are defined by a nonidentically zero function φ such that

   $$-\Delta\varphi = \lambda\varphi \quad \text{in } \Omega$$

   and $\varphi(x, y) = 0$ for $(x, y) \in \partial\Omega$. For which values α and β is $\varphi(x, y) = \sin(\alpha x) \sin(\beta y)$ an eigenfunction? What is the corresponding eigenvalue? Plot on the unit square the first four eigenfunctions of the Laplacian, that is, the eigenfunctions corresponding to the smallest eigenvalues. Interpret the curves of the previous question.

6.8. Let A be the matrix defined by A=Laplacian2dD(5), and $A = LU$ its LU factorization given by LUFacto. Use the function spy to display the matrices L and U. Explain.

6.9. Let A be a band matrix of order n and half bandwidth p. For $n \gg p \gg 1$ compute the number of operations $N_{op}(n, p)$ required for the LU factorization (having in mind Proposition 6.2.1).

6.10. The goal of this exercise is to program the so-called incomplete LU factorization of a matrix A, which is defined as the approximate factorization $A \approx \tilde{L}\tilde{U}$, where $\tilde{L}$ and $\tilde{U}$ are computed by the program LUFacto modified as follows: the entries $\tilde{L}_{i,j}$ and $\tilde{U}_{i,j}$ are computed if and only if the entry $A_{i,j}$ is not zero. If this entry is zero, we set $\tilde{L}_{i,j} = 0$ and $\tilde{U}_{i,j} = 0$.

1. Write a program ILUfacto computing the incomplete LU factorization of a matrix. Because of rounding errors, the condition $A_{i,j} = 0$ has to be replaced by $|A_{i,j}| < \varepsilon$, where $\varepsilon > 0$ is a prescribed small threshold.
2. For A=Laplacian2dD(10), compute $\text{cond}_2(A)$ and $\text{cond}_2(\tilde{U}^{-1}\tilde{L}^{-1}A)$. Explain.

7

Least Squares Problems

7.1 Motivation

The origin of the least squares data-fitting problem is the need of a notion of "generalized solutions" for a linear system $Ax = b$ that has no solution in the classical sense (that is, b does not belong to the range of A). The idea is then to look for a vector x such that Ax is "the closest possible" to b. Several norms are at hand to measure the distance between Ax and b, but the simplest choice (which corresponds to the denomination "least squares") is the Euclidean vector norm. In other words, a least squares problem amounts to finding the solution (possibly nonunique) $x \in \mathbb{R}^p$ to the following minimization problem:

$$\|b - Ax\|_n = \min_{y \in \mathbb{R}^p} \|b - Ay\|_n, \tag{7.1}$$

where $A \in \mathcal{M}_{n,p}(\mathbb{R})$ is a matrix with n rows and p columns, b is a vector of $\mathbb{R}^n$, and $\| \cdot \|_n$ denotes the Euclidean norm in $\mathbb{R}^n$.

In the square case $p = n$, if the matrix A is nonsingular, then there exists a unique minimizer $x = A^{-1}b$, and the minimum is equal to zero. In such a case, a least squares problem is equivalent to solving a linear system. If A is singular or if $p \neq n$, the notion of least squares yields a generalization of a linear system solving to nonsquare or singular matrices. If a solution of the linear system $Ax = b$ exists, then it is also a solution of the least squares problem. The converse is not true, as we shall see in the following geometrical argument.

The least squares problem (7.1) has a geometrical interpretation as finding the orthogonal projection of b on the range of A. Indeed, Ax is the closest vector in $\text{Im}(A)$ to b. A well-known property of the orthogonal projection is that $b - Ax$ is actually orthogonal to $\text{Im}(A)$. We display in Figure 7.1 a vector b and its orthogonal projection Ax onto the vector subspace $\text{Im}(A)$. It is therefore clear that (7.1) always admits at least one solution x (such that Ax is the orthogonal projection of b,) although the linear system $Ay = b$ may have no solution if b does not belong to $\text{Im}(A)$.

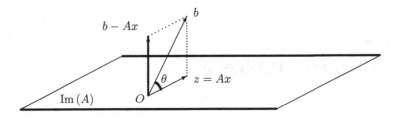

Fig. 7.1. Least squares problem: projection of b onto $\mathrm{Im}\,(A)$.

Finally, let us recall that one of the main motivations of least squares problems is data-fitting (see Section 1.2).

7.2 Main Results

We consider the least squares problem (7.1): find $x \in \mathbb{R}^p$ that minimizes $\|b - Ay\|_n$ over $\mathbb{R}^p$, where $A \in \mathcal{M}_{n,p}(\mathbb{R})$ is a matrix with n rows and p columns, $b \in \mathbb{R}^n$, and $\|\cdot\|_n$ denotes the Euclidean norm in $\mathbb{R}^n$.

Lemma 7.2.1. *A vector $x \in \mathbb{R}^p$ is a solution to the least squares problem (7.1) if and only if it satisfies the so-called normal equation*

$$A^*Ax = A^*b. \tag{7.2}$$

(Observe that A^*A is a square matrix of size p.)

Proof. Let $x \in \mathbb{R}^p$ be a solution of (7.1), i.e.,

$$\|b - Ax\|_n^2 \le \|b - Ay\|_n^2, \quad \forall y \in \mathbb{R}^p.$$

For any $z \in \mathbb{R}^p$ and any $t \in \mathbb{R}$, set $y = x + tz$. Then

$$\|b - Ax\|_n^2 \le \|b - Ax\|_n^2 + 2t\langle Ax - b, Az\rangle + t^2\|Az\|_n^2.$$

We infer

$$0 \le 2\,\mathrm{sign}\,(t)\langle Ax - b, Az\rangle + |t|\|Az\|_n^2,$$

which implies that as t tends to 0 (from above and then from below),

$$\langle Ax - b, Az\rangle = 0, \quad \forall z \in \mathbb{R}^p.$$

Thus we deduce that $A^*Ax - A^*b = 0$. Conversely, if x is a solution of the normal equation (7.2), then

$$\langle Ax - b, Az\rangle = 0, \quad \forall z \in \mathbb{R}^p.$$

Thus

$$\|b - Ax\|_n^2 \le \|b - Ay\|_n^2, \quad \forall y = x + tz \in \mathbb{R}^p.$$

That is, x is also a solution of (7.1). □

Theorem 7.2.1. *For any matrix $A \in M_{n,p}(\mathbb{R})$, there always exists at least one solution of the normal equation (7.2). Furthermore, this solution is unique if and only if $\operatorname{Ker} A = \{0\}$.*

Proof. If A^*A is nonsingular, there exists, of course, a unique solution to the normal equation. If it is singular, we now show that there still exists a solution that is not unique. Let us prove that $A^*b \in \operatorname{Im} A^*A$, or more generally, that $\operatorname{Im} A^* \subset \operatorname{Im} A^*A$. The opposite inclusion, $\operatorname{Im} A^*A \subset \operatorname{Im} A^*$, is obvious, as well as $\operatorname{Ker} A \subset \operatorname{Ker} A^*A$. On the other hand, the relation $\operatorname{Im} A^* = (\operatorname{Ker} A)^\perp$ implies that

$$\mathbb{R}^p = \operatorname{Ker} A \oplus \operatorname{Im} A^*. \tag{7.3}$$

Moreover, A^*A is real symmetric, so is diagonalizable in an orthonormal basis of eigenvectors. Since the range and the kernel of a diagonalizable matrix are in direct sum, we deduce

$$\mathbb{R}^p = \operatorname{Ker} A^*A \oplus \operatorname{Im} A^*A. \tag{7.4}$$

If we can show that $\operatorname{Ker} A^*A \subset \operatorname{Ker} A$ (and thereby that $\operatorname{Ker} A = \operatorname{Ker} A^*A$), then (7.3) and (7.4), together with the relation $\operatorname{Im} A^*A \subset \operatorname{Im} A^*$, imply that $\operatorname{Im} A^* = \operatorname{Im} A^*A$, which is the desired result. Let us prove that $\operatorname{Ker} A^*A \subset \operatorname{Ker} A$. If $x \in \operatorname{Ker} A^*A$, then

$$A^*Ax = 0 \Rightarrow \langle A^*Ax, x \rangle = 0 \Leftrightarrow \|Ax\| = 0 \Leftrightarrow Ax = 0,$$

and thus $x \in \operatorname{Ker} A$. This proves the existence of at least one solution. Clearly two solutions of the normal equation differ by a vector in $\operatorname{Ker} A^*A$, which is precisely equal to $\operatorname{Ker} A$. $\qquad\square$

A particular solution of the normal equation can be expressed in terms of the pseudoinverse $A^\dagger$ of A (see Definition 2.7.2).

Proposition 7.2.1. *The vector $x_b = A^\dagger b$ is a solution of the least squares problem (7.1). When (7.1) has several solutions, x_b is the unique solution with minimal norm, i.e., for all $x \neq x_b$ such that $\|Ax_b - b\|_2 = \|Ax - b\|_2$, we have*

$$\|x_b\|_2 < \|x\|_2.$$

Proof. For any $x \in \mathbb{R}^p$, we decompose $Ax - b$ as follows:

$$Ax - b = A(x - x_b) - (I - AA^\dagger)b.$$

This decomposition is orthogonal since $A(x - x_b) \in \operatorname{Im} A$ and $(I - AA^\dagger)b \in (\operatorname{Im} A)^\perp$, because $AA^\dagger$ is the orthogonal projection matrix of $\mathbb{C}^m$ onto $\operatorname{Im} A$; see Exercise 2.29. We deduce from this decomposition that

$$\|Ax - b\|_2^2 = \|Ax - Ax_b\|_2^2 + \|Ax_b - b\|_2^2 \geq \|Ax_b - b\|_2^2, \tag{7.5}$$

which proves that x_b is a solution of (7.1). In addition, if $\|Ax_b - b\|_2 = \|Ax - b\|_2$, then (7.5) shows that $Ax = Ax_b$ and $z = x - x_b \in \operatorname{Ker} A$. We obtain thus a decomposition of x into $x = z + x_b$. This decomposition is orthogonal, since $z \in \operatorname{Ker} A$ and $x_b = A^{\dagger} b \in (\operatorname{Ker} A)^{\perp}$ (by definition of $A^{\dagger}$ in Exercise 2.29). Hence, if $x \neq x_b$, we have

$$\|x\|_2^2 = \|z\|_2^2 + \|x_b\|_2^2 > \|x_b\|_2^2.$$

$\square$

Remark 7.2.1. The vector $x_b = A^{\dagger} b$ has a simple geometric characterization: it is the unique vector of $(\operatorname{Ker} A)^{\perp}$ whose image under the matrix A is equal to the projection of b onto $\operatorname{Im} A$; for more details see Exercise 2.29.

7.3 Numerical Algorithms

Before introducing efficient numerical methods for solving problem (7.2), we first study the sensitivity of the solution to variations of the data.

7.3.1 Conditioning of Least Squares Problems

In this section we assume that system (7.2) has a unique solution, that is, $\operatorname{Ker} A$ is reduced to the zero vector. Note that this is possible only if $p \leq n$. In this case, the square matrix $A^* A$ is nonsingular, and by Theorem 7.2.1, the least squares problem has a unique solution, equal to $A^{\dagger} b$.

Sensitivity of the Solution to Variations of b.

For a given $b_0 \in \mathbb{R}^n$, we call $x_0 = A^{\dagger} b_0$ the solution to the least squares problem

$$\min_{y \in \mathbb{R}^p} \|Ay - b_0\|. \tag{7.6}$$

Similarly for a given $b_1 \in \mathbb{R}^n$, we call $x_1 = A^{\dagger} b_1$ the solution to the least squares problem

$$\min_{y \in \mathbb{R}^p} \|Ay - b_1\|. \tag{7.7}$$

Before analyzing the variations of x in terms of the variations of b, let us first observe that only the projection of the vector b onto $\operatorname{Im} A$ counts; it is the point of the next remark.

Remark 7.3.1. As already explained in Section 7.1, a solution x of the least squares problem can be obtained by taking Ax as the orthogonal projection of b onto $\operatorname{Im}(A)$. Therefore, if we modify the vector b without changing its orthogonal projection onto $\operatorname{Im}(A)$, we preserve the same solution of the least squares problem. In other words, if b_0 and b_1 have the same projection z onto $\operatorname{Im}(A)$, we have $Ax_0 = Ax_1 = z$. Since the kernel of A is $\operatorname{Ker} A = \{0\}$, we clearly obtain equality between the two solutions, $x_0 = x_1$.

Bound from above of the absolute variation

We have a direct upper bound of the variations of the solution:

$$\|x_1 - x_0\|_2 = \|A^\dagger(b_1 - b_0)\|_2 \leq \frac{1}{\mu_p}\|b_1 - b_0\|_2,$$

where μ_p is the smallest nonzero singular value of the matrix A; see Remark 5.3.4.

Bound from above of the relative variation

Assuming that x_0 and b_0 are nonzero, the relative error on x can be bounded in terms of the relative error on b.

Proposition 7.3.1. *Assume that* $\mathrm{Ker}\, A = \{0\}$. *Let* b_0, b_1 *be the vectors defined in (7.6), (7.7), and* x_0, x_1 *their corresponding solutions. They satisfy*

$$\frac{\|x_1 - x_0\|_2}{\|x_0\|_2} \leq C_b \frac{\|b_1 - b_0\|_2}{\|b_0\|_2}, \tag{7.8}$$

where

$$C_b = \|A^\dagger\|_2 \frac{\|b_0\|}{\|x_0\|}. \tag{7.9}$$

The constant C_b is a measure of the amplification of the relative error on the solution x with respect to the relative error on the right-hand side b. This constant is the product of several quantities:

$$C_b = \frac{\mathrm{cond}(A)}{\eta \cos\theta},$$

where

✗ $\mathrm{cond}(A) = \|A\|_2\|A^\dagger\|_2$ is the generalized conditioning of A,
✗ θ denotes the angle formed by the vectors Ax_0 and b_0 (see Figure 7.1), i.e., $\cos\theta = \|Ax_0\|_2/\|b_0\|_2$,
✗ $\eta = \|A\|_2\,\|x_0\|_2/\|Ax_0\|_2$ indicates the gap between the norm of the vector Ax_0 and the maximal value that can be taken by this norm (it always satisfies $\eta \geq 1$).

We single out the following particular cases:

✗ if $b_0 \in \mathrm{Im}\, A$, then $\theta = 0$ and since $\eta \geq 1$, the amplification constant C_b is at most equal to $\mathrm{cond}_2(A)$;
✗ if $b_0 \in (\mathrm{Im}\, A)^\perp$, then $\theta = \pi/2$ and $z_0 = 0 = Ax_0$. The amplification constant C_b is infinite in this case.

Sensitivity of the Solution to Variations of A.

We now vary the matrix A. Let A_0 be a reference matrix with $\operatorname{Ker} A_0 = \{0\}$, and let

$$A_\varepsilon = A_0 + \varepsilon B, \quad B \in \mathcal{M}_{n,p}(\mathbb{C}) \tag{7.10}$$

be the matrix of the perturbed problem. For ε small enough, $A_\varepsilon^* A_\varepsilon$ is nonsingular, and the respective solutions to the reference and perturbed problems are denoted by x_0 and x_ε. They satisfy

$$A_0^* A_0 x_0 = A_0^* b \quad \text{and} \quad A_\varepsilon^* A_\varepsilon x_\varepsilon = A_\varepsilon^* b.$$

We perform a Taylor expansion for ε close to 0:

$$\begin{aligned}
x_\varepsilon &= (A_\varepsilon^* A_\varepsilon)^{-1} A_\varepsilon^* b = \left[(A_0^* + \varepsilon B^*)(A_0 + \varepsilon B) \right]^{-1} (A_0^* + \varepsilon B^*) b \\
&= \left[A_0^* A_0 + \varepsilon(A_0^* B + B^* A_0) + \mathcal{O}(\varepsilon^2) \right]^{-1} (A_0^* b + \varepsilon B^* b) \\
&= \left[I + \varepsilon(A_0^* A_0)^{-1}(A_0^* B + B^* A_0) + \mathcal{O}(\varepsilon^2) \right]^{-1} (A_0^* A_0)^{-1}(A_0^* b + \varepsilon B^* b) \\
&= \left[I - \varepsilon(A_0^* A_0)^{-1}(A_0^* B + B^* A_0) + \mathcal{O}(\varepsilon^2) \right] \left[x_0 + \varepsilon(A_0^* A_0)^{-1} B^* b \right].
\end{aligned}$$

Therefore, we deduce that

$$x_\varepsilon - x_0 = (A_0^* A_0)^{-1}(\varepsilon B^*)(b - A_0 x_0) - (A_0^* A_0)^{-1} A_0^*(\varepsilon B) x_0 + \mathcal{O}(\varepsilon^2).$$

Setting $\Delta A_0 = A_\varepsilon - A_0$, we get the following upper bound:

$$\frac{\|x_\varepsilon - x_0\|_2}{\|x_0\|_2} \leq \|(A_0^* A_0)^{-1}\|_2 \, \|\Delta A_0\|_2 \frac{\|b - A_0 x_0\|_2}{\|A_0 x_0\|_2} + \|A_0^\dagger\|_2 \, \|\Delta A_0\|_2 + \mathcal{O}(\varepsilon^2).$$

On the other hand, we have $\tan \theta = \|b - z_0\|_2 / \|z_0\|_2$ and

$$\|(A^* A)^{-1}\|_2 = \frac{1}{\sigma} = \frac{1}{\min_{\lambda \in \sigma(A^* A)} |\lambda|} = \frac{1}{\min_i \mu_i^2} = \|A^\dagger\|_2^2,$$

where σ is the smallest singular value of $A^* A$, μ_i are the singular values of A, and we have used the fact that $A^* A$ is normal. Hence, we have proved the following result:

Proposition 7.3.2. *Assume that $\operatorname{Ker} A_0 = \{0\}$. Let x_0 and x_ε be the solutions to the least squares problems associated with the matrices A_0 and A_ε respectively, with the same right-hand side b. The following upper bound holds as ε tends to 0:*

$$\frac{\|x_\varepsilon - x_0\|_2}{\|x_0\|_2} \leq C_{A_0} \frac{\|A_\varepsilon - A_0\|_2}{\|A_0\|_2} + \mathcal{O}(\varepsilon^2), \tag{7.11}$$

where $C_A = \operatorname{cond}_2(A) + \dfrac{\tan \theta}{\eta} \operatorname{cond}_2(A)^2$.

Therefore, the relative error on x can be amplified by the factor C_A. We single out the following particular cases:

- ✗ if $b_0 \in \text{Im } A$, then $C_A = \text{cond}_2(A)$;
- ✗ if $b_0 \in (\text{Im } A)^{\perp}$, the amplification factor C_A is infinite;
- ✗ in all other cases, C_A is usually of the same order as $\text{cond}_2(A)^2$. Of course, if $(\tan \theta)/\eta$ is very small (much smaller than $\text{cond}_2(A)^{-1}$), C_A is of the order of $\text{cond}_2(A)$, while if $(\tan \theta)/\eta$ is very large, C_A is larger than $\text{cond}_2(A)^2$.

7.3.2 Normal Equation Method

Lemma 7.2.1 tells us that the solution to the least squares problem is also a solution of the normal equation defined by

$$A^* A x = A^* b.$$

Since $A^* A$ is a square matrix of size p, we can apply to this linear system the methods for solving linear systems, as seen in Chapter 6. If $\text{Ker } A = \{0\}$, then the matrix $A^* A$ is even symmetric and positive definite, so we can apply the most efficient algorithm, that is, the Cholesky method.

Operation Count

As usual, we count only multiplications and we give an equivalent for n and p large. When applying the Cholesky algorithm to the normal equation, the following operations are performed:

- Multiplication of A^* by A: it is the product of a $p \times n$ matrix by an $n \times p$ matrix. The matrix $A^* A$ is symmetric, so only the upper part has to be computed. The number of operations N_{op} is exactly

$$N_{op} = \frac{np(p+1)}{2}.$$

- Cholesky factorization:

$$N_{op} \approx \frac{p^3}{6}.$$

- Computing the right-hand side: the matrix-vector product $A^* b$ costs

$$N_{op} = pn.$$

- Substitutions: solving two triangular linear systems costs

$$N_{op} \approx p^2.$$

In general, n is much larger than p, which makes the cost of the Cholesky factorization marginal with respect to the cost of the matrix product A^*A. However, this method is not recommended if p is large and the conditioning of A is also large. Actually, the amplification of rounding errors, while solving the normal equation, is governed by the conditioning of A^*A, which is in general of the order of the square of the conditioning of A. We shall see other methods where the conditioning is simply equal to that of A.

7.3.3 QR Factorization Method

The main idea of the QR factorization method is to reduce the problem to a least squares problem with a triangular matrix. We thus factorize A as the product of a triangular matrix R and an orthogonal (unitary) matrix Q. We recall that the multiplication by an orthogonal matrix preserves the Euclidean norm of a vector:

$$\|Qz\|_n = \|z\|_n, \quad \forall z \in \mathbb{R}^n \text{ if } Q^{-1} = Q^*.$$

Let $A \in \mathcal{M}_{n,p}(\mathbb{R})$ be a (not necessarily square) matrix. We determine $R \in \mathcal{M}_{n,p}(\mathbb{R})$ such that $r_{i,j} = 0$ if $i < j$, and $Q \in \mathcal{M}_{n,n}(\mathbb{R})$ such that $Q^{-1} = Q^*$, satisfying

$$A = QR.$$

The original least squares problem (7.1) is then equivalent to the following triangular problem:

$$\|Q^*b - Rx\|_n = \min_{y \in \mathbb{R}^p} \|Q^*b - Ry\|_n,$$

which is easily solved by a simple back substitution. We first study this method based on the Gram–Schmidt procedure (in the next section, we shall see another more powerful algorithm). We distinguish three cases.

Case $n = p$

If the matrix A is nonsingular, then we know that the solution to the least squares problem is unique and equal to the solution of the linear system $Ax = b$. We have seen in Chapter 6 how the QR method is applied to such a system. If the matrix A is singular, we need to slightly modify the previous QR method. Let $a_1, \ldots, a_n$ be the column vectors of A. Since these vectors are linearly dependent, there exists i such that $a_1, \ldots, a_i$ are linearly independent, and a_{i+1} is generated by $a_1, \ldots, a_i$. The Gram–Schmidt procedure (see Theorem 2.1.1) would stop at the $(i+1)$th step, because

$$\tilde{a}_{i+1} = a_{i+1} - \sum_{k=1}^{i} \langle q_k, a_{i+1}\rangle q_k = 0.$$

Indeed, since the subspaces $span\{q_1,\dots,q_i\}$ and $span\{a_1,\dots,a_i\}$ are equal and a_{i+1} belongs to the latter one, the orthogonal projection of a_{i+1} onto $span\{q_1,\dots,q_i\}$ is equal to a_{i+1}, and $\tilde{a}_{i+1}$ vanishes. To avoid this difficulty, we first swap the columns of A to bring into the first positions the linearly independent columns of A. In other words, we multiply A by a permutation matrix P such that the $\mathrm{rk}\,(A)$ first columns of AP are linearly independent and the $n - \mathrm{rk}\,(A)$ last columns of AP are spanned by the $\mathrm{rk}\,(A)$ first ones. This permutation can be carried out simultaneously with the Gram–Schmidt procedure: if a norm is zero at step $i + 1$, we perform a circular permutation from the $(i + 1)$th column to the nth. Permutation matrices are orthogonal, so the change of variable $z = P^t y$ yields

$$\|b - Ay\|_n = \|b - APP^t y\|_n = \|b - (AP)z\|_n.$$

We apply the Gram–Schmidt procedure to the matrix AP up to step $\mathrm{rk}\,(A)$ (we cannot go further). Hence, we obtain orthonormal vectors $q_1,\dots,q_{\mathrm{rk}\,(A)}$, to which we can add vectors $q_{\mathrm{rk}\,(A)+1},\dots,q_n$ in order to obtain an orthonormal basis of $\mathbb{R}^n$. We call Q the matrix formed by these column vectors. We have

$$q_i = \frac{a_i - \sum_{k=1}^{i-1}\langle q_k, a_i\rangle q_k}{\|a_i - \sum_{k=1}^{i-1}\langle q_k, a_i\rangle q_k\|}, \quad 1 \le i \le \mathrm{rk}\,(A),$$

and $a_i \in span\{a_1,\dots,a_{\mathrm{rk}\,(A)}\} = span\{q_1,\dots,q_{\mathrm{rk}\,(A)}\}$ if $\mathrm{rk}\,(A)+1 \le i \le n$. Therefore, we infer that there exist scalars $r_{k,i}$ such that

$$\begin{cases} a_i = \sum_{k=1}^{i} r_{k,i}q_k, & \text{with } r_{ii} > 0 \text{ if } 1 \le i \le \mathrm{rk}\,(A), \\ a_i = \sum_{k=1}^{\mathrm{rk}\,(A)} r_{k,i}q_k & \text{if } \mathrm{rk}\,(A) + 1 \le i \le n. \end{cases} \tag{7.12}$$

We set $r_{k,i} = 0$ if $k > i$, and call R the upper triangular matrix with entries $(r_{k,i})$:

$$R = \begin{pmatrix} R_{1,1} & R_{1,2} \\ 0 & 0 \end{pmatrix}, \quad \text{with} \quad R_{1,1} = \begin{pmatrix} r_{1,1} & \cdots & r_{1,\mathrm{rk}\,(A)} \\ & \ddots & \vdots \\ 0 & & r_{\mathrm{rk}\,(A),\mathrm{rk}\,(A)} \end{pmatrix}.$$

Relations (7.12) are simply written $AP = QR$. Let $z = (z_1, z_2)$ with z_1 the vector of the first $\mathrm{rk}\,(A)$ entries, and z_2 that of the last $n - \mathrm{rk}\,(A)$. We have

$$\|b-APz\|_n^2 = \|Q^*b - Rz\|_n^2 = \|(Q^*b)_1 - R_{1,1}z_1 - R_{1,2}z_2\|_{rg(A)}^2 + \|(Q^*b)_2\|_{n-\mathrm{rk}\,(A)}^2.$$

Since $R_{1,1}$ is upper triangular and nonsingular, by a simple back substitution, and whatever the vector z_2 is, we can compute a solution:

$$z_1 = R_{1,1}^{-1}\left((Q^*b)_1 - R_{1,2}z_2\right). \tag{7.13}$$

Consequently, the value of the minimum is

$$\|(Q^*b)_2\|_{n-\operatorname{rk}(A)} = \min_{y \in \mathbb{R}^p} \|b - Ay\|_n.$$

Since z_2 is not prescribed, there is an infinite number of solutions (a vector space of dimension $n - \operatorname{rk}(A)$) to the least squares problem.

Case $n < p$

In this case, we always have $\operatorname{Ker} A \neq \{0\}$. Therefore, there is an infinity of solutions. For simplicity, we assume that the rank of A is maximal, i.e., equal to n. Otherwise, we have to slightly modify the argument that follows. Let $a_1, \ldots, a_p \in \mathbb{R}^n$ be the columns of A. Since $\operatorname{rk}(A) = n$, possibly after permuting the columns, the first n columns of A are linearly independent in $\mathbb{R}^n$ and we can apply the Gram–Schmidt procedure to them. We thus obtain an orthogonal matrix Q of size n, with columns $q_1, \ldots, q_n$ satisfying

$$q_i = \frac{a_i - \sum_{k=1}^{i-1}\langle q_k, a_i\rangle q_k}{\|a_i - \sum_{k=1}^{i-1}\langle q_k, a_i\rangle q_k\|}, \quad 1 \le i \le n.$$

On the other hand, $a_{n+1}, \ldots, a_p$ are spanned by $q_1, \ldots, q_n$, which is a basis of $\mathbb{R}^n$. That is, there exist entries $r_{k,i}$ such that

$$\begin{cases} a_i = \sum_{k=1}^{i} r_{k,i} q_k, \text{ with } r_{ii} > 0 \text{ if } 1 \le i \le n, \\ a_i = \sum_{k=1}^{n} r_{k,i} q_k \qquad\qquad \text{if } n+1 \le i \le p. \end{cases}$$

Set $r_{k,i} = 0$ if $k > i$, and call R the $n \times p$ upper triangular matrix with entries $(r_{k,i})$:

$$R = \left(R_{1,1} \ R_{1,2} \right), \quad \text{with} \quad R_{1,1} = \begin{pmatrix} r_{1,1} & \cdots & r_{1,n} \\ & \ddots & \vdots \\ 0 & & r_{n,n} \end{pmatrix},$$

and $R_{1,2}$ is an $n \times (p-n)$ matrix. Set $z = (z_1, z_2)$ with z_1 the vector formed by the first n entries, and z_2 by the last $p - n$. We have

$$\|b - Az\|_n = \|Q^*b - R_{1,1}z_1 - R_{1,2}z_2\|_n.$$

Since $R_{1,1}$ is upper triangular and nonsingular, by a simple back substitution, and for any choice of z_2, we can compute a solution:

$$z_1 = R_{1,1}^{-1}\left((Q^*b) - R_{1,2}z_2\right). \tag{7.14}$$

As a consequence, the minimum value is

$$0 = \min_{y \in \mathbb{R}^p} \|b - Ay\|_n.$$

Since z_2 is not prescribed, there is an infinite number of solutions (a vector space of dimension $p - n$) to the least squares problem.

Case $n > p$

This is the most widespread case in practice, that is, there are more equations than unknowns. For simplicity, we assume that $\operatorname{Ker} A = \{0\}$ (which is equivalent to $\operatorname{rk}(A) = p$), so the least squares fitting problem has a unique solution. If $\operatorname{Ker} A \neq \{0\}$, then what follows should be modified as in the case $n = p$.

We apply the Gram–Schmidt procedure to the (linearly independent) columns $a_1, \ldots, a_p$ of A in order to obtain orthonormal vectors $q_1, \ldots, q_p$. We complement this set of vectors by $q_{p+1}, \ldots, q_n$ to get an orthonormal basis of $\mathbb{R}^n$. We call Q the matrix formed by these column vectors. We have

$$a_i = \sum_{k=1}^{i} r_{k,i} q_k, \quad \text{with} \quad r_{ii} > 0 \text{ if } 1 \leq i \leq p.$$

Set $r_{k,i} = 0$ if $k > i$, and call R the $n \times p$ upper triangular matrix with entries $(r_{k,i})$:

$$R = \begin{pmatrix} R_{1,1} \\ 0 \end{pmatrix} \quad \text{with} \quad R_{1,1} = \begin{pmatrix} r_{1,1} & \cdots & r_{1,p} \\ & \ddots & \vdots \\ 0 & & r_{p,p} \end{pmatrix}.$$

Denoting by $(Q^*b)_p$ (respectively, $(Q^*b)_{n-p}$) the vector of the first p (respectively, the last $n - p$) entries of Q^*b, we write

$$\|b - Az\|_n^2 = \|Q^*b - Rz\|_n^2 = \|(Q^*b)_p - R_{1,1}z\|_p^2 + \|(Q^*b)_{n-p}\|_{n-p}^2.$$

Since $R_{1,1}$ is upper triangular and nonsingular, by a simple back substitution we can compute the solution

$$z = R_{1,1}^{-1}(Q^*b)_p. \tag{7.15}$$

Consequently, the minimum value is

$$\|(Q^*b)_{n-p}\|_{n-p} = \min_{y \in \mathbb{R}^p} \|b - Ay\|_n.$$

Note that in this case, there is a unique solution to the least squares problem given by formula (7.15).

Operation Count

We compute the number of multiplications required by the Gram–Schmidt algorithm when $n > p$.

- Orthonormalization: at each step $1 \leq i \leq p$, we compute $i - 1$ scalar products of vectors of $\mathbb{R}^n$, and $i - 1$ vector-scalar products. The number of operations N_{op} is therefore

$$N_{op} \approx \sum_{i=1}^{p} 2(i - 1)n \approx np^2.$$

- Updating the right-hand side: the cost of the matrix-vector product Q^*b can be reduced by remarking that in (7.15), only the first p entries of Q^*b are required; hence

$$N_{op} \approx pn.$$

- Substitution: solving a triangular linear system of size p costs

$$N_{op} \approx p^2/2.$$

For large n, the QR method with the Gram–Schmidt algorithm is less efficient than the normal equation method if we compare the number of operations. The triangular system $Rx = Q^*b$ (in the case $n = p$) is, however, better conditioned. Indeed, $\mathrm{cond}_2(R) = \mathrm{cond}_2(A)$, since R and A differ by the multiplication of an orthogonal matrix (for the normal equation method, it is the conditioning of A^*A that matters). Nevertheless, in practice, this algorithm is not recommended for large matrices A (the following Householder algorithm shall be preferred). Indeed, the Gram–Schmidt algorithm is numerically unstable in the sense that for large values of p, the columns of Q are no longer perfectly orthogonal, so Q^{-1} is numerically no longer equal to Q^*.

7.3.4 Householder Algorithm

The Householder algorithm is an implementation of the QR method that does not rely on the Gram–Schmidt algorithm. It amounts to multiplying the matrix A by a sequence of very simple orthogonal matrices (the so-called Householder matrices) so as to shape A progressively into an upper triangular matrix.

Definition 7.3.1. *Let $v \in \mathbb{R}^n$ be a nonzero vector. The Householder matrix associated with the vector v, denoted by $H(v)$, is defined by*

$$H(v) = I - 2\frac{vv^t}{\|v\|^2}.$$

We set $H(0) = I$; the identity is thus considered as a Householder matrix.

Remark 7.3.2. The product vv^t of an $n \times 1$ matrix by a $1 \times n$ matrix is indeed a square matrix of order n, which is equivalently denoted by $v \otimes v$. It easy to check by associativity that the product $(vv^t)x$ is equal to $\langle v, x \rangle v$.

Householder matrices feature interesting properties that are described in the following result; see also Figure 7.2.

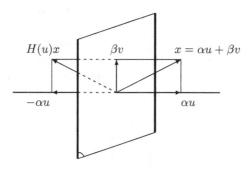

Fig. 7.2. Householder transformation $H(u)$ = orthogonal symmetry with respect to the hyperplane that is orthogonal to u.

Lemma 7.3.1. *Let $H(v)$ be a Householder matrix.*

(i) $H(v)$ is symmetric and orthogonal.
(ii) Let e be a unitary vector. $\forall v \in \mathbb{R}^n$, we have

$$H(v + \|v\|e)v = -\|v\|e \tag{7.16}$$

and

$$H(v - \|v\|e)v = +\|v\|e. \tag{7.17}$$

Proof. Obviously $H(v)^t = H(v)$. On the other hand, we have

$$H^2 = I - 4\frac{vv^t}{\|v\|^2} + 4\frac{(vv^t)(vv^t)}{\|v\|^4} = I,$$

since $(vv^t)(vv^t) = \|v\|^2(vv^t)$. Hence $H(v)$ is also orthogonal. Without loss of generality, we can assume that e is the first vector of the canonical basis $(e_i)_{1 \le i \le n}$. Let $w = v + \|v\|e$. If w is the null vector, then $H(w)v = v = -\|v\|e$, and relation (7.16) holds. For $w \ne 0$,

$$H(w)v = v - 2\frac{ww^t}{\|w\|^2}v = v - 2\frac{(\|v\|^2 + \|v\|v_1)(v + \|v\|e)}{(v_1 + \|v\|)^2 + \sum_{k \ne 1}|v_k|^2}$$

$$= v - \frac{2(\|v\|^2 + \|v\|v_1)(v + \|v\|e)}{2\|v\|^2 + 2v_1\|v\|} = v - (v + \|v\|e) = -\|v\|e.$$

A similar computation gives $H(v - \|v\|e)v = \|v\|e$. $\square$

We describe the Householder algorithm in the case $n \geq p$, that is, the matrix $A \in \mathcal{M}_{n,p}(\mathbb{R})$ has more rows than columns. The Householder algorithm defines a sequence of Householder matrices $H^k \in \mathcal{M}_n(\mathbb{R})$ and a sequence of matrices $A^{k+1} \in \mathcal{M}_{n,p}(\mathbb{R})$ for $1 \leq k \leq p$ satisfying

$$A^1 = A, \quad A^{k+1} = H^k A^k, \quad A^{p+1} = R,$$

where R is upper triangular. Each matrix A^k has zeros below the diagonal in its first $(k-1)$ columns, and the Householder matrix H^k is built in such a way that it reduces to zero the kth column below the diagonal in A^{k+1}.

Step 1 We set $A^1 = A$. Let a^1 be the first column of A^1. If we have

$$a^1 = \begin{pmatrix} a_{1,1} \\ 0 \\ \vdots \\ 0 \end{pmatrix},$$

then we are done by taking $H^1 = I$. Otherwise, we set

$$H^1 = H(a^1 + \|a^1\|e_1),$$

and define $A^2 = H^1 A^1$. By virtue of Lemma 7.3.1, the first column of A^2 is

$$A^2 e_1 = H^1 a^1 = \begin{pmatrix} -\|a^1\| \\ 0 \\ \vdots \\ 0 \end{pmatrix},$$

which is the desired result at the first step. We could also have taken $H^1 = H(a^1 - \|a^1\|e_1)$; we choose the sign according to numerical stability criteria.

Step k Assume that the first $k - 1$ columns of A^k have zeros below the diagonal:

$$A^k = \begin{pmatrix} a^k_{1,1} \cdots & \cdots & \cdots\cdots & a^k_{1,p} \\ 0 & \ddots & & \vdots \\ \vdots & \ddots & a^k_{k-1,k-1} \; \times \cdots \times \\ \vdots & & 0 & \vdots & \vdots \\ \vdots & & \vdots & \vdots & \vdots \\ 0 \cdots & & 0 & \times \cdots a^k_{n,p} \end{pmatrix}.$$

Let a^k be the vector of size $(n+1-k)$ made of the last $(n+1-k)$ entries of the kth column of A^k. If we have

$$a^k = \begin{pmatrix} a^k_{k,k} \\ 0 \\ \vdots \\ 0 \end{pmatrix},$$

then we choose $H^k = I$. Otherwise, we set

$$H^k = \left(\begin{array}{c|c} I_{k-1} & 0 \\ \hline 0 & H(a^k + \|a^k\|e_1) \end{array} \right),$$

where I_{k-1} is the identity matrix of order $k - 1$, and $H(a^k + \|a^k\|e_1)$ is a Householder matrix of order $(n + 1 - k)$. We define $A^{k+1} = H^k A^k$. By virtue of Lemma 7.3.1, the kth column of A^{k+1} is

$$A^{k+1}e_k = H^k \begin{pmatrix} a^k_{1,k} \\ \vdots \\ a^k_{k-1,k} \\ \hline a^k \end{pmatrix} = \begin{pmatrix} a^k_{1,k} \\ \vdots \\ a^k_{k-1,k} \\ \hline H(a^k + \|a^k\|e_1)a^k \end{pmatrix} = \begin{pmatrix} a^k_{1,k} \\ \vdots \\ a^k_{k-1,k} \\ -\|a^k\| \\ 0 \\ \vdots \\ 0 \end{pmatrix}.$$

Furthermore, in view of the structure of H^k, the first $(k-1)$ columns of A^{k+1} are exactly the same as those of A^k. Consequently,

$$A^{k+1} = \begin{pmatrix} a^{k+1}_{1,1} & \cdots & \cdots & \cdots\cdots & a^{k+1}_{1,p} \\ 0 & \ddots & & & \vdots \\ \vdots & & a^{k+1}_{k,k} & \times \cdots & \times \\ \vdots & & 0 & \vdots & \vdots \\ \vdots & & \vdots & \vdots & \vdots \\ 0 & \cdots & 0 & \times \cdots & a^{k+1}_{n,p} \end{pmatrix},$$

which is the desired result at the kth step. We could also have taken $H(a^k - \|a^k\|e_1)$ in H^k; we choose the sign according to numerical stability criteria. **After p steps** we have thus obtained an upper triangular matrix A^{p+1} such that

$$R = A^{p+1} = \begin{pmatrix} a_{1,1}^{p+1} & \cdots & a_{1,p}^{p+1} \\ 0 & \ddots & \vdots \\ \vdots & \ddots & a_{p,p}^{p+1} \\ \vdots & & 0 \\ \vdots & & \vdots \\ 0 & \cdots & 0 \end{pmatrix}.$$

Setting $Q = H^1 \cdots H^p$, which is an orthogonal matrix, we have indeed obtained that $A = QR$.

Remark 7.3.3. The QR factorization by the Householder algorithm is possible even if the matrix A is singular. It is an advantage with respect to the Gram–Schmidt algorithm, where column permutations are required if A is singular. The Householder algorithm is numerically stable, so it is the one used in practice. The algorithm still works if $n < p$ with obvious modifications.

Operation Count

At each step k, the vector $a^k + \|a^k\| e_1$ is computed as well as the product of the matrices H^k and A^k. Due to the special shape of the matrix H^k, this matrix product is equivalent to running $(p + 1 - k)$ operations of the type

$$\frac{(vv^t)a}{\|v\|^2} = \frac{\langle v, a \rangle v}{\|v\|^2},$$

where $v = a^k + \|a^k\| e_1$, and a is successively the vector containing the last $(n + 1 - k)$ entries of the last $(p + 1 - k)$ column vectors of A^k. Hence, there are mainly $(p + 1 - k)$ scalar products and vector-scalar multiplications (we neglect all lower-order terms such as, for example, the computation of $\|v\|$). Finally, we obtain

$$N_{op} \approx \sum_{k=1}^{p} 2(p + 1 - k)(n + 1 - k) \approx np^2 - \frac{1}{3}p^3.$$

This number of operations is smaller than that for the Gram–Schmidt procedure. In the case $n = p$, this method can be used to solve a linear system, and the number of operations is of order $\frac{2}{3}n^3$, which makes the Householder algorithm twice slower than the Gauss elimination algorithm. We shall use Householder matrices again in Chapter 10, concerning eigenvalue computations.

7.4 Exercises

7.1. Define a matrix A=reshape(1:28,7,4) and vectors b1=ones(7,1), and b2=[1;2;3;4;3;2;1].

1. Compute the solutions x_1 and x_2, of minimal norm, for the least squares problems

$$\min_{x \in \mathbb{R}^3} \|Ax - b_1\|_2 \quad \text{and} \quad \min_{x \in \mathbb{R}^3} \|Ax - b_2\|_2,$$

as well as the corresponding minimum values. Comment.
2. How can the other solutions of these two minimization problem be computed?

7.2. Define a matrix A by

```
A=reshape(1:6,3,2);A=[A eye(size(A)); -eye(A) -A];
```

1. Define a vector b_0 by b0=[2 4 3 -2 -4 -3]'. Compute the solution x_0, of minimal norm, of the least squares problem associated to the matrix A and b_0. Let b be a (small) variation of b_0 defined by e=1.E-2;b=b0+e*rand(6,1). Compute the solution x associated with b. Compute the relative errors $\|x - x_0\|_2/\|x_0\|$ and $\|b - b_0\|_2/\|b_0\|$. Compute the amplification coefficient C_b defined by equality (7.9).
2. Same questions for the vector b_1 defined by b1=[3 0 -2 -3 0 2]'. Display both vectors x_1 and x. What do you observe?
3. For i varying from $1/100$ to 1 by a step size of $1/100$, compute the amplification coefficient $C_b(i)$ associated with the vector $b_2 = ib_0 + (1 - i)b_1$. Plot the results on a graph. Comment.

7.3. The goal of this exercise is to approximate a smooth function f defined on the interval $(0, 1)$ by a polynomial $p \in \mathbb{P}_{n-1}$ in the least squares sense, i.e.,

$$\int_0^1 |f(x) - p(x)|^2 dx = \min_{q \in \mathbb{P}_{n-1}} \int_0^1 |f(x) - q(x)|^2 \, dx. \qquad (7.18)$$

Writing $p = \sum_{i=1}^n a_i \varphi_i$ in a basis $(\varphi_i)_{i=1}^n$ of $\mathbb{P}_{n-1}$, the unknowns of the problem are thus the coefficients a_i. Problem (7.18) is equivalent to determining the minimum of the function E from $\mathbb{R}^n$ into $\mathbb{R}$, defined by

$$E(\alpha_1, \ldots, \alpha_n) = \int_0^1 \left| f(x) - \sum_{i=1}^n \alpha_i \varphi_i(x) \right|^2 dx,$$

which admits a unique solution $(a_1, \ldots, a_n)$ characterized by the relations

$$\frac{\partial E}{\partial \alpha_k}(a_1, \ldots, a_n) = 0, \quad k = 1, \ldots, n.$$

1. Show that the vector $a = (a_1, \ldots, a_n)^t$ is the solution of a linear system $Aa = b$ whose matrix and right-hand side are to be specified.
2. Take $\varphi_i(x) = x^{i-1}$ and show that the matrix A is the Hilbert matrix; see Exercise 2.2.

3. Prove, using the Gram–Schmidt procedure, that there exists a basis $(\varphi_i)_{i=1}^n$ of $\mathbb{P}_{n-1}$ such that

$$\int_0^1 \varphi_i(x)\varphi_j(x)dx = \delta_{i,j}, \quad 1 \leq i, j \leq n.$$

What is the point in using this basis to compute p?

7.4. Define a matrix A by `A=MatRank(300,100,100)` (see Exercise 2.7), and `b=rand(300,1)`.

1. Compare (in terms of computational time) the following three methods for solving the least squares problem

$$\min_{x \in \mathbb{R}^p} \|Ax - b\|_2. \tag{7.19}$$

(a) The Cholesky method for solving the normal equations. Use the function `chol`.
(b) The QR factorization method.
(c) The SVD method where A is factorized as $A = V\Sigma U^*$ (see Theorem 2.7.1). *Hint:* the solutions of (7.19) are given by $x = Uy$, where $y \in \mathbb{R}^n$ can be determined explicitly in terms of the singular values of A and the vector b.

Compute the solutions x of (7.19), the minima $\|Ax - b\|$, and the computational time. Conclude.

2. Now define A and b by

```
e=1.e-5;P=[1 1 0;0 1 -1; 1 0 -1]
A=P*diag([e,1,1/e])*inv(P);b=ones(3,1)
```

Compare the solutions of the least squares problem obtained by the Cholesky method and the QR method. Explain.

7.5 (∗). Program a function `Householder` to execute the QR factorization of a matrix by the Householder algorithm. Compare the results with the factorization obtained by the Gram–Schmidt method.

8

Simple Iterative Methods

8.1 General Setting

This chapter is devoted to solving the linear system

$$Ax = b$$

by means of iterative methods. In the above equation, $A \in \mathcal{M}_n(\mathbb{R})$ is a nonsingular square matrix, $b \in \mathbb{R}^n$ is the right-hand side, and x is the unknown vector. A method for solving the linear system $Ax = b$ is called iterative if it is a numerical method computing a sequence of approximate solutions x_k that converges to the exact solution x as the number of iterations k goes to $+\infty$.

In this chapter, we consider only iterative methods whose sequence of approximate solutions is defined by a simple induction relation, that is, x_{k+1} is a function of x_k only and not of the previous iterations $x_{k-1}, \dots, x_1$.

Definition 8.1.1. *Let A be a nonsingular matrix. A pair of matrices (M, N) with M nonsingular (and easily invertible in practice) satisfying*

$$A = M - N$$

is called a splitting (or regular decomposition) of A. An iterative method based on the splitting (M, N) is defined by

$$\begin{cases} x_0 \text{ given in } \mathbb{R}^n, \\ M x_{k+1} = N x_k + b \quad \forall k \geq 1. \end{cases} \tag{8.1}$$

In the iterative method (8.1), the task of solving the linear system $Ax = b$ is replaced by a sequence of several linear systems $M\tilde{x} = \tilde{b}$ to be solved. Therefore, M has to be much easier to invert than A.

Remark 8.1.1. If the sequence of approximate solutions x_k converges to a limit x as k tends to infinity, then by taking the limit in the induction relation (8.1) we obtain

$$(M - N)x = Ax = b.$$

Accordingly, should the sequence of approximate solutions converge, its limit is necessarily the solution of the linear system.

From a practical viewpoint, a convergence criterion is required to decide when to terminate the iterations, that is, when x_k is sufficiently close to the unknown solution x. We will address this issue at the end of the chapter.

Definition 8.1.2. *An iterative method is said to converge if for any choice of the initial vector $x_0 \in \mathbb{R}^n$, the sequence of approximate solutions x_k converges to the exact solution x.*

Definition 8.1.3. *We call the vector $r_k = b - Ax_k$ (respectively, $e_k = x_k - x$) residual (respectively, error) at the kth iteration.*

Obviously, an iterative method converges if and only if e_k converges to 0, which is equivalent to $r_k = Ae_k$ converging to 0. In general, we have no knowledge of e_k because x is unknown! However, it is easy to compute the residuals r_k, so convergence is detected on the residual in practice.

The sequence defined by (8.1) is also equivalently given by

$$x_{k+1} = M^{-1}Nx_k + M^{-1}b. \tag{8.2}$$

The matrix $M^{-1}N$ is called an iteration matrix or amplification matrix of the iterative method. Theorem 8.1.1 below shows that the convergence of the iterative method is linked to the spectral radius of $M^{-1}N$.

Theorem 8.1.1. *The iterative method defined by (8.1) converges if and only if the spectral radius of $M^{-1}N$ satisfies*

$$\varrho(M^{-1}N) < 1.$$

Proof.   The error e_k is given by the induction relation

$$e_k = x_k - x = (M^{-1}Nx_{k-1} + M^{-1}b) - (M^{-1}Nx + M^{-1}b)$$
$$= M^{-1}N(x_{k-1} - x) = M^{-1}Ne_{k-1}.$$

Hence $e_k = (M^{-1}N)^k e_0$, and by Lemma 3.3.1 we infer that $\lim_{k \to +\infty} e_k = 0$, for any e_0, if and only if $\varrho(M^{-1}N) < 1$. □

Example 8.1.1. To solve a linear system $Ax = b$, we consider Richardson's iterative method (also called gradient method)

$$x_{k+1} = x_k + \alpha(b - Ax_k),$$

where α is a real number. It corresponds to the splitting (8.1) with $M = \alpha^{-1}I$ and $N = \alpha^{-1}I - A$. The eigenvalues of the iteration matrix $B_\alpha = I - \alpha A$ are $(1 - \alpha\lambda_i)$, where $(\lambda_i)_i$ are the eigenvalues of A. Richardson's method converges if and only if $|1 - \alpha\lambda_i| < 1$ for any eigenvalue λ_i. If the eigenvalues of A satisfy $0 < \lambda_1 \leq \cdots \leq \lambda_n \equiv \varrho(A)$, the latter condition is equivalent to $\alpha \in (0, 2/\varrho(A))$; see Figure 9.1.

In some cases, it is not necessary to compute the spectral radius of $M^{-1}N$ to prove convergence, as shown in the following theorem.

Theorem 8.1.2. *Let A be a Hermitian positive definite matrix. Consider a splitting of $A = M - N$ with M nonsingular. Then the matrix $(M^* + N)$ is Hermitian. Furthermore, if $(M^* + N)$ is also positive definite, we have*

$$\varrho(M^{-1}N) < 1.$$

Proof. First of all, $M^* + N$ is indeed Hermitian since it is the sum of two Hermitian matrices

$$M^* + N = (M^* - N^*) + (N^* + N) = A^* + (N^* + N).$$

Since A is positive definite, we define the following vector norm:

$$|x|_A = \sqrt{\langle Ax, x \rangle}, \quad \forall x \in \mathbb{R}^n.$$

We denote by $\|.\|$ the matrix norm subordinate to $|.|_A$. Let us show that $\|M^{-1}N\| < 1$, which yields the desired result thanks to Proposition 3.1.4. By Proposition 3.1.1, there exists v depending on $M^{-1}N$ such that $|v|_A = 1$ and satisfying

$$\|M^{-1}N\|^2 = \max_{|x|_A=1} |M^{-1}Nx|_A^2 = |M^{-1}Nv|_A^2.$$

Since $N = M - A$, setting $w = M^{-1}Av$, we get

$$
\begin{aligned}
|M^{-1}Nv|_A^2 &= \langle AM^{-1}Nv, M^{-1}Nv \rangle = \langle AM^{-1}(M - A)v, M^{-1}(M - A)v \rangle \\
&= \langle (Av - AM^{-1}Av), (I - M^{-1}A)v \rangle \\
&= \langle Av, v \rangle - \langle AM^{-1}Av, v \rangle + \langle AM^{-1}Av, M^{-1}Av \rangle - \langle Av, M^{-1}Av \rangle \\
&= 1 - \langle w, Mw \rangle + \langle Aw, w \rangle - \langle Mw, w \rangle = 1 - \langle (M^* + N)w, w \rangle.
\end{aligned}
$$

By assumption, $(M^* + N)$ is positive definite and $w \neq 0$, since A and M are nonsingular. Thus $\langle (M^* + N)w, w \rangle > 0$. As a result, $\|M^{-1}N\|^2 = 1 - \langle (M^* + N)w, w \rangle < 1$. □

Iterative methods for solving linear systems may require a large number of iterations to converge. Thus, one might think that the accumulation of rounding errors during the iterations completely destroys the convergence of these methods on computers (or even worse, makes them converge to wrong solutions). Fortunately enough, this is not the case, as is shown by the following result.

Theorem 8.1.3. *Consider a splitting of $A = M - N$ with A and M nonsingular. Let $b \in \mathbb{R}^n$ be the right-hand side, and let $x \in \mathbb{R}^n$ be the solution of $Ax = b$. We assume that at each step k the iterative method is tainted by an error $\varepsilon_k \in \mathbb{R}^n$, meaning that x_{k+1} is not exactly given by (8.1) but rather by*

$$x_{k+1} = M^{-1}Nx_k + M^{-1}b + \varepsilon_k.$$

We assume that $\varrho(M^{-1}N) < 1$, and that there exist a vector norm and a positive constant ε such that for all $k \geq 0$,

$$\|\varepsilon_k\| \leq \varepsilon.$$

Then, there exists a constant K, which depends on $M^{-1}N$ but not on ε, such that

$$\limsup_{k \to +\infty} \|x_k - x\| \leq K\varepsilon.$$

Proof. The error $e_k = x_k - x$ satisfies $e_{k+1} = M^{-1}Ne_k + \varepsilon_k$, so that

$$e_k = \left(M^{-1}N\right)^k e_0 + \sum_{i=0}^{k-1} \left(M^{-1}N\right)^i \varepsilon_{k-i-1}. \tag{8.3}$$

By virtue of Proposition 3.1.4, there exists a subordinate matrix norm $\|\cdot\|_s$ such that $\|M^{-1}N\|_s < 1$, since $\varrho(M^{-1}N) < 1$. We use the same notation for the associated vector norm. Now, all vector norms on $\mathbb{R}^n$ are equivalent, so there exists a constant $C \geq 1$, which depends only on $M^{-1}N$, such that

$$C^{-1}\|y\| \leq \|y\|_s \leq C\|y\|, \quad \forall y \in \mathbb{R}^n.$$

Bounding (8.3) from above yields

$$\|e_k\|_s \leq \|M^{-1}N\|_s^k \|e_0\|_s + \sum_{i=0}^{k-1} \|M^{-1}N\|_s^i C\varepsilon \leq \|M^{-1}N\|_s^k \|e_0\|_s + \frac{C\varepsilon}{1 - \|M^{-1}N\|_s}.$$

Letting k go to infinity leads to the desired result with $K = C^2/(1 - \|M^{-1}N\|_s)$. $\qquad\square$

Iterative methods are often used with sparse matrices. A matrix is said to be sparse if it has relatively few nonzero entries. Sparse matrices arise, for example, in the discretization of partial differential equations by the finite difference or finite element method. A simple instance is given by tridiagonal matrices.

Storage of sparse matrices. The idea is to keep track only of nonzero entries of a matrix A, thereby saving considerable memory in practice for matrices of large size. We introduce sparse or Morse storage through the following illustrative example; for more details we refer to [5], [14].

Example 8.1.2. Define a matrix

$$A = \begin{pmatrix} 9 & 0 & -3 & 0 \\ 7 & -1 & 0 & 4 \\ 0 & 5 & 2 & 0 \\ 1 & 0 & -1 & 2 \end{pmatrix}. \tag{8.4}$$

The entries of A are stored in a vector array STOCKA. We define another array BEGINL, which indicates where the rows of A are stored in STOCKA.

More precisely, STOCKA(BEGINL(i)) contains the first nonzero entry of row i. We also need a third array INDICC that gives the column of every entry stored in STOCKA. If $a_{i,j}$ is stored in STOCKA(k), then INDICC(k) $= j$. The number of nonzero entries of A is equal to the size of the vectors INDICC and STOCKA. The vector BEGINL has size $(n+1)$, where n is the number of rows of A, because its last entry BEGINL($n+1$) is equal to the size of INDICC and STOCKA plus 1. This is useful in computing the product $z = Ay$ with such a storage: each entry $z(i)$ of z is given by

$$z(i) = \sum_{k=k_i}^{k_{i+1}-1} \text{STOCKA}(k)\, y(\text{INDICC}(k)),$$

where $k_i =$ BEGINL(i) and $(k_{n+1} - 1)$ is precisely the size of INDICC and STOCKA (see Table 8.1 for its application to A).

STOCKA	INDICC	BEGINL
2	4	
−1	3	
1	1	
2	3	
5	2	
4	4	11
−1	2	8
7	1	6
−3	3	3
9	1	1

Table 8.1. Morse storage of the matrix A in (8.4).

8.2 Jacobi, Gauss–Seidel, and Relaxation Methods

8.2.1 Jacobi Method

For any matrix $A = (a_{i,j})_{1 \le i,j \le n}$, its diagonal D is defined as

$$D = \text{diag}\,(a_{1,1}, \ldots, a_{n,n}).$$

Definition 8.2.1. *The Jacobi method is the iterative method defined by the splitting*

$$M = D, \quad N = D - A.$$

The iteration matrix of this method is denoted by $\mathcal{J} = M^{-1}N = I - D^{-1}A$.

Remark 8.2.1.
1. The Jacobi method is well defined if the diagonal matrix is nonsingular.
2. If A is Hermitian, the Jacobi method converges if A and $2D - A$ are positive definite (by virtue of Theorem 8.1.2).
3. There exists a block Jacobi method; see Section 8.6.

8.2.2 Gauss–Seidel Method

For any matrix $A = (a_{i,j})_{1 \leq i,j \leq n}$, consider the decomposition $A = D - E - F$, where D is the diagonal, $-E$ is the lower part, and $-F$ is the upper part of A. Namely,

$$
\begin{cases}
d_{i,j} = a_{i,j}\delta_{i,j}; \\
e_{i,j} = -a_{i,j} & \text{if } i > j, \text{ and } 0 \text{ otherwise;} \\
f_{i,j} = -a_{i,j} & \text{if } i < j, \text{ and } 0 \text{ otherwise.}
\end{cases}
$$

Definition 8.2.2. *The Gauss–Seidel method is the iterative method defined by the splitting*

$$
M = D - E, \quad N = F.
$$

The iteration matrix of this method is denoted by $\mathcal{G}_1 = M^{-1}N = (D-E)^{-1}F$.

Remark 8.2.2.
1. The Gauss–Seidel method is well defined if the matrix $D - E$ is nonsingular, which is equivalent to asking that D be nonsingular.
2. The matrix $(D - E)$ is easy to invert, since it is triangular.
3. If A is Hermitian and positive definite, then $M^* + N = D$ is also Hermitian and positive definite, so Gauss–Seidel converges (by virtue of Theorem 8.1.2).
4. There exists a block Gauss–Seidel method; see Section 8.6.

Comparison between the Jacobi method and the Gauss–Seidel method.

In the Jacobi method, we successively compute the entries of x^{k+1} in terms of all the entries of x^k:

$$
x_i^{k+1} = \frac{1}{a_{i,i}}[-a_{i,1}x_1^k - \cdots - a_{i,i-1}x_{i-1}^k - a_{i,i+1}x_{i+1}^k - \cdots - a_{i,n}x_n^k + b_i].
$$

In the Gauss–Seidel method, we use the information already computed in the $(i - 1)$ first entries. Namely,

$$
x_i^{k+1} = \frac{1}{a_{i,i}}[-a_{i,1}x_1^{k+1} - \cdots - a_{i,i-1}x_{i-1}^{k+1} - a_{i,i+1}x_{i+1}^k - \cdots - a_{i,n}x_n^k + b_i].
$$

From a practical point of view, two vectors of size n are required to store x^k and x^{k+1} separately in the Jacobi method, while only one vector is required in the Gauss–Seidel method (the entries of x^{k+1} progressively override those of x^k).

8.2.3 Successive Overrelaxation Method (SOR)

The successive overrelaxation method (SOR) can be seen as an extrapolation of the Gauss–Seidel method.

Definition 8.2.3. *Let $\omega \in \mathbb{R}^+$. The iterative method relative to the splitting*

$$M = \frac{D}{\omega} - E, \quad N = \frac{1 - \omega}{\omega} D + F$$

is called relaxation method for the parameter ω. We denote by $\mathcal{G}_\omega$ the iteration matrix

$$\mathcal{G}_\omega = M^{-1}N = \left(\frac{D}{\omega} - E \right)^{-1} \left(\frac{1 - \omega}{\omega} D + F \right).$$

Remark 8.2.3.
1. The relaxation method is well defined if the diagonal D is invertible.
2. If $\omega = 1$, we recover the Gauss–Seidel method.
3. If $\omega < 1$, we talk about an under-relaxation method.
4. If $\omega > 1$, we talk about an over-relaxation method.
5. The idea behind the relaxation method is the following. If the efficiency of an iterative method is measured by the spectral radius of its iteration matrix $M^{-1}N$, then, since $\varrho(\mathcal{G}_\omega)$ is continuous with respect to ω, we can find an optimal ω that produces the smallest spectral radius possible. Accordingly, the associated iterative method is more efficient than Gauss–Seidel. We shall see that in general, $\omega_{\mathrm{opt}} > 1$, hence the name SOR (over-relaxation).
6. A block relaxation method is defined in Section 8.6.
7. A relaxation approach for the Jacobi method is discussed in Exercise 8.7.

Theorem 8.2.1. *Let A be a Hermitian positive definite matrix. Then for any $\omega \in]0, 2[$, the relaxation method converges.*

Proof. Since A is definite positive, so is D. As a result, $\frac{D}{\omega} - E$ is nonsingular. Moreover,

$$M^* + N = \frac{D}{\omega} - E^* + \frac{1 - \omega}{\omega} D + F = \frac{2 - \omega}{\omega} D,$$

since $E^* = F$. We conclude that $M^* + N$ is positive definite if and only if $0 < \omega < 2$. Theorem 8.1.2 yields the result. $\qquad\square$

Theorem 8.2.2. *For any matrix A, we always have*

$$\varrho(\mathcal{G}_\omega) \geq |1 - \omega|, \quad \forall \omega \neq 0.$$

Consequently, the relaxation method can converge only if $0 < \omega < 2$.

Proof. The determinant of $\mathcal{G}_\omega$ is equal to

$$\det(\mathcal{G}_\omega) = \det\left((\frac{1-\omega}{\omega}D + F\right) / \det\left(\frac{D}{\omega} - E\right) = (1-\omega)^n.$$

We deduce that

$$\varrho(\mathcal{G}_\omega)^n \geq \prod_{i=1}^n |\lambda_i(\mathcal{G}_\omega)| = |\det(\mathcal{G}_\omega)| = |1-\omega|^n,$$

where $\lambda_i(\mathcal{G}_\omega)$ are the eigenvalues of $\mathcal{G}_\omega$. This yields the result. □

8.3 The Special Case of Tridiagonal Matrices

We compare the Jacobi, Gauss–Seidel, and relaxation methods in the special case of tridiagonal matrices.

Theorem 8.3.1. *Let A be a tridiagonal matrix. We have*

$$\varrho(\mathcal{G}_1) = \varrho(\mathcal{J})^2,$$

so the Jacobi and Gauss–Seidel methods converge or diverge simultaneously, but Gauss–Seidel always converges faster than Jacobi.

Theorem 8.3.2. *Let A be a tridiagonal Hermitian positive definite matrix. Then all three methods converge. Moreover, there exists a unique optimal parameter ω_{opt} in the sense that*

$$\varrho(\mathcal{G}_{\omega_{opt}}) = \min_{0<\omega<2} \varrho(\mathcal{G}_\omega),$$

where

$$\omega_{opt} = \frac{2}{1 + \sqrt{1 - \varrho(\mathcal{J})^2}},$$

and

$$\varrho(\mathcal{G}_{\omega_{opt}}) = \omega_{opt} - 1.$$

Remark 8.3.1. Theorem 8.3.2 shows that in the case of a tridiagonal Hermitian positive definite matrix, we have $\omega_{opt} \geq 1$. Therefore, it is better to perform overrelaxation than underrelaxation.

To prove the above theorems, we need a technical lemma.

Lemma 8.3.1. *For any nonzero real number $\mu \neq 0$, we define a tridiagonal matrix $A(\mu)$ by*

$$A(\mu) = \begin{pmatrix} b_1 & \mu^{-1}c_1 & & 0 \\ \mu a_2 & \ddots & \ddots & \\ & \ddots & \ddots & \mu^{-1}c_{n-1} \\ 0 & & \mu a_n & b_n \end{pmatrix},$$

where a_i, b_i, c_i are given real numbers. The determinant of $A(\mu)$ is independent of μ. In particular, $\det A(\mu) = \det A(1)$.

Proof. The matrices $A(\mu)$ and $A(1)$ are similar, since $A(\mu)Q(\mu) = Q(\mu)A(1)$ with $Q(\mu) = \operatorname{diag}(\mu, \mu^2, \ldots, \mu^n)$, which yields the result.   □

Proof of Theorem 8.3.1.   The eigenvalues of A are the roots of its characteristic polynomial $P_A(\lambda) = \det(A - \lambda I)$. We have

$$P_{\mathcal{J}}(\lambda) = \det(-D^{-1}) \det(\lambda D - E - F)$$

and

$$P_{\mathcal{G}_1}(\lambda^2) = \det(E - D)^{-1} \det(\lambda^2 D - \lambda^2 E - F).$$

We define a matrix $A(\mu)$ by

$$A(\mu) = \lambda^2 D - \mu\lambda^2 E - \frac{1}{\mu}F.$$

By Lemma 8.3.1, we get $\det A(\frac{1}{\lambda}) = \det A(1)$. Hence

$$P_{\mathcal{G}_1}(\lambda^2) = (-1)^n \lambda^n P_{\mathcal{J}}(\lambda).$$

As a consequence, for any $\lambda \neq 0$, we deduce that λ is an eigenvalue of $\mathcal{J}$ if and only if λ^2 is an eigenvalue of $\mathcal{G}_1$. Thus, $\varrho(\mathcal{G}_1) = \varrho(\mathcal{J})^2$.   □

Proof of Theorem 8.3.2.   Since A is Hermitian, positive definite, we already know by Theorem 8.2.1 that the relaxation method converges for $0 < \omega < 2$. In particular, the Gauss–Seidel method converges. Now, Theorem 8.3.1 states that $\varrho(\mathcal{J})^2 = \varrho(\mathcal{G}_1) < 1$. Therefore, the Jacobi method converges too. It remains to determine ω_{opt}. Let $A(\mu)$ be the matrix defined by

$$A(\mu) = \frac{\lambda^2 + \omega - 1}{\omega}D - \mu\lambda^2 E - \frac{1}{\mu}F.$$

By Lemma 8.3.1, we know that $\det A(\frac{1}{\lambda}) = \det A(1)$. Accordingly,

$$\det\left(\frac{\lambda^2 + \omega - 1}{\omega}D - \lambda^2 E - F\right) = \lambda^n \det\left(\frac{\lambda^2 + \omega - 1}{\lambda\omega}D - E - F\right).$$

Observing that

$$P_{\mathcal{G}_\omega}(\lambda^2) = \det\left(E - \frac{D}{\omega}\right)^{-1}\det\left(\frac{\lambda^2 + \omega - 1}{\omega}D - \lambda^2 E - F\right),$$

we deduce that there exists a constant c (independent of λ) such that

$$P_{\mathcal{G}_\omega}(\lambda^2) = c\lambda^n P_{\mathcal{J}}\left(\frac{\lambda^2 + \omega - 1}{\lambda\omega}\right).$$

In other words, for any $\lambda \neq 0$, λ^2 is an eigenvalue of $\mathcal{G}_\omega$ if and only if $(\lambda^2 + \omega - 1)/(\lambda\omega)$ is an eigenvalue of $\mathcal{J}$. For an eigenvalue α of $\mathcal{J}$, we denote by $\lambda^\pm(\alpha)$ the two roots of the following equation:

$$\frac{\lambda^2 + \omega - 1}{\lambda\omega} = \alpha,$$

that is,

$$\lambda^\pm(\alpha) = \frac{\alpha\omega \pm \sqrt{\alpha^2\omega^2 - 4(\omega - 1)}}{2}.$$

We have just proved that $\mu^+(\alpha) = \lambda^+(\alpha)^2$ and $\mu^-(\alpha) = \lambda^-(\alpha)^2$ are eigenvalues of $\mathcal{G}_\omega$. Now, if α is an eigenvalue of $\mathcal{J}$, so is $-\alpha$. Then, $\lambda^+(\alpha) = -\lambda^-(-\alpha)$. Hence,

$$\varrho(\mathcal{G}_\omega) = \max_{\alpha \in \sigma(\mathcal{J})} |\mu^+(\alpha)|,$$

with

$$|\mu^+(\alpha)| = \left|\frac{1}{2}(\alpha^2\omega^2 - 2\omega + 2) + \frac{\alpha\omega}{2}\sqrt{\alpha^2\omega^2 - 4(\omega - 1)}\right|. \qquad (8.5)$$

In order to compute ω_{opt}, we have to maximize (8.5) over all eigenvalues of $\mathcal{J}$. Let us first show that the eigenvalues of $\mathcal{J}$ are real. Denote by α and $v \neq 0$ an eigenvalue and its corresponding eigenvector of $\mathcal{J}$. By definition,

$$\mathcal{J}v = \alpha v \Leftrightarrow (E + F)v = \alpha Dv \Leftrightarrow Av = (1 - \alpha)Dv.$$

Taking the scalar product with v yields

$$\langle Av, v \rangle = (1 - \alpha)\langle Dv, v \rangle,$$

which implies that $(1 - \alpha)$ is a positive real number, since A and D are Hermitian positive definite. The next step amounts to computing explicitly $|\mu^+(\alpha)|$. Note that $\mu^+(\alpha)$ may be complex, because the polynomial $\alpha^2\omega^2 - 4\omega + 4 = 0$ has two roots:

$$\omega^+(\alpha) = \frac{2}{1 + \sqrt{1 - \alpha^2}} < \omega^-(\alpha) = \frac{2}{1 - \sqrt{1 - \alpha^2}},$$

and may therefore be negative. Since $|\alpha| \leq \varrho(\mathcal{J}) < 1$, we get

$$1 < \omega^+(\alpha) < 2 < \omega^-(\alpha).$$

If $\omega^+(\alpha) < \omega < 2$, then $\mu^+(\alpha)$ is complex, and a simple computation shows that $|\mu^+(\alpha)| = |\omega - 1| = \omega - 1$. Otherwise, since $\omega \in (0, 2)$, we have $0 < \omega < \omega^+(\alpha)$, and $\mu^+(\alpha)$ is real. Thus

$$|\mu^+(\alpha)| = \begin{cases} \omega - 1 & \text{if } \omega^+(\alpha) < \omega < 2, \\ \lambda^+(\alpha)^2 & \text{if } 0 < \omega < \omega^+(\alpha). \end{cases}$$

When $\mu^+(\alpha)$ is real, we have $\lambda^+(\alpha) \geq \lambda^+(-\alpha)$ if $\alpha > 0$. Furthermore, $\omega^+(\alpha) = \omega^+(-\alpha)$, so

$$|\mu^+(\alpha)| \geq |\mu^+(-\alpha)|, \quad \text{if } \alpha > 0.$$

In other words, we can restrict the maximization of $|\mu^+(\alpha)|$ to positive α. Moreover, for $\alpha > 0$ we have

$$\frac{d}{d\alpha}\left(\lambda^+(\alpha)^2\right) = \lambda^+(\alpha)\left(\omega + \frac{\omega^2\alpha}{\sqrt{\alpha^2\omega^2 - 4(\omega - 1)}}\right) > 0.$$

Accordingly, for a fixed ω, the maximum is attained at $\alpha = \varrho(\mathcal{J})$:

$$\varrho(\mathcal{G}_\omega) = |\mu^+(\varrho(\mathcal{J}))| = \max_{\alpha \in \sigma(\mathcal{J})} |\mu^+(\alpha)|.$$

From now on, we replace α by the maximizer $\varrho(\mathcal{J})$ and we eventually minimize with respect to ω. The derivative is

$$\frac{d}{d\omega}\left(\lambda^+(\varrho(\mathcal{J}))^2\right) = 2\lambda^+(\varrho(\mathcal{J}))\frac{d}{d\omega}\lambda^+(\varrho(\mathcal{J}))$$

$$= 2\lambda^+(\varrho(\mathcal{J}))\left(\frac{\varrho(\mathcal{J})}{2} + \frac{2\varrho(\mathcal{J})^2\omega - 4}{4\sqrt{\varrho(\mathcal{J})^2\omega^2 - 4(\omega - 1)}}\right)$$

$$= \lambda^+(\varrho(\mathcal{J}))\frac{2(\varrho(\mathcal{J})\lambda^+(\varrho(\mathcal{J})) - 1)}{\sqrt{\varrho(\mathcal{J})^2\omega^2 - 4(\omega - 1)}}.$$

Since $0 < \varrho(\mathcal{J}) < 1$ and $\lambda^+(\varrho(\mathcal{J})) \leq \varrho(\mathcal{G}_\omega) < 1$, we deduce

$$\frac{d}{d\omega}\left(\lambda^+(\varrho(\mathcal{J}))^2\right) < 0,$$

and the minimum of $\lambda^+(\varrho(\mathcal{J}))^2$ on $[0, \omega^+(\varrho(\mathcal{J}))]$ is attained at $\omega^+(\varrho(\mathcal{J}))$. Likewise, the minimum of $\omega - 1$ on $[\omega^+(\varrho(\mathcal{J})), 2]$ is attained at $\omega^+(\varrho(\mathcal{J}))$. We deduce that as ω varies in $]0, 2[$, the minimum of $\varrho(\mathcal{G}_\omega)$ is attained at $\omega^+(\varrho(\mathcal{J}))$ and we obtain (see Figure 8.1) $\min_{0 < \omega < 2} \varrho(\mathcal{G}_\omega) = \omega_{\text{opt}} - 1$, and $\omega_{\text{opt}} = \omega^+(\varrho(\mathcal{J}))$. $\qquad\square$

Remark 8.3.2. If only a rough approximation of the optimal parameter ω_{opt} is available, it is better to overevaluate it than to underevaluate it, since (see Figure 8.1)

$$\lim_{\omega \to \omega_{\text{opt}}^-} \frac{d\varrho(\mathcal{G}_\omega)}{d\omega} = -\infty \quad \text{and} \quad \lim_{\omega \to \omega_{\text{opt}}^+} \frac{d\varrho(\mathcal{G}_\omega)}{d\omega} = 1.$$

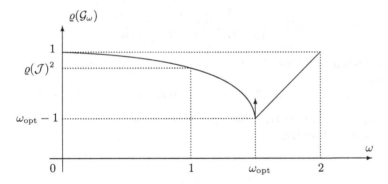

$$\varrho(\mathcal{G}_\omega)$$

Fig. 8.1. Spectral radius of $\mathcal{G}_\omega$ in terms of ω.

8.4 Discrete Laplacian

We revisit the finite difference discretization of the Laplacian (see Sections 1.1 and 5.3.3), which leads to the linear system $A_n x = b$, with a tridiagonal symmetric positive definite matrix $A_n \in \mathcal{M}_{n-1}(\mathbb{R})$, defined by (5.12), and $b \in \mathbb{R}^{n-1}$. The eigenvalues of A_n are (see Section 5.3.3)

$$\lambda_k = 4n^2 \sin^2\left(k\frac{\pi}{2n}\right), \quad k = 1, \ldots, n-1.$$

We now compare some iterative methods for solving the corresponding linear system. As usual $(x_k)_k$ denotes the sequence of vector iterates and $e_k = x_k - x$ is the error.

✓ *Jacobi method.* According to our notations we have

$$M = 2n^2 I_{n-1}, \quad N = 2n^2 I_{n-1} - A_n, \quad \mathcal{J} = M^{-1}N = I_{n-1} - \frac{1}{2n^2}A_n.$$

The eigenvalues of the Jacobi matrix $\mathcal{J}$ are therefore $\mu_k = 1 - \lambda_k/(2n^2)$, with λ_k eigenvalue of A_n, and

$$\varrho(\mathcal{J}) = \max_{1 \le k \le n-1} \left|\cos k\frac{\pi}{n}\right| = \max_{k \in \{1, n-1\}} \left|\cos k\frac{\pi}{n}\right| = \cos\frac{\pi}{n}.$$

Since $\varrho(\mathcal{J}) < 1$, the Jacobi method converges, and as $n \to +\infty$,

$$\varrho(\mathcal{J}) = 1 - \frac{1}{2}\frac{\pi^2}{n^2} + \mathcal{O}(n^{-4}).$$

The matrix $\mathcal{J} = I - \frac{1}{2n^2}A_n$ is symmetric, and therefore normal. Thus, $\|\mathcal{J}^k\|_2 = \varrho(\mathcal{J})^k$ and $\|e_k\|_2 \le \varrho(\mathcal{J})^k\|e_0\|_2$.

Let ε be a given error tolerance. What is the minimal number of iterations k_0 such that the error after k_0 iterations is reduced from a factor ε? In mathematical terms we want $\|e_k\|_2 \leq \varepsilon \|e_0\|_2$, for $k \geq k_0$. We compute an upper bound by looking for k_0 such that $\varrho(\mathcal{J})^{k_0} \leq \varepsilon$. We find

$$k_0 = \frac{\ln \varepsilon}{\ln \varrho(\mathcal{J})} \approx Cn^2, \quad \text{with} \quad C = -2\frac{\ln \varepsilon}{\pi^2}.$$

✓ *Gauss–Seidel method.* Since A_n is tridiagonal, we have

$$\varrho(\mathcal{G}_1) = \varrho(\mathcal{J})^2 = \cos^2\frac{\pi}{n} < 1,$$

and the Gauss–Seidel method converges too. As $n \to +\infty$,

$$\varrho(\mathcal{G}_1) = 1 - \frac{\pi^2}{n^2} + \mathcal{O}(n^{-4}).$$

For all $k \geq k_1$, we have $\|e_k\|_2 \leq \varepsilon\|e_0\|_2$, where k_1 is defined by

$$k_1 = \frac{\ln \varepsilon}{\ln \varrho(\mathcal{G}_1)} \approx Cn^2, \quad \text{with} \quad C = -\frac{\ln \varepsilon}{\pi^2}.$$

Note that $k_1 \approx k_0/2$. Put differently, for large values of n, the Gauss–Seidel method takes half as many iterations as the Jacobi method to meet some prescribed convergence criterion.

✓ *Relaxation method.* Since A_n is tridiagonal, symmetric, and positive definite, the relaxation method converges if and only if $\omega \in (0,2)$, and the optimal value of the parameter ω (see Theorem 8.3.2) is

$$\omega_{\text{opt}} = \frac{2}{1 + \sqrt{1 - \varrho(\mathcal{J})^2}} = \frac{2}{1 + \sin\frac{\pi}{n}}.$$

As $n \to +\infty$, we have $\omega_{\text{opt}} = 2(1 - \frac{\pi}{n}) + \mathcal{O}(n^{-2})$, and

$$\varrho(\mathcal{G}_{\omega_{\text{opt}}}) = \omega_{\text{opt}} - 1 = 1 - \frac{2\pi}{n} + \mathcal{O}(n^{-2}).$$

With the choice $\omega = \omega_{\text{opt}}$, we get $\|e_k\|_2 \leq \varepsilon\|e_0\|_2$ for $k \geq k_2$ satisfying

$$k_2 = \frac{\ln \varepsilon}{\ln \varrho(\mathcal{L}_{\omega_{\text{opt}}})} \approx Cn, \quad \text{with } C = -\frac{\ln \varepsilon}{2\pi}.$$

Note that $k_2 \approx \frac{\pi}{2n}k_1$, so the convergence is much faster than for Jacobi or Gauss–Seidel, since we save one order of magnitude in n.

As a conclusion, to achieve a given fixed error ε, the Gauss–Seidel method is (asymptotically) twice as fast as the Jacobi method. The speedup is all the more considerable as we move from the Jacobi and Gauss–Seidel methods to the relaxation method (with optimal parameter), since we save a factor n. For instance, for $n = 100$ and $\varepsilon = 0.1$, we approximately obtain:

- $k_0 = 9342$ iterations for the Jacobi method,
- $k_1 = 4671$ iterations for the Gauss–Seidel method,
- $k_2 = 75$ iterations for the optimal relaxation method.

8.5 Programming Iterative Methods

We first define a convergence criterion for an iterative method, that is, a test that if satisfied at the kth iteration allows us to conclude that x_{k+1} is an approximation of x and to have control over the error $e_{k+1} = x_{k+1} - x$ made at this approximation.

Convergence criterion

Since we do not know x, we cannot decide to terminate the iterations as soon as $\|x - x_k\| \leq \varepsilon$, where ε is the desired accuracy or tolerance. However, we know Ax (which is equal to b), and a simpler convergence criterion is $\|b - Ax_k\| \leq \varepsilon$. Nevertheless, if the norm of A^{-1} is large, this criterion may be misleading, since $\|x - x_k\| \leq \|A^{-1}\| \|b - Ax_k\| \leq \varepsilon \|A^{-1}\|$, which may not be small. Therefore in practice, a relative criterion is preferred:

$$\frac{\|b - Ax_k\|}{\|b - Ax_0\|} \leq \varepsilon. \tag{8.6}$$

Another simple (yet dangerous!) criterion that is sometimes used to detect the convergence of x_k is $\|x_{k+1} - x_k\| \leq \varepsilon$. This criterion is dangerous because it is a necessary, albeit not sufficient, condition for convergence (a notorious counterexample is the scalar sequence $x_k = \sum_{i=1}^{k} 1/i$, which goes to infinity although $|x_{k+1} - x_k|$ goes to zero).

pseudolanguage algorithm

We recall that the simple iterative method $Mx_{k+1} = Nx_k + b$ can also be written as follows:

$$x_{k+1} = x_k + M^{-1}r_k, \tag{8.7}$$

where $r_k = b - Ax_k$ is the residual at the kth iteration. Formula (8.7) is the induction relation that we shall program.

- ✓ The initial guess is usually chosen as $x_0 = 0$, unless we have some information about the exact solution that we shall exploit by choosing x_0 close to x.
- ✓ At step 2 of the algorithm, the variable $\mathbf{x}$ is equal to x_k as the input and to x_{k+1} as the output. Same for r at step 3, which is equal to r_k as the input and to r_{k+1} as the output.
- ✓ Step 3 of the algorithm is based on the relation

$$r_{k+1} = b - Ax_{k+1} = b - A(x_k + M^{-1}r_k) = r_k - AM^{-1}r_k.$$

- ✓ The algorithm stops as soon as $\|r\|_2 < \varepsilon\|b\|$, which is the relative criterion (8.6) for $x_0 = 0$.
- ✓ In practice, we cut down the number of iterations by adding to the condition of the "while" loop an additional condition $k \leq k_{\max}$, where $k_{\max}$ is the maximum authorized number of iterations. The iteration number k is incremented inside the loop.

Data: A, b. Output: $\mathbf{x}$ (approximation of x)
> *Initialization:*
>> choose $\mathbf{x} \in \mathbb{R}^n$.
>> compute $r = b - A\mathbf{x}$.
>
> **While** $\|r\|_2 > \varepsilon\|b\|_2$
>> 1. compute $y \in \mathbb{R}^n$ solution of
>>> $My = r$
>>
>> 2. Update the solution
>>> $\mathbf{x} = \mathbf{x} + y$
>>
>> 3. Compute the residual
>>> $r = r - Ay$
>
> **End While**

Algorithm 8.1: Iterative method for a splitting $A = M - N$.

Computational complexity.

For iterative methods, we compute the number of operations per iteration.

✓ Convergence criterion. It takes n elementary operations since the test can be done on $\|r\|_2^2$ in order to avoid computing a square root.

✓ Step 1. Computing y requires n operations for the Jacobi method (M is diagonal), and $n^2/2$ operations for the Gauss–Seidel and relaxation methods (M is triangular).

✓ Step 3. The product Ay requires n^2 operations.

The number of operations per iteration is at most $\frac{3}{2}n^2$. This is very favorable compared to direct methods if the total number of iterations is sensibly smaller than n.

8.6 Block Methods

We can extend the Jacobi, Gauss–Seidel, and relaxation methods to the case of block matrices. Figure 8.2 shows an example of block decomposition of a matrix A. We always have $A = D - E - F$ where D is a block diagonal matrix, $-E$ is block lower triangular, and $-F$ is block upper triangular. Assume that the size of the matrix is $n \times n$, and let $A_{i,j}$ ($1 \le i, j \le p$) be the blocks constituting this matrix. Each block $A_{i,j}$ has size $n_i \times n_j$. In particular, each diagonal block $A_{i,i}$ is square, of size $n_i \times n_i$ (note that $n = \sum_{i=1}^{p} n_i$). The block decomposition of A suggests the following decomposition, for any $b \in \mathbb{R}^n$:

$$b = (b^1, \ldots, b^p)^t, \quad b^i \in \mathbb{R}^{n_i}, \quad 1 \le i \le p.$$

If the diagonal blocks are nonsingular, the Jacobi, Gauss–Seidel, and relaxation methods are well defined.

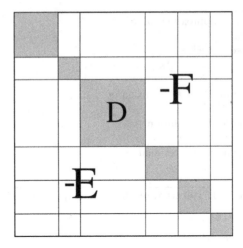

Fig. 8.2. Example of a matrix decomposition by blocks.

Block–Jacobi method.

An iteration of the Jacobi method reads

$$Dx_{k+1} = (E + F)x_k + b. \tag{8.8}$$

Writing the vectors $x_k \in \mathbb{R}^n$ as $x_k = (x_k^1, \ldots, x_k^p)^t$, with $x_k^i \in \mathbb{R}^{n_i}$, $1 \leq i \leq p$, equation (8.8) becomes

$$A_{1,1}x_{k+1}^1 = b^1 - \sum_{j=2}^{p} A_{1,j}x_k^j, \tag{8.9}$$

$$A_{i,i}x_{k+1}^i = b^i - \sum_{j\neq i} A_{i,j}x_k^j, \quad \text{for } 2 \leq i \leq p-1, \tag{8.10}$$

$$A_{p,p}x_{k+1}^p = b^p - \sum_{j=1}^{p-1} A_{p,j}x_k^j. \tag{8.11}$$

Since all diagonal blocks $A_{i,i}$ are nonsingular, we can compute each vectorpiece x_{k+1}^i (for $i = 1, \ldots, p$), thereby determining x_{k+1} completely. Since at each iteration of the algorithm we have to invert the same matrices, it is wise to compute once and for all the LU or Cholesky factorizations of the blocks $A_{i,i}$.

To count the number of operations required for one iteration, we assume for simplicity that all blocks have the same size $n_i = n/p$. Computing x_{k+1} from x_k requires $p(p-1)$ block-vector multiplications, each having a cost of $(n/p)^2$, and p back-and-forth substitution on the diagonal blocks, with cost $(n/p)^2$ again. Each iteration has a total cost on the order on $n^2 + n^2/p$.

Likewise, we can define the block-Gauss–Seidel and block-relaxation methods. Theorems 8.2.1, 8.3.1, and 8.3.2 apply to block methods as well (see [3] for proofs).

8.7 Exercises

8.1. Let A=[1 2 2 1;-1 2 1 0;0 1 -2 2;1 2 1 2]. Check that this matrix is nonsingular. In order to solve the system $Ax = (1, 1, 1, 1)^t$, we decompose $A = M - N$ (with M nonsingular) and then build a sequence of vectors by (8.1).

1. Let M=diag(diag(A));N=M-A. What is the method thus defined? Compute x_{100}, x_{200}, ... Does the sequence x_k converge?
2. Same questions for M=tril(A).
3. Is the conclusion the same for M=2*tril(A)?

8.2. Write a function JacobiCvg(A) returning the value 1 if the Jacobi method applied to the matrix A is well defined and converges, and returning 0 if it diverges. It is not asked to program the method. Same question for a function GaussSeidelCvg(A) informing about the convergence of the Gauss–Seidel method. For each of the matrices

$$A_1 = \begin{pmatrix} 1 & 2 & 3 & 4 \\ 4 & 5 & 6 & 7 \\ 4 & 3 & 2 & 0 \\ 0 & 2 & 3 & 4 \end{pmatrix}, \quad A_2 = \begin{pmatrix} 2 & 4 & -4 & 1 \\ 2 & 2 & 2 & 0 \\ 2 & 2 & 1 & 0 \\ 2 & 0 & 0 & 2 \end{pmatrix},$$

do the Jacobi and Gauss–Seidel methods converge? Comment.

8.3. For different values of n, define the matrix A=DiagDomMat(n,n) (see Exercise 2.23). Do the Jacobi and Gauss–Seidel methods converge when applied to this matrix? Justify your answers.

8.4. Let A, M_1, and M_2 be the matrices defined by

$$A = \begin{pmatrix} 5 & 1 & 1 & 1 \\ 0 & 4 & -1 & 1 \\ 2 & 1 & 5 & 1 \\ -2 & 1 & 0 & 4 \end{pmatrix}, \quad M_1 = \begin{pmatrix} 3 & 0 & 0 & 0 \\ 0 & 3 & 0 & 0 \\ 2 & 1 & 3 & 0 \\ -2 & 1 & 0 & 4 \end{pmatrix}, \quad M_2 = \begin{pmatrix} 4 & 0 & 0 & 0 \\ 0 & 4 & 0 & 0 \\ 2 & 1 & 4 & 0 \\ -2 & 1 & 0 & 4 \end{pmatrix}.$$

Set $N_i = M_i - A$ and $b = A(8, 4, 9, 3)^t$. Compute the first 20 terms of the sequences defined by $x_0 = (0, 0, 0, 0)^t$ and $M_i x_{k+1} = N_i x_k + b$. For each sequence, compare x_{20} with the exact solution. Explain.

8.5. Program a function [x,iter]=Jacobi(A,b,tol,iterMax,x0). The input arguments are:

- a matrix A and a right-hand side b;
- the initial vector x0;
- tol, the tolerance ε for the convergence criterion;
- iterMax, the maximum number of iterations.

The output arguments are:

- the approximate solution x;
- iter, the number of performed iterations.

Hint: use the command nargin to determine the number of input arguments.

- If it is equal to 4, set x0=zeros(b);
- If it is equal to 3, set x0=zeros(b);iterMax=200;
- If it is equal to 2, set x0=zeros(b);iterMax=200;tol=1.e-4.

For $n = 20$, define A=Laplacian1dD(n);xx=(1:n)'/(n+1);b=xx.*sin(xx), and sol=A\b. For different values of the parameter tol= $10^{-s}, s = 2, 3, \ldots$, compute the approximate solution x=Jacobi(A,b,tol,1000) and compare norm(x-sol) and norm(inv(A))*tol. Comment.

8.6 (∗). Write a function [x, iter]=Relax(A,b,w,tol,iterMax,x0) programming the relaxation method with parameter ω equal to w. For the same matrix A as in Exercise 8.5 and the same vector b, plot the curve that gives the number of iterations carried out by the relaxation method in terms of $\omega = i/10$ $(i = 1, 2, \ldots, 20)$. Take iterMAx = 1000, tol = 10^{-6}, and x0 = 0. Find the value of ω that yields the solution with a minimal number of iterations. Compare with the theoretical value given in Theorem 8.3.2.

8.7. Let $A \in \mathcal{M}_n(\mathbb{R})$ be a nonsingular square matrix for which the Jacobi method is well defined. To solve the system $Ax = b$, we consider the following iterative method, known as the relaxed Jacobi method (ω is a nonzero real parameter):

$$\frac{D}{\omega} x_{k+1} = \left(\frac{1-\omega}{\omega} D + E + F\right) x_k + b, \quad k \geq 1, \tag{8.12}$$

and $x_0 \in \mathbb{R}^n$ is a given initial guess.

1. Program this algorithm (function RelaxJacobi). As in Exercise 8.6, find the optimal value of ω for which the solution is obtained with a minimal number of iterations. Take $n = 10$, iterMAx = 1000, tol = 10^{-4}, x0 = 0, and vary ω between 0.1 and 2 with a step size equal to 0.1. Compute the norms as well as the residuals of the solutions obtained for a value of $\omega < 1$ and a value $\omega > 1$. Explain.
2. Theoretical study. We assume that A is symmetric, positive definite, and tridiagonal, and we denote by $\mathcal{J}_\omega$ the iteration matrix associated with algorithm (8.12).
   (a) Find a relationship between the eigenvalues of the matrix $\mathcal{J}_\omega$ and those of $\mathcal{J}_1$.
   (b) Prove that the relaxed Jacobi method converges if and only if ω belongs to an interval I to be determined.
   (c) Find the value $\bar{\omega}$ ensuring the fastest convergence, i.e., such that

$$\varrho(\mathcal{J}_{\bar{\omega}}) = \inf_{\omega \in I} \varrho(\mathcal{J}_\omega).$$

Compute $\varrho(\mathcal{J}_{\bar{\omega}})$, and conclude.

8.8. The goal of this exercise is to study a process of acceleration of convergence for any iterative method. To solve the linear system $Ax = b$ of order n, we have at our disposal an iterative method that reads $x_{k+1} = Bx_k + c$, where B is a matrix of order n and $c \in \mathbb{R}^n$. We assume that this iterative method converges, i.e., the sequence x_k converges to the unique solution x. The convergence acceleration of this iterative method amounts to building a new sequence of vectors $(x'_j)_{j \geq 0}$ that converges faster to x than the original sequence $(x_k)_{k \geq 0}$. The sequence $(x'_j)_j$ is defined by

$$x'_j = \sum_{k=0}^{j} \alpha_k^j x_k, \qquad (8.13)$$

where the coefficients $\alpha_k^j \in \mathbb{R}$ are chosen in order to ensure the fastest convergence rate. We set $e_k = x_k - x$, and $e'_j = x'_j - x$. We shall use `Matlab` only at the very last question of this exercise.

1. Explain why the condition (which shall be assumed in the sequel)

$$\sum_{k=0}^{j} \alpha_k^j = 1 \qquad (8.14)$$

is necessary.
2. Show that

$$e'_{j+1} = p_j(B)e_0, \qquad (8.15)$$

   where $p_j \in \mathbb{P}_j$ is defined by $p_j(t) = \sum_{k=0}^{j} \alpha_k^j t^k$.
3. We assume that B is a normal matrix and that $\|e_0\|_2 = 1$. Prove that

$$\|e'_{j+1}\|_2 \leq \|p_j(D)\|_2, \qquad (8.16)$$

   where D is a diagonal matrix made up of the eigenvalues of B, denoted by $\sigma(B)$. Deduce that

$$\|e'_{j+1}\|_2 \leq \max_{\lambda \in \sigma(B)} |p_j(\lambda)|. \qquad (8.17)$$

4. Show that the eigenvalues of B belong to an interval $[-\alpha, \alpha]$ with $0 < \alpha < 1$.
5. Clearly the fastest convergence rate of the acceleration process is obtained if the polynomial p_j is chosen such that the right-hand side of (8.17) is minimal. However, since we do not know the eigenvalues of B, we substitute the search for p_j making minimal the maximum of the right-hand side in (8.17) by the search for p_j making minimal

$$\max_{\lambda \in [-\alpha, \alpha]} |p_j(\lambda)|. \qquad (8.18)$$

   Observing that $p_j(1) = 1$, determine the solution to this problem using Proposition 9.5.3 on Chebyshev polynomials.

6. Use relation (9.12) to establish the following induction relation between three consecutive vectors of the sequence $(x'_j)_{j \geq 1}$:

$$x'_{j+1} = \mu_j(Bx'_j + c - x'_{j-1}) + x'_{j-1}, \qquad \forall j \geq 1, \qquad (8.19)$$

where μ_j is a real number to be determined. This relation allows for computing x'_j directly without having previously computed the vectors x_k, provided that the real sequence $(\mu_j)_j$ can be computed.

7. Compute μ_0 and μ_1. Express $\frac{1}{\mu_{j+1}}$ in terms of μ_j. Prove that $\mu_j \in (1,2)$ for all $j \geq 1$. Check that the sequence $(\mu_j)_j$ converges.

8. Programming the method. We consider the Laplacian in two space dimensions with a right-hand side $f(x,y) = \cos(x)\sin(y)$; see Exercise 6.7. Compare the Jacobi method and the accelerated Jacobi method for solving this problem. Take $n = 10, \alpha = 0.97$, and plot on the same graph the errors (assuming that the exact solution is A\b) of both methods in terms of the iteration number (limited to 50).

9

Conjugate Gradient Method

From a practical viewpoint, all iterative methods considered in the previous chapter have been supplanted by the conjugate gradient method, which is actually a direct method used as an iterative one. For simplicity, we will restrict ourselves, throughout this chapter, to real symmetric matrices. The case of complex self-adjoint matrices is hardly more difficult. However, that of non-self-adjoint matrices is relatively more delicate to handle.

9.1 The Gradient Method

Often called the "steepest descent method," the gradient method is also known as Richardson's method. It is a classical iterative method with a particular choice of regular decomposition. We recall its definition already presented in Example 8.1.1.

Definition 9.1.1. *The iterative method, known as the gradient method, is defined by the following regular decomposition:*

$$M = \frac{1}{\alpha}I_n \quad and \quad N = \left(\frac{1}{\alpha}I_n - A\right),$$

where α is a real nonzero parameter. In other words, the gradient method consists in computing the sequence x_k defined by

$$\begin{cases} x_0 \text{ given in } \mathbb{R}^n, \\ x^{k+1} = x^k + \alpha(b - Ax^k), \quad \forall k \geq 1. \end{cases}$$

For the implementation of this method, see Algorithm 9.1.

Theorem 9.1.1. *Let A be a matrix with eigenvalues $\lambda_1 \leq \lambda_2 \leq \cdots \leq \lambda_n$.*

(i) If $\lambda_1 \leq 0 \leq \lambda_n$, then the gradient method does not converge for any value of α.

Data: A, b. Output: $\mathbf{x}$ (approximation of x)
    Initialization:
        choose α.
        choose $\mathbf{x} \in \mathbb{R}^n$.
        compute $r = b - A\mathbf{x}$.
    While $\|r\|_2 > \varepsilon \|b\|_2$
        $\mathbf{x} = \mathbf{x} + \alpha r$
        $r = b - A\mathbf{x}$
    End While

Algorithm 9.1: Gradient algorithm.

(ii)If $0 < \lambda_1 \leq \cdots \leq \lambda_n$, then the gradient method converges if and only if $0 < \alpha < 2/\lambda_n$. In this case, the optimal parameter α, which minimizes $\varrho(M^{-1}N)$, is

$$\alpha_{\text{opt}} = \frac{2}{\lambda_1 + \lambda_n} \quad and \quad \min_\alpha \varrho(M^{-1}N) = \frac{\lambda_n - \lambda_1}{\lambda_n + \lambda_1} = \frac{\text{cond}_2(A) - 1}{\text{cond}_2(A) + 1}.$$

Remark 9.1.1. Note that if the matrix A is diagonalizable with eigenvalues $\lambda_1 \leq \cdots \leq \lambda_n < 0$, then a symmetric result of *(ii)* occurs by changing α into $-\alpha$. On the other hand, the conditioning of a normal invertible matrix A is $\text{cond}_2(A) = \lambda_n/\lambda_1$. Thus, for the optimal parameter $\alpha = \alpha_{\text{opt}}$, the spectral radius of the iteration matrix, $\varrho(M^{-1}N) = \frac{\text{cond}_2(A)-1}{\text{cond}_2(A)+1}$, is an increasing function of $\text{cond}_2(A)$. In other words, the better the matrix A is conditioned, the faster the gradient method converges.

Proof of Theorem 9.1.1.   According to Theorem 8.1.1, we know that the gradient method converges if and only if $\varrho(M^{-1}N) < 1$. Here, $M^{-1}N = (I_n - \alpha A)$; hence

$$\varrho(M^{-1}N) < 1 \iff |1 - \alpha\lambda_i| < 1 \iff -1 < 1 - \alpha\lambda_i < 1 , \quad \forall i.$$

This implies that $\alpha\lambda_i > 0$ for all $1 \leq i \leq n$. As a result, all eigenvalues of A have to be nonzero and bear the same sign as α. Therefore, the gradient method does not converge if two eigenvalues have opposite signs, whatever the value of α. If, on the other hand, we have $0 < \lambda_1 \leq \cdots \leq \lambda_n$, then we deduce

$$-1 < 1 - \alpha\lambda_n \Rightarrow \alpha < \frac{2}{\lambda_n}.$$

To compute the optimal parameter α_{opt}, note that the function $\lambda \to |1 - \alpha\lambda|$ is decreasing on $]-\infty, 1/\alpha]$ and then increasing on $[1/\alpha, +\infty[$; see Figure 9.1.

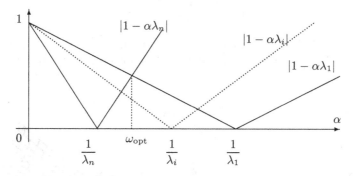

Fig. 9.1. Eigenvalues of the matrix $B_\alpha = I - \alpha A$.

Thus

$$\varrho(M^{-1}N) = \max\{|1 - \alpha\lambda_1|, |1 - \alpha\lambda_n|\}.$$

The piecewise affine function $\alpha \in [0, 2/\lambda_n] \mapsto \varrho(M^{-1}N)$ attains its minimum at the point α_{opt} defined by $1 - \alpha_{opt}\lambda_1 = \alpha_{opt}\lambda_n - 1$, i.e., $\alpha_{opt} = \frac{2}{\lambda_1 + \lambda_n}$. At this point we check that $\varrho(M^{-1}N) = \frac{\lambda_n - \lambda_1}{\lambda_n + \lambda_1}$.   □

9.2 Geometric Interpretation

This section provides an explanation of the name of the gradient method. To this end, we introduce several technical tools.

Definition 9.2.1. *Let f be a function from $\mathbb{R}^n$ into $\mathbb{R}$. We call the vector of partial derivatives at the point x the gradient (or differential) of the function f at x, which we denote by*

$$\nabla f(x) = \left(\frac{\partial f}{\partial x_1}(x), \ldots, \frac{\partial f}{\partial x_n}(x)\right)^t.$$

We recall that the partial derivative $\frac{\partial f}{\partial x_i}(x)$ is computed by differentiating $f(x)$ with respect to x_i while keeping the other entries x_j, $j \neq i$, constant.

We consider the problem of minimizing quadratic functions from $\mathbb{R}^n$ into $\mathbb{R}$ defined by

$$f(x) = \frac{1}{2}\langle Ax, x\rangle - \langle b, x\rangle = \frac{1}{2}\sum_{i,j=1}^{n} a_{ij}x_i x_j - \sum_{i=1}^{n} b_i x_i, \qquad (9.1)$$

where A is a symmetric matrix in $\mathcal{M}_n(\mathbb{R})$ and b is a vector in $\mathbb{R}^n$. The function $f(x)$ is said to have a minimum (or attains its minimum) at x_0 if $f(x) \geq f(x_0)$ for all $x \in \mathbb{R}^n$. Figure 9.2 shows the surfaces $x \in \mathbb{R}^2 \mapsto \langle A_i x, x\rangle$ for each of the symmetric matrices

$$A_1 = \begin{pmatrix} 5 & -3 \\ -3 & 5 \end{pmatrix}, \quad A_2 = \begin{pmatrix} 1 & 3 \\ 3 & 1 \end{pmatrix}. \tag{9.2}$$

The matrix A_1 has eigenvalues 2 and 8; it is therefore positive definite. The matrix A_2 has eigenvalues -2 and 4, so it is not positive (or negative!).

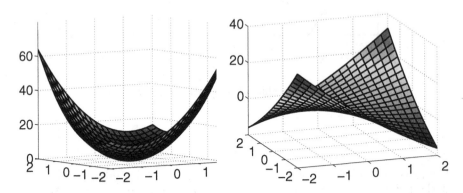

Fig. 9.2. Surfaces $x \in \mathbb{R}^2 \mapsto \langle Ax, x \rangle$, for A_1 (left) and A_2 (right), see (9.2).

Proposition 9.2.1. *The gradient of the function $f(x)$ defined by (9.1) is*

$$\nabla f(x) = Ax - b.$$

Moreover,

(i) *if A is positive definite, then f admits a unique minimum at x_0 that is a solution of the linear system $Ax = b$;*

(ii) *if A is positive indefinite and if b belongs to the range of A, then f attains its minimum at all vectors x_0 that solve the linear system $Ax = b$ and at these vectors only;*

(iii) *in all other cases, that is, if A is not positive or if b does not belong to the range of A, f does not have a minimum, i.e., its infimum is $-\infty$.*

Proof. We compute the kth partial derivative of f,

$$\frac{\partial f}{\partial x_k}(x) = a_{kk}x_k + \frac{1}{2}\sum_{i \neq k} a_{ik}x_i + \frac{1}{2}\sum_{i \neq k} a_{ki}x_i - b_k = \sum_i a_{ik}x_i - b_k = (Ax - b)_k,$$

thereby easily deducing that $\nabla f(x) = Ax - b$. Since A is real symmetric, we can diagonalize it. Let (λ_i) and $(\hat{e}_i)$ be its eigenvalues and eigenvectors. Setting $x = \sum_i \hat{x}_i \hat{e}_i$ and $b = \sum_i \hat{b}_i \hat{e}_i$, we have

$$f(x) = \frac{1}{2}\sum_{i=1}^n \lambda_i \hat{x}_i^2 - \sum_{i=1}^n \hat{b}_i \hat{x}_i = \frac{1}{2}\sum_{i=1}^n \left(\lambda_i (\hat{x}_i - \frac{\hat{b}_i}{\lambda_i})^2 - \frac{\hat{b}_i^2}{\lambda_i} \right).$$

If A is positive definite, then $\lambda_i > 0$ for all $1 \le i \le n$. In this case, we minimize $f(x)$ by minimizing each of the squared terms. There exists a unique minimum x whose entries are $\hat{x}_i = \frac{\hat{b}_i}{\lambda_i}$, and this vector is the unique solution to the system $Ax = b$.

If A is only positive indefinite, then there exists at least one index i_0 such that $\lambda_{i_0} = 0$. In this case, if $\hat{b}_{i_0} \ne 0$, taking $x = -t\hat{b}_{i_0}\hat{e}_{i_0}$, we obtain

$$\lim_{t \to +\infty} f(x) = -\infty,$$

so f is not bounded from below. If $\hat{b}_i = 0$ for all indices i such that $\lambda_i = 0$, then $b \in \mathrm{Im}\,(A)$. Let $\tilde{A}$ be the restriction of A to $\mathrm{Im}\,(A)$. Hence the minimum of f is attained at all points $x = \tilde{A}^{-1}b + y$, where y spans $\mathrm{Ker}\,(A)$.

Finally, if A admits an eigenvalue $\lambda_{i_0} < 0$, then, taking x parallel to $\hat{e}_{i_0}$, we easily see that f is not bounded from below. $\qquad\square$

Corollary 9.2.1. *Let A be a real positive definite symmetric matrix. Let $f(x)$ be the function defined by (9.1). Let F be a vector subspace of $\mathbb{R}^n$. There exists a unique vector $x_0 \in F$ such that*

$$f(x_0) \le f(x), \quad \forall x \in F.$$

Furthermore, x_0 is the unique vector of F such that

$$\langle Ax_0 - b, y \rangle = 0, \quad \forall y \in F.$$

Proof. Denote by P the orthogonal projection on the vector subspace F in $\mathbb{R}^n$. We also denote by P its matrix representation in the canonical basis. By definition of the orthogonal projection we have $P^* = P$, $P^2 = P$, and the mapping P is onto from $\mathbb{R}^n$ into F; consequently,

$$\min_{x \in F} f(x) = \min_{y \in \mathbb{R}^n} f(Py).$$

We compute $f(Py) = \frac{1}{2}\langle P^*APy, y \rangle - \langle P^*b, y \rangle$. Proposition 9.2.1 can be applied to this function. Since A is positive definite, we easily see that P^*AP is nonnegative. If P^*b did not belong to the range of (P^*AP), then we would infer that the infimum of $f(Py)$ is $-\infty$, which is impossible, since

$$\inf_{y \in \mathbb{R}^n} f(Py) = \inf_{x \in F} f(x) \ge \min_{x \in \mathbb{R}^n} f(x) > -\infty,$$

because A is positive definite. Thus $P^*b \in \mathrm{Im}\,(P^*AP)$ and the minimum of $f(Py)$ is attained by all solutions of $P^*APx = P^*b$. Let us prove that this equation has a unique solution in F. Let x_1 and x_2 be two solutions in F of $P^*APx = P^*b$. We have

$$P^*AP(x_1 - x_2) = 0 \Rightarrow \langle A(Px_1 - Px_2), (Px_1 - Px_2) \rangle = 0 \Rightarrow Px_1 = Px_2.$$

Since x_1 and x_2 belong to F, we have $x_1 = Px_1 = Px_2 = x_2$. Multiplying the equation $P^*APx_0 = P^*b$ by $y \in F$ and using the fact that $Px_0 = x_0$ and $Py = y$, because $x_0, y \in F$, yields the relation $\langle Ax_0 - b, y \rangle = 0$ for all y in F. $\square$

Theorem 9.2.1. *Let A be a positive definite symmetric matrix, and let f be the function defined on $\mathbb{R}^n$ by (9.1).*

(i) $x \in \mathbb{R}^n$ is the minimum of f if and only if $\nabla f(x) = 0$.
(ii) Let $x \in \mathbb{R}^n$ be such that $\nabla f(x) \neq 0$. Then $\forall \alpha \in \big(0, 2/\varrho(A)\big)$, we have

$$f(x - \alpha \nabla f(x)) < f(x).$$

Remark 9.2.1. From this theorem we infer an iterative method for minimizing f. We construct a sequence of points $(x_k)_k$ such that the sequence $(f(x_k))_k$ is decreasing:

$$x_{k+1} = x_k - \alpha \nabla f(x_k) = x_k + \alpha(b - Ax_k).$$

This is exactly the gradient method that we have studied in the previous section. In other words, we have shown that solving a linear system whose matrix is symmetric and positive definite is equivalent to minimizing a quadratic function. This is a very important idea that we shall use in the sequel of this chapter.

Proof of Theorem 9.2.1.   We already know that f attains its minimum at a unique point x that is a solution of $Ax = b$. At this point x, we therefore have $\nabla f(x) = 0$. Next, let x be a point such that $\nabla f(x) \neq 0$ and set $\delta = -\alpha(Ax-b)$. Since A is symmetric, we compute

$$f(x + \delta) = \frac{1}{2}\langle A(x+\delta), (x+\delta) \rangle - \langle b, x+\delta \rangle = f(x) + \frac{1}{2}\langle A\delta, \delta \rangle + \langle Ax - b, \delta \rangle.$$

Now, $\langle A\delta, \delta \rangle \leq \|A\|_2 \|\delta\|^2$, and for a symmetric real matrix, we have $\|A\|_2 = \varrho(A)$. Thus

$$f(x + \delta) \leq f(x) + \big(\alpha^2 \varrho(A)/2 - \alpha\big) \|Ax - b\|^2.$$

As a consequence, $f(x + \delta) < f(x)$ if $0 < \alpha < 2/\varrho(A)$. $\square$

9.3 Some Ideas for Further Generalizations

We can improve the gradient method with constant step size, which we have just discussed, by choosing at each iterative step a different coefficient α_k that minimizes $f(x_k - \alpha \nabla f(x_k))$.

Definition 9.3.1. *The following iterative method for solving the linear system* $Ax = b$ *is called the gradient method with variable step size:*

$$\begin{cases} x_0 \text{ given in } \mathbb{R}^n, \\ x^{k+1} = x^k + \alpha_k(b - Ax^k) \quad \forall k \geq 1, \end{cases}$$

where α_k *is chosen as the minimizer of the function*

$$g(\alpha) = f\left(x_k - \alpha \nabla f(x_k)\right),$$

with $f(x)$ *defined by (9.1).*

The optimal value α_k is given by the following lemma, the proof of which is left to the reader as an easy exercise.

Lemma 9.3.1. *Let A be a positive definite symmetric matrix. For the gradient method with variable step size, there exists a unique optimal step size given by*

$$\alpha_k = \frac{\|Ax_k - b\|^2}{\langle A(Ax_k - b), (Ax_k - b)\rangle}.$$

We observe that α_k is always well defined except when $Ax_k - b = 0$, in which case the method has already converged! The gradient method with optimal step size is implemented in Algorithm 9.2.

Data A, b. Output: $\mathbf{x}$ (approximation of x)
    Initialization:
        choose $\mathbf{x} \in \mathbb{R}^n$.
        compute $r = b - A\mathbf{x}$.
        $\alpha = \frac{\|r\|_2^2}{\langle Ar, r\rangle}$
    While $\|r\|_2 > \varepsilon\|b\|_2$
        $\mathbf{x} = \mathbf{x} + \alpha r$
        $r = b - A\mathbf{x}$
        $\alpha = \frac{\|r\|_2^2}{\langle Ar, r\rangle}$ *optimal step size*
    End While

Algorithm 9.2: Gradient algorithm with variable step size.

The gradient method with variable step size is a bit more complex than the one with constant step size (more operations are needed to compute α_k), and in practice, it is not much more efficient than the latter. In order to improve the gradient method and construct the conjugate gradient method, we introduce the important notion of Krylov space.

Definition 9.3.2. *Let* r *be a vector in* $\mathbb{R}^n$. *We call the vector subspace of* $\mathbb{R}^n$ *spanned by the* $k + 1$ *vectors* $\{r, Ar, \ldots, A^k r\}$. *The Krylov space associated with the vector* r *(and matrix A), denoted by* $K_k(A, r)$ *or simply* K_k.

The Krylov spaces $(K_k)_{k\geq 0}$ form by inclusion an increasing sequence of vector subspaces. Since $K_k \subset \mathbb{R}^n$, this sequence becomes stationary from a certain k. Namely, we prove the following result.

Lemma 9.3.2. *The sequence of Krylov spaces $(K_k)_{k\geq 0}$ is increasing,*

$$K_k \subset K_{k+1} \quad \forall k \geq 0.$$

Moreover, for all vectors $r_0 \neq 0$, there exists $k_0 \in \{0, 1, \ldots, n-1\}$ such that

$$\begin{cases} \dim K_k = k+1 & \text{if } 0 \leq k \leq k_0, \\ \dim K_k = k_0 + 1 & \text{if } k_0 \leq k. \end{cases}$$

This integer k_0 is called the Krylov critical dimension.

Proof. It is clear that $\dim K_k \leq k+1$ and $\dim K_0 = 1$. Since $\dim K_k \leq n$ for any k, there exists k_0 that is the greatest integer such that $\dim K_k = k+1$ for any $k \leq k_0$. By definition of k_0, the dimension of K_{k_0+1} is strictly smaller than $k_0 + 2$. However, since $K_{k_0} \subset K_{k_0+1}$, we necessarily have $\dim K_{k_0+1} = k_0 + 1$, and the vector $A^{k_0+1}r_0$ is a linear combination of the vectors $(r_0, Ar_0, \ldots, A^{k_0}r_0)$. Thereby we infer by a simple induction argument that all vectors $A^k r_0$ for $k \geq k_0 + 1$ are also linear combinations of the vectors $(r_0, Ar_0, \ldots, A^{k_0}r_0)$. Accordingly, $K_k = K_{k_0}$ for all $k \geq k_0$.   □

Proposition 9.3.1. *We consider the gradient method (with constant or variable step size)*

$$\begin{cases} x_0 \in \mathbb{R}^n \text{ initial choice,} \\ x_{k+1} = x_k + \alpha_k(b - Ax_k). \end{cases}$$

The vector $r_k = b - Ax_k$, called the residual, satisfies the following properties:

1. *r_k belongs to the Krylov space K_k corresponding to the initial residual r_0.*
2. *x_{k+1} belongs to the affine space $[x_0 + K_k]$ defined as the collection of vectors x such that $x - x_0$ belongs to the vector subspace K_k.*

Proof. By definition we have $x_{k+1} = x_k + \alpha_k r_k$, which, by multiplication by A and subtraction to b, yields

$$r_{k+1} = r_k - \alpha_k A r_k. \tag{9.3}$$

By induction on (9.3), we easily infer that $r_k \in span\{r_0, Ar_0, \ldots, A^k r_0\}$. Then, a similar induction on $x_{k+1} = x_k + \alpha_k r_k$ shows that $x_{k+1} \in [x_0 + K_k]$. □

Lemma 9.3.3. *Let $(x_k)_{k\geq 0}$ be a sequence in $\mathbb{R}^n$. Let K_k be the Krylov space relative to the vector $r_0 = b - Ax_0$. If $x_{k+1} \in [x_0 + K_k]$, then $r_{k+1} = (b - Ax_{k+1}) \in K_{k+1}$.*

Proof. If $x_{k+1} \in [x_0 + K_k]$, there exist coefficients $(\alpha_i)_{0 \le i \le k}$ such that

$$x_{k+1} = x_0 + \sum_{i=0}^{k} \alpha_i A^i r_0.$$

We multiply this equation by A and we subtract to b, which yields

$$r_{k+1} = r_0 - \sum_{i=0}^{k} \alpha_i A^{i+1} r_0.$$

Hence $r_{k+1} \in K_{k+1}$. $\square$

9.4 Theoretical Definition of the Conjugate Gradient Method

We now assume that all matrices considered here are symmetric and positive definite. To improve the gradient method, we forget, from now on, the induction relation that gives x_{k+1} in terms of x_k, and we keep as the starting point only the relation provided by Proposition 9.3.1, namely

$$x_{k+1} \in [x_0 + K_k],$$

where K_k is the Krylov space relative to the initial residual $r_0 = b - Ax_0$ (x_0 is some initial choice). Of course, there exists an infinity of possible choices for x_{k+1} in the affine space $[x_0 + K_k]$. To determine x_{k+1} in a unique fashion, we put forward two simple criteria:

- **1st definition (orthogonalization principle).** We choose $x_{k+1} \in [x_0 + K_k]$ such that $r_{k+1} \perp K_k$.
- **2nd definition (minimization principle).** We choose $x_{k+1} \in [x_0 + K_k]$ that minimizes in $[x_0 + K_k]$

$$f(x) = \frac{1}{2} \langle Ax, x \rangle - \langle b, x \rangle.$$

Theorem 9.4.1. *Let A be a symmetric positive definite matrix. For the two above definitions, there exists indeed a unique vector $x_{k+1} \in [x_0 + K_k]$. Both definitions correspond to the same algorithm in the sense that they lead to the same value of x_{k+1}. Furthermore, this algorithm converges to the solution of the linear system $Ax = b$ in at most n iterations. We call this method the "conjugate gradient method."*

Remark 9.4.1. The previous theorem shows that the conjugate gradient algorithm that we have devised as an iterative method is in fact a direct method, since it converges in a finite number of iterations (precisely $k_0 + 1$, where k_0 is the critical Krylov dimension defined in Lemma 9.3.2). However, in practice, we use it like an iterative method that (hopefully) converges numerically in fewer than $k_0 + 1 \leq n$ iterations.

Intuitively, it is easy to see why the conjugate gradient improves the simple gradient method. Actually, instead of merely decreasing $f(x)$ at each iteration (cf. Theorem 9.2.1), we minimize $f(x)$ on an increasing sequence of affine subspaces.

Remark 9.4.2. Introducing $r = b - Ax$, we have

$$h(r) = \langle A^{-1}r, r \rangle /2 = f(x) + \langle A^{-1}b, b \rangle /2.$$

Thanks to Lemma 9.3.3, the second definition is equivalent to finding $x_{k+1} \in [x_0 + K_k]$ such that its residual $r_{k+1} = b - Ax_{k+1}$ minimizes the function $h(r)$ in K_{k+1}.

Proof. First, let us prove that the two suggested definitions are identical and uniquely define x_{k+1}. For any $y \in K_k$, we set

$$g(y) = f(x_0 + y) = \frac{1}{2}\langle Ay, y \rangle - \langle r_0, y \rangle + f(x_0).$$

Minimizing f on $[x_0 + K_k]$ is equivalent to minimizing g on K_k. Now, by Corollary 9.2.1, $g(y)$ admits a unique minimum in K_k, which we denote by $(x_{k+1} - x_0)$. As a consequence, the second definition gives a unique x_{k+1}. Furthermore, Corollary 9.2.1 also states that

$$\langle Ax_{k+1} - b, y \rangle = 0 \quad \forall y \in K_k,$$

which is nothing else but the definition of $r_{k+1} \perp K_k$. The two algorithms are therefore identical and yield the same unique value of x_{k+1}.

If the critical dimension of the Krylov space k_0 is equal to $n - 1$, then $\dim K_{k_0} = n$, and the affine subspace $[x_0 + K_{k_0}]$ coincides with $\mathbb{R}^n$ entirely. Accordingly, $x_{k_0+1} = x_n$ is the minimum of f on $\mathbb{R}^n$ that satisfies, by Proposition 9.2.1, $Ax_n = b$. The gradient method has thereby converged in n iterations.

If $k_0 < n - 1$, then by virtue of Lemma 9.3.2, for all $k \geq k_0$, $\dim K_k = k_0 + 1$. In particular, $A^{k_0+1}r_0 \in K_{k_0}$, which means that $A^{k_0+1}r_0$ is a linear combination of vectors $(r_0, Ar_0, \ldots, A^{k_0}r_0)$,

$$A^{k_0+1}r_0 = \sum_{i=0}^{k_0} \alpha_i A^i r_0.$$

The coefficient α_0 is inevitably nonzero. As a matter of fact, if this were not true, we could multiply the above equation by A^{-1} (we recall that A is nonsingular, since it is positive definite) and show that $A^{k_0}r_0$ is a linear combination of vectors $(r_0, Ar_0, \ldots, A^{k_0-1}r_0)$, which would imply that $K_{k_0} = K_{k_0-1}$, contradicting the definition of the critical dimension k_0. Since $r_0 = b - Ax_0$, we get

$$A \left(\frac{1}{\alpha_0} A^{k_0} r_0 - \sum_{i=1}^{k_0} \frac{\alpha_i}{\alpha_0} A^{i-1} r_0 + x_0 \right) = b.$$

It turns out from this equation that the solution of $Ax = b$ belongs to the affine space $[x_0 + K_{k_0}]$. Hence, at the $(k_0 + 1)$th iteration, the minimum of $f(x)$ on $[x_0 + K_{k_0}]$ happens to be the minimum on all of $\mathbb{R}^n$, and the iterate x_{k_0+1} is nothing but the exact solution. The conjugate gradient method has thus converged in exactly $k_0 + 1 \leq n$ iterations. $\square$

The definition of the conjugate gradient method that we have just given is purely theoretical. Actually, no practical algorithm has been provided either to minimize $f(x)$ on $[x_0 + K_k]$ or to construct r_{k+1} orthogonal to K_k. It remains to show how, in practice, we compute x_{k+1}. To do so, we will make use of an additional property of the conjugate gradient method.

Proposition 9.4.1. *Let A be a symmetric, positive definite matrix. Let $(x_k)_{0 \leq k \leq n}$ be the sequence of approximate solutions obtained by the conjugate gradient method. Set*

$$r_k = b - Ax_k \quad and \quad d_k = x_{k+1} - x_k.$$

Then

(i) the Krylov space K_k, defined by $K_k = span\{r_0, Ar_0, \ldots, A^k r_0\}$, satisfies

$$K_k = span\{r_0, \ldots, r_k\} = span\{d_0, \ldots, d_k\},$$

(ii)   the sequence $(r_k)_{0 \leq k \leq n-1}$ is orthogonal, i.e.,

$$\langle r_k, r_l \rangle = 0 \text{ for all } 0 \leq l < k \leq n - 1,$$

(iii)   the sequence $(d_k)_{0 \leq k \leq n-1}$ is "conjugate" with respect to A, or "A-conjugate," i.e.,

$$\langle Ad_k, d_l \rangle = 0 \text{ for all } 0 \leq l < k \leq n - 1.$$

Remark 9.4.3. A conjugate sequence with respect to a matrix A is in fact orthogonal for the scalar product defined by $\langle Ax, y \rangle$ (we recall that A is symmetric and positive definite). This property gave its name to the conjugate gradient method.

Proof of Proposition 9.4.1.   Let us first note that the result is independent of the critical dimension k_0, defined by Lemma 9.3.2. When $k \geq k_0 + 1$, that is, when the conjugate gradient method has already converged, we have $r_k = 0$ and $x_k = x_{k_0+1}$; thus $d_k = 0$. In this case, the sequence r_k is indeed orthogonal and d_k is indeed conjugate for $k \geq k_0 + 1$. When $k \leq k_0$, the first definition of the conjugate gradient implies that $r_{k+1} \in K_{k+1}$ and $r_{k+1} \perp K_k$. Now, $K_k \subset K_{k+1}$ and $\dim K_k = k + 1$ for $k \leq k_0$. Therefore the family $(r_k)_{0 \leq k \leq k_0}$

is free and orthogonal. In particular, this entails that $K_k = span\{r_0, \ldots, r_k\}$ for all $k \geq 0$.

On the other hand, $d_k = x_{k+1} - x_k$, with $x_k \in [x_0 + K_{k-1}]$ and $x_{k+1} \in [x_0 + K_k]$, implies that d_k belongs to K_k for all $k \geq 1$. As a consequence,

$$span\{d_0, \ldots, d_k\} \subset K_k.$$

Let us show that the family $(d_k)_{1 \leq k \leq k_0}$ is conjugate with respect to A. For $l < k$, we have

$$\langle Ad_k, d_l \rangle = \langle Ax_{k+1} - Ax_k, d_l \rangle = \langle r_k - r_{k+1}, d_l \rangle = 0,$$

since $d_l \in K_l = span\{r_0, \ldots, r_l\}$ and the family (r_k) is orthogonal. We deduce that the family $(d_k)_{0 \leq k \leq k_0}$ is orthogonal for the scalar product $\langle Ax, y \rangle$. Now, $d_k \neq 0$ for $k \leq k_0$, because otherwise, we would have $x_{k+1} = x_k$ and accordingly $r_{k+1} = r_k \neq 0$, which is not possible since $r_{k+1} \in K_{k+1}$ and $r_{k+1} \perp K_k$. An orthogonal family of nonzero vectors is free, which implies $K_k = span\{d_0, \ldots, d_k\}$.   □

9.5 Conjugate Gradient Algorithm

In order to find a practical algorithm for the conjugate gradient method, we are going to use Proposition 9.4.1, namely that the sequence $d_k = x_{k+1} - x_k$ is conjugate with respect to A. Thanks to the Gram–Schmidt procedure, we will easily construct a sequence (p_k) conjugate with respect to A that will be linked to the sequence (d_k) by the following result of "almost uniqueness" of the Gram–Schmidt orthogonalization procedure.

Lemma 9.5.1. *Let $(a_i)_{1 \leq i \leq p}$ be a family of linearly independent vectors of $\mathbb{R}^n$, and let $(b_i)_{1 \leq i \leq p}$ and $(c_i)_{1 \leq i \leq p}$ be two orthogonal families for the same scalar product on $\mathbb{R}^n$ such that for all $1 \leq i \leq p$,*

$$span\{a_1, \ldots, a_i\} = span\{b_1, \ldots, b_i\} = span\{c_1, \ldots, c_i\}.$$

Then each vector b_i is parallel to c_i for $1 \leq i \leq p$.

Proof. By the Gram–Schmidt orthonormalization procedure (cf. Theorem 2.1.1), there exists a unique orthonormal family $(d_i)_{1 \leq i \leq p}$ (up to a sign change) such that

$$span\{a_1, \ldots, a_i\} = span\{d_1, \ldots, d_i\}, \quad \forall 1 \leq i \leq p.$$

Since the family (a_i) is linearly independent, the vectors b_i and c_i are never zero. We can thus consider the orthonormal families $(b_i/\|b_i\|)$ and $(c_i/\|c_i\|)$, which must then coincide up to a sign change. This proves that each b_i and c_i differ only by a multiplicative constant.   □

We will apply this result to the scalar product $\langle A\cdot,\cdot\rangle$ and to the family $(d_k)_{0\leq k\leq k_0}$, which is orthogonal with respect to this scalar product. Accordingly, if we can produce another sequence $(p_k)_{0\leq k\leq k_0}$ that would also be orthogonal with respect to $\langle A\cdot,\cdot\rangle$ (for instance, by the Gram–Schmidt procedure), then for all $0\leq k\leq k_0$ there would exist a scalar α_k such that

$$x_{k+1}=x_k+\alpha_k p_k.$$

More precisely, we obtain the following fundamental result.

Theorem 9.5.1. *Let A be a symmetric, positive definite matrix. Let (x_k) be the sequence of approximate solutions of the conjugate gradient method. Let $(r_k=b-Ax_k)$ be the associated residual sequence. Then there exists an A-conjugate sequence (p_k) such that*

$$p_0=r_0=b-Ax_0, \text{ and for } 0\leq k\leq k_0, \quad \begin{cases} x_{k+1}=x_k+\alpha_k p_k, \\ r_{k+1}=r_k-\alpha_k Ap_k, \\ p_{k+1}=r_{k+1}+\beta_k p_k, \end{cases} \quad (9.4)$$

with

$$\alpha_k=\frac{\|r_k\|^2}{\langle Ap_k,p_k\rangle} \quad and \quad \beta_k=\frac{\|r_{k+1}\|^2}{\|r_k\|^2}.$$

Conversely, let (x_k,r_k,p_k) be three sequences defined by the induction relations (9.4). Then (x_k) is nothing but the sequence of approximate solutions of the conjugate gradient method.

Proof. We start by remarking that should (x_k) be the sequence of approximate solutions of the conjugate gradient method, then for all $k\geq k_0+1$ we would have $x_k=x_{k_0+1}$ and $r_k=0$. Reciprocally, let (x_k,r_k,p_k) be three sequences defined by the induction relations (9.4). We denote by k_0 the smallest index such that $r_{k_0+1}=0$. We then easily check that for $k\geq k_0+1$ we have $x_k=x_{k_0+1}$ and $r_k=p_k=0$. As a consequence, in the sequel we restrict indices to $k\leq k_0$ for which $r_k\neq 0$ (for both methods).

Consider (x_k), the sequence of approximate solutions of the conjugate gradient method. Let us show that it satisfies the induction relations (9.4). We construct a sequence (p_k), orthogonal with respect to the scalar product $\langle A\cdot,\cdot\rangle$, by applying the Gram–Schmidt procedure to the family (r_k). We define, in this fashion, for $0\leq k\leq k_0$,

$$p_k=r_k+\sum_{j=0}^{k-1}\beta_{j,k}p_j,$$

with

$$\beta_{j,k}=-\frac{\langle Ar_k,p_j\rangle}{\langle Ap_j,p_j\rangle}. \quad (9.5)$$

Applying Lemma 9.5.1, we deduce that, since the sequences (p_k) and $(d_k = x_{k+1} - x_k)$ are both conjugate with respect to A, for all $0 \leq k \leq k_0$ there exists a scalar α_k such that

$$x_{k+1} = x_k + \alpha_k p_k.$$

We deduce

$$\langle Ar_k, p_j \rangle = \langle r_k, Ap_j \rangle = \left\langle r_k, \frac{A(x_{j+1} - x_j)}{\alpha_j} \right\rangle = \frac{1}{\alpha_j} \langle r_k, r_j - r_{j+1} \rangle, \quad (9.6)$$

on the one hand. On the other hand, we know that the sequence r_k is orthogonal (for the canonical scalar product). Therefore (9.5) and (9.6) imply

$$\beta_{j,k} = \begin{cases} 0 & \text{if } 0 \leq j \leq k-2, \\ \dfrac{\|r_k\|^2}{\alpha_{k-1} \langle Ap_{k-1}, p_{k-1} \rangle} & \text{if } j = k-1. \end{cases}$$

In other words, we have obtained

$$p_k = r_k + \beta_{k-1} p_{k-1} \text{ with } \beta_{k-1} = \beta_{k-1,k}.$$

Moreover, the relation $x_{k+1} = x_k + \alpha_k p_k$ implies that

$$r_{k+1} = r_k - \alpha_k Ap_k.$$

We have hence derived the three induction relations (9.4). It remains to find the value of α_k. To do so, we use the fact that r_{k+1} is orthogonal to r_k:

$$0 = \|r_k\|^2 - \alpha_k \langle Ap_k, r_k \rangle.$$

Now, $r_k = p_k - \beta_{k-1} p_{k-1}$ and $\langle Ap_k, p_{k-1} \rangle = 0$, and thus

$$\alpha_k = \frac{\|r_k\|^2}{\langle Ap_k, p_k \rangle}$$

and

$$\beta_{k-1} = \frac{\|r_k\|^2}{\|r_{k-1}\|^2}.$$

Conversely, consider three sequences (x_k, r_k, p_k) defined by the induction relations (9.4). It is easy to show by induction that the relations

$$r_0 = b - Ax_0 \quad \text{and} \quad \begin{cases} r_{k+1} = r_k - \alpha_k Ap_k, \\ x_{k+1} = x_k + \alpha_k p_k, \end{cases}$$

imply that the sequence r_k is indeed the residual sequence, namely $r_k = b - Ax_k$. Another elementary induction argument shows that the relations

$$r_0 = p_0 \quad \text{and} \quad \begin{cases} r_k = r_{k-1} - \alpha_{k-1} Ap_{k-1}, \\ p_k = r_k + \beta_{k-1} p_{k-1}, \end{cases}$$

imply that p_k and r_k belong to the Krylov space K_k, for all $k \geq 0$. As a result, the induction relation $x_{k+1} = x_k + \alpha_k p_k$ entails that x_{k+1} belongs to the affine space $[x_0 + K_k]$. To conclude that the sequence (x_k) is indeed that of the conjugate gradient method, it remains to show that r_{k+1} is orthogonal to K_k (first definition of the conjugate gradient method). To this end, we now prove by induction that r_{k+1} is orthogonal to r_j, for all $0 \leq j \leq k$, and that p_{k+1} is conjugate to p_j, for all $0 \leq j \leq k$. At order 0 we have

$$\langle r_1, r_0 \rangle = \|r_0\|^2 - \alpha_0 \langle Ap_0, r_0 \rangle = 0,$$

since $p_0 = r_0$, and

$$\langle Ap_1, p_0 \rangle = \langle (r_1 + \beta_0 p_0), Ap_0 \rangle = \alpha_0^{-1} \langle (r_1 + \beta_0 r_0), (r_0 - r_1) \rangle = 0.$$

Assume that up to the kth order, we have

$$\langle r_k, r_j \rangle = 0 \text{ for } 0 \leq j \leq k-1, \text{ and } \langle Ap_k, p_j \rangle = 0 \text{ for } 0 \leq j \leq k-1.$$

Let us prove that this still holds at the $(k+1)$th order. Multiplying the definition of r_{k+1} by r_j leads to

$$\langle r_{k+1}, r_j \rangle = \langle r_k, r_j \rangle - \alpha_k \langle Ap_k, r_j \rangle,$$

which together with the relation $r_j = p_j - \beta_{j-1} p_{j-1}$ implies

$$\langle r_{k+1}, r_j \rangle = \langle r_k, r_j \rangle - \alpha_k \langle Ap_k, p_j \rangle + \alpha_k \beta_{j-1} \langle Ap_k, p_{j-1} \rangle.$$

By the induction assumption, we easily infer that $\langle r_{k+1}, r_j \rangle = 0$ if $j \leq k-1$, whereas the formula for α_k implies that $\langle r_{k+1}, r_k \rangle = 0$. Moreover,

$$\langle Ap_{k+1}, p_j \rangle = \langle p_{k+1}, Ap_j \rangle = \langle r_{k+1}, Ap_j \rangle + \beta_k \langle p_k, Ap_j \rangle,$$

and since $Ap_j = (r_j - r_{j+1})/\alpha_j$ we have

$$\langle Ap_{k+1}, p_j \rangle = \alpha_j^{-1} \langle r_{k+1}, (r_j - r_{j+1}) \rangle + \beta_k \langle p_k, Ap_j \rangle.$$

For $j \leq k-1$, the induction assumption and the orthogonality of r_{k+1} (which we have just obtained) prove that $\langle Ap_{k+1}, p_j \rangle = 0$. For $j = k$, we infer $\langle Ap_{k+1}, p_k \rangle = 0$ from the formulas giving α_k and β_k. This ends the induction.

Since the family $(r_k)_{0 \leq k \leq k_0}$ is orthogonal, it is linearly independent as long as $r_k \neq 0$. Now, $r_k \in K_k$, which entails that $K_k = span\{r_0, \ldots, r_k\}$, since these two spaces have the same dimension. Thus, we have derived the first definition of the conjugate gradient, namely

$$x_{k+1} \in [x_0 + K_k] \quad \text{and} \quad r_{k+1} \perp K_k.$$

Hence, the sequence x_k is indeed that of the conjugate gradient. $\square$

Remark 9.5.1. The main relevance of Theorem 9.5.1 is that it provides a practical algorithm for computing the sequence of approximate solutions (x_k). It actually suffices to apply the three induction relations to (x_k, r_k, p_k) to derive x_{k+1} starting from x_k. It is no longer required to orthogonalize r_{k+1} with respect to K_k, nor to minimize $\frac{1}{2}\langle Ax, x \rangle - \langle b, x \rangle$ on the space $[x_0 + K_k]$ (cf. the two theoretical definitions of the conjugate gradient method).

Instead of having a single induction on x_k (as in the simple gradient method), there are three inductions on (x_k, r_k, p_k) in the conjugate gradient method (in truth, there are only two, since r_k is just an auxiliary for the computations we can easily get rid of). In the induction on p_k, if we substitute p_k with $(x_{k+1} - x_k)/\alpha_k$, we obtain a three-term induction for x_k:

$$x_{k+1} = x_k + \alpha_k (b - Ax_k) + \frac{\alpha_k \beta_{k-1}}{\alpha_{k-1}}(x_k - x_{k-1}).$$

This three-term induction is more complex than the simple inductions we have studied so far in the iterative methods (see Chapter 8). This partly accounts for the higher efficiency of the conjugate gradient method over the simple gradient method.

9.5.1 Numerical Algorithm

In practice, the conjugate gradient algorithm is implemented following Algorithm 9.3. As soon as $r_k = 0$, the algorithm has converged. That is, x_k is the

---

Data: A and b. Output: $\mathbf{x}$ (approximation of x)

    Initialization:

        choose $\mathbf{x} \in \mathbb{R}^n$

        compute $r = b - A\mathbf{x}$

        set $p = r$

        compute $\gamma = \|r\|^2$

    While $\|\gamma\| > \varepsilon$

        $y = Ap$

        $\alpha = \frac{\gamma}{\langle y, p \rangle}$

        $\mathbf{x} = \mathbf{x} + \alpha p$

        $r = r - \alpha y$

        $\beta = \frac{\|r\|^2}{\gamma}$

        $\gamma = \|r\|^2$

        $p = r + \beta p$

    End While

Algorithm 9.3: Conjugate gradient algorithm.

---

solution to the system $Ax = b$. We know that the convergence is achieved after $k_0 + 1$ iterations, where $k_0 \leq n - 1$ is the critical dimension of the Krylov spaces (of which we have no knowledge a priori). However, in practice, computer calculations are always prone to errors due to rounding, so we do not find exactly $r_{k_0+1} = 0$. That is why we introduce a small parameter ε (for instance, 10^{-4} or 10^{-8} according to the desired precision), and we decide that the algorithm has converged as soon as

$$\frac{\|r_k\|}{\|r_0\|} \le \varepsilon.$$

Moreover, for large systems (for which n and k_0 are large, orders of magnitude ranging from 10^4 to 10^6), the conjugate gradient method is used as an iterative method, i.e., it converges in the sense of the above criterion after a number of iterations much smaller than $k_0 + 1$ (cf. Proposition 9.5.1 below).

Remark 9.5.2.

1. In general, if we have no information about the solution, we choose to initialize the conjugate gradient method with $x_0 = 0$. If we are solving a sequence of problems that bear little difference between them, we can initialize x_0 to the previous solution.
2. At each iteration, a single matrix-vector product is computed, namely Ap_k, since r_k is computed by the induction formula and not through the relation $r_k = b - Ax_k$.
3. To implement the conjugate gradient method, it is not necessary to store the matrix A in an array if we know how to compute the product Ay for any vector y. For instance, for the Laplacian matrix in one space dimension, we have

$$A = \begin{pmatrix} 2 & -1 & & 0 \\ -1 & \ddots & \ddots & \\ & \ddots & \ddots & -1 \\ 0 & & -1 & 2 \end{pmatrix}, \quad \begin{cases} (Ay)_1 = 2y_1 - y_2, \\ (Ay)_n = 2y_n - y_{n-1}, \\ \text{and for } i = 2, \ldots, n-1, \\ \quad (Ay)_i = 2y_i - y_{i+1} - y_{i-1}. \end{cases}$$

4. The conjugate gradient method is very efficient. It has many variants and generalizations, in particular to the case of nonsymmetric positive definite matrices.

9.5.2 Number of Operations

If we consider the conjugate gradient method as a direct method, we can count the number of operations necessary to solve a linear system in the most unfavorable case, $k_0 = n - 1$. At each iteration, on the one hand, a matrix-vector product is computed (n^2 products), and on the other hand, two scalar products and three linear combinations of vectors such as $x + \alpha y$ (of the order of magnitude of n products) have to be carried out. After n iterations, we obtain a number of operations equal, to first order, to

$$N_{op} \approx n^3.$$

It is hence less efficient than the Gauss and Cholesky methods. But one must recall that it is being used as an iterative method and that generally convergence occurs after fewer than n iterations.

9.5.3 Convergence Speed

Recall that Remark 9.1.1 tells us that the gradient method reduces the error at each iteration by a factor of $(\text{cond}_2(A) - 1)/(\text{cond}_2(A) + 1)$. We shall now prove that the convergence speed of the conjugate gradient is much faster, since it reduces the error at each iteration by a smaller factor $(\sqrt{\text{cond}_2(A)} - 1)/(\sqrt{\text{cond}_2(A)} + 1)$.

Proposition 9.5.1. *Let A be a symmetric real and positive definite matrix. Let x be the exact solution to the system $Ax = b$. Let $(x_k)_k$ be the sequence of approximate solutions produced by the conjugate gradient method. We have*

$$\|x_k - x\|_2 \leq 2\sqrt{\text{cond}_2(A)} \left(\frac{\sqrt{\text{cond}_2(A)} - 1}{\sqrt{\text{cond}_2(A)} + 1} \right)^k \|x_0 - x\|_2.$$

Proof. Recall that x_k can be derived as the minimum on the space $[x_0 + K_{k-1}]$ of the function f defined by

$$z \in \mathbb{R}^n \longmapsto f(z) = \frac{1}{2}\langle Az, z \rangle - \langle b, z \rangle = \frac{1}{2}\|x - z\|_A^2 - \frac{1}{2}\langle Ax, x \rangle,$$

where $\|y\|_A^2 = \langle Ay, y \rangle$. Computing x_k is thus equivalent to minimizing $\|x - z\|_A^2$, that is, the error in the $\|.\|_A$ norm, on the space $[x_0 + K_{k-1}]$. Now we compute $\|e_k\|_A$. Relations (9.4) show that

$$x_k = x_0 + \sum_{j=0}^{k-1} \alpha_j p_j,$$

with each p_j equal to a polynomial in A (of degree $\leq j$) applied to p_0, so there exists a polynomial $q_{k-1} \in \mathbb{P}_{k-1}$ such that

$$x_k = x_0 + q_{k-1}(A)p_0.$$

Since $p_0 = r_0 = b - Ax_0 = A(x - x_0)$, the errors e_k satisfy the relations

$$e_k = x_k - x = e_0 + q_{k-1}(A)p_0 = Q_k(A)e_0,$$

where Q_k is the polynomial defined, for all $t \in \mathbb{R}$, by $Q_k(t) = 1 - q_{k-1}(t)t$. We denote by u_j an orthonormal basis of eigenvectors of A, i.e., $Au_j = \lambda_j u_j$, and by $e_{0,j}$ the entries of the initial error e_0 in this basis: $e_0 = \sum_{j=1}^{n} e_{0,j} u_j$. We have

$$\|e_0\|_A^2 = \langle e_0, Ae_0 \rangle = \sum_{j=1}^n \lambda_j |e_{0,j}|^2$$

and

$$\|e_k\|_A^2 = \|Q_k(A)e_0\|_A^2 = \sum_{j=1}^n \lambda_j |e_{0,j}Q_k(\lambda_j)|^2.$$

Since the conjugate gradient method minimizes $\|e_k\|_A$, the polynomial Q_k must satisfy

$$\|e_k\|_A^2 = \sum_{j=1}^n \lambda_j |e_{0,j}Q_k(\lambda_j)|^2 = \min_{Q \in \mathbb{P}_k^0} \sum_{j=1}^n \lambda_j |e_{0,j}Q(\lambda_j)|^2, \qquad (9.7)$$

where minimization is carried in $\mathbb{P}_k^0$, the set of polynomials $Q \in \mathbb{P}_k$ that satisfy $Q(0) = 1$ (indeed, we have $Q_k \in \mathbb{P}_k^0$). From (9.7) we get an easy upper bound

$$\frac{\|e_k\|_A^2}{\|e_0\|_A^2} \le \left(\min_{Q \in \mathbb{P}_k^0} \max_{1 \le j \le n} |Q_k(\lambda_j)|^2 \right) \le \left(\min_{Q \in \mathbb{P}_k^0} \max_{\lambda_1 \le x \le \lambda_n} |Q_k(x)|^2 \right). \qquad (9.8)$$

The last min–max problem in the right-hand side of (9.8) is a classical and celebrated polynomial approximation problem: *Find a polynomial $p \in \mathbb{P}_k$ minimizing the quantity $\max_{x \in [a,b]} |p(x)|$.* To avoid the trivial zero solution, we impose an additional condition on p, for instance, $p(\beta) = 1$ for a number $\beta \notin [a, b]$. This problem is solved at the end of this chapter, Section 9.5.5. Its unique solution (see Proposition 9.5.3) is the polynomial

$$\frac{T_k \left(\frac{2x-(a+b)}{b-a} \right)}{T_k \left(\frac{2\beta-(a+b)}{b-a} \right)}, \qquad (9.9)$$

where T_k is the kth-degree Chebyshev polynomial defined for all $t \in [-1, 1]$ by $T_k(t) = \cos(k \arccos t)$. The maximum value reached by (9.9) on the interval $[a, b]$ is

$$\frac{1}{\left| T_k \left(\frac{2\beta-(a+b)}{b-a} \right) \right|}.$$

In our case (9.7), we have $a = \lambda_1, b = \lambda_n, \beta = 0$. The matrix A being positive definite, $\beta \notin [a, b]$, we conclude that

$$\min_{Q \in \mathbb{P}_k^0} \max_{\lambda_1 \le x \le \lambda_n} |Q(x)| = \frac{1}{\left| T_k \left(\frac{\lambda_n+\lambda_1}{\lambda_n-\lambda_1} \right) \right|}. \qquad (9.10)$$

Observe now that

- A being symmetric, we have $\frac{\lambda_n+\lambda_1}{\lambda_n-\lambda_1} = \frac{\kappa+1}{\kappa-1}$, with $\kappa = \text{cond}_2(A) = \lambda_n/\lambda_1$;

- Chebyshev polynomials satisfy (9.16), which we recall here:

$$2T_k(x) = [x + \sqrt{x^2 - 1}]^k + [x - \sqrt{x^2 - 1}]^k,$$

for all $x \in (-\infty, -1] \cup [1, +\infty)$. Therefore, we infer the following lower bound:

$$\left(2T_k \left(\frac{\lambda_n + \lambda_1}{\lambda_n - \lambda_1} \right) \right)^{1/k} \geq \frac{\kappa + 1}{\kappa - 1} + \sqrt{\left(\frac{\kappa + 1}{\kappa - 1} \right)^2 - 1}$$

$$\geq \frac{1}{\kappa - 1} \left((\kappa + 1) + 2\sqrt{\kappa} \right) = \frac{\sqrt{\kappa} + 1}{\sqrt{\kappa} - 1}.$$

Combining these results, equality (9.10), and the upper bound (9.8), we obtain

$$\|e_k\|_A^2 \leq 4 \left(\frac{\sqrt{\kappa} + 1}{\sqrt{\kappa} - 1} \right)^{2k} \|e_0\|_A^2.$$

Finally, the proposition follows immediately from the equivalence between the norms $\|.\|_2$ and $\|.\|_A$: $\lambda_1 \|x\|_2^2 \leq \|x\|_A^2 \leq \lambda_n \|x\|_2^2$. □

It is possible to improve further Proposition 9.5.1 by showing that the convergence speed is indeed faster when the eigenvalues of A are confined to a reduced number of values.

We draw from Proposition 9.5.1 three important consequences. First of all, the conjugate gradient method works indeed as an iterative method. In fact, even if we do not perform the requisite n iterations to converge, the more we iterate, the more the error between x and x_k diminishes. Furthermore, the convergence speed depends on the square root of the conditioning of A, and not on the conditioning itself as for the simple gradient method. Hence, the conjugate gradient method converges faster than the simple gradient one (we say that the convergence is quadratic instead of linear). Lastly, as usual, the closer $\text{cond}_2(A)$ is to 1, the greater the speed of convergence, which means that the matrix A has to be well conditioned for a quick convergence of the conjugate gradient method.

9.5.4 Preconditioning

Definition 9.5.1. *Let $Ax = b$ be the linear system to be solved. We call a matrix C that is easy to invert and such that $\text{cond}_2(C^{-1}A)$ is smaller than $\text{cond}_2(A)$ a preconditioning of A. We call the equivalent system $C^{-1}Ax = C^{-1}b$ a preconditioned system.*

The idea of preconditioning is that if $\text{cond}_2(C^{-1}A) < \text{cond}_2(A)$, the conjugate gradient method converges faster for the preconditioned system than for the original one. Of course, the price to pay for this faster convergence is the requirement of inverting C. Nevertheless, we recall that it is not necessary to form the matrix $(C^{-1}A)$; we merely successively multiply matrices A and C^{-1} by vectors. The problem here is to choose a matrix C that is an easily invertible approximation of A. In practice, the preconditioning technique is very efficient and there is a large literature on this topic.

Observe that it is not obvious whether we can choose C in such a way that on the one hand, $\mathrm{cond}(C^{-1}A) \ll \mathrm{cond}(A)$, and on the other hand, $C^{-1}A$ remains symmetric positive definite (a necessary condition for the application of the conjugate gradient method). For this reason, we introduce a "symmetric" preconditioning. Let C be a symmetric positive definite matrix. We write $C = BB^t$ (Cholesky decomposition) and we substitute the original system $Ax = b$, whose matrix A is badly conditioned, by the equivalent system

$$\tilde{A}\tilde{x} = \tilde{b}, \text{ where } \tilde{A} = B^{-1}AB^{-t}, \ \tilde{b} = B^{-1}b, \text{ and } \tilde{x} = B^t x.$$

Since the matrix $\tilde{A}$ is symmetric positive definite, we can use Algorithm 9.3 of the conjugate gradient method to solve this problem, i.e.,

> Initialization
> > choice of $\tilde{x}_0$
> > $\tilde{r}_0 = \tilde{p}_0 = \tilde{b} - \tilde{A}\tilde{x}_0$
> Iterations
> > For $k \geq 1$ and $\tilde{r}_k \neq 0$
> > > $\tilde{\alpha}_{k-1} = \dfrac{\|\tilde{r}_{k-1}\|^2}{\langle \tilde{A}\tilde{p}_{k-1}, \tilde{p}_{k-1}\rangle}$
> > > $\tilde{x}_k = \tilde{x}_{k-1} + \tilde{\alpha}_{k-1}\tilde{p}_{k-1}$
> > > $\tilde{r}_k = \tilde{r}_{k-1} - \tilde{\alpha}_{k-1}\tilde{A}\tilde{p}_{k-1}$
> > > $\tilde{\beta}_{k-1} = \dfrac{\|\tilde{r}_k\|^2}{\|\tilde{r}_{k-1}\|^2}$
> > > $\tilde{p}_k = \tilde{r}_k + \tilde{\beta}_{k-1}\tilde{p}_{k-1}$
> End

We can compute $\tilde{x}$ in this fashion and next compute x by solving the upper triangular system $B^t x = \tilde{x}$. In practice, we do not proceed this way in order to cut down the cost of the computations.

Note that residuals are linked by the relation $\tilde{r}_k = \tilde{b} - \tilde{A}\tilde{x}_k = B^{-1}r_k$, the new conjugate directions are $\tilde{p}_k = B^t p_k$, and

$$\langle \tilde{A}\tilde{p}_k, \tilde{p}_j\rangle = \langle Ap_k, p_j\rangle.$$

We thereby obtain the relations

$$\tilde{\alpha}_{k-1} = \frac{\langle C^{-1}r_{k-1}, r_{k-1}\rangle}{\langle Ap_{k-1}, p_{k-1}\rangle},$$

$$\tilde{\beta}_{k-1} = \frac{\langle C^{-1}r_k, r_k\rangle}{\langle C^{-1}r_{k-1}, r_{k-1}\rangle}.$$

We thus see that the only new operation (which is also the most costly one) is the computation of $z_k = C^{-1}r_k$, namely solving the linear system $Cz_k = r_k$. Knowing the Cholesky decomposition of C, the calculation of each z_k can be carried out in n^2 operations at most. From this observation, we infer a preconditioned version of the conjugate gradient algorithm (see Algorithm 9.4), in which the precise knowledge of the Cholesky factorization of C is no longer required. This is Algorithm 9.4, which is used in numerical practice.

Data: A and b. Output: x_k
$\quad$ Initialization
$\quad\quad$ choice of x_0
$\quad\quad$ $r_0 = b - Ax_0$
$\quad\quad$ $z_0 = C^{-1}r_0$ $\qquad\qquad\qquad\qquad\qquad$ *preconditioning*
$\quad\quad$ $p_0 = z_0$
$\quad$ Iterations
$\quad\quad$ **For** $k \geq 1$ and $r_k \neq 0$
$\quad\quad\quad$ $\alpha_{k-1} = \dfrac{\langle z_{k-1}, r_{k-1} \rangle}{\langle Ap_{k-1}, p_{k-1} \rangle}$
$\quad\quad\quad$ $x_k = x_{k-1} + \alpha_{k-1}p_{k-1}$
$\quad\quad\quad$ $r_k = r_{k-1} - \alpha_{k-1}Ap_{k-1}$
$\quad\quad\quad$ $z_k = C^{-1}r_k$ $\qquad\qquad\qquad\qquad$ *preconditioning*
$\quad\quad\quad$ $\beta_{k-1} = \dfrac{\langle z_k, r_k \rangle}{\langle z_{k-1}, r_{k-1} \rangle}$
$\quad\quad\quad$ $p_k = z_k + \beta_{k-1}p_{k-1}$
$\quad$ **End**

Algorithm 9.4: Preconditioned conjugate gradient method.

SSOR preconditioning

To illustrate the interest in preconditioning the conjugate gradient method, we consider again the matrix given by the finite difference discretization of the Laplacian, which is a tridiagonal, symmetric, positive definite matrix $A_n \in \mathcal{M}_{n-1}(\mathbb{R})$, defined by (5.12). For simplicity, in the sequel we drop the subscript n and write A instead of A_n. We denote by D the diagonal of A and by $-E$ its strictly lower triangular part, $A = D - E - E^t$. For $\omega \in (0, 2)$, the symmetric matrix

$$C_\omega = \frac{\omega}{2 - \omega} \left(\frac{D}{\omega} - E \right) D^{-1} \left(\frac{D}{\omega} - E^t \right)$$

is positive definite. Indeed, for $x \neq 0$, we have

$$\langle C_\omega x, x \rangle = \frac{\omega}{2 - \omega} \left\langle D^{-1} \left(\frac{D}{\omega} - E \right)^t x, \left(\frac{D}{\omega} - E \right)^t x \right\rangle,$$

and since D^{-1} is symmetric positive definite and $\frac{D}{\omega} - E$ is nonsingular, we have $\langle C_\omega x, x \rangle > 0$ if and only if $\frac{\omega}{2-\omega} > 0$. We now precondition the system $Ax = b$ by the matrix C_ω. We denote by $B_\omega = \sqrt{\frac{\omega}{2-\omega}}(\frac{D}{\omega} - E)D^{-1/2}$ the

Cholesky factor of C_ω. The matrices $\tilde{A}_\omega = B_\omega^{-1} A B_\omega^{-t}$ and $C_\omega^{-1} A$ are similar, since

$$C_\omega^{-1} A = B_\omega^{-t} B_\omega^{-1} A = B_\omega^{-t} \left(B_\omega^{-1} A B_\omega^{-t} \right) B_\omega^t = B_\omega^{-t} \tilde{A}_\omega B_\omega^t.$$

Hence

$$\mathrm{cond}_2(C_\omega^{-1} A) = \mathrm{cond}_2(\tilde{A}_\omega) = \frac{\lambda_{\max}(\tilde{A}_\omega)}{\lambda_{\min}(\tilde{A}_\omega)}.$$

To evaluate the performance of the preconditioning we determine an upper bound for $\lambda_{\max}(\tilde{A}_\omega)$ and a lower bound for $\lambda_{\min}(\tilde{A}_\omega)$. For any $x \neq 0$, we have

$$\frac{\langle \tilde{A}_\omega x, x \rangle}{\langle x, x \rangle} = \frac{\langle B_\omega^{-1} A B_\omega^{-t} x, x \rangle}{\langle x, x \rangle} = \frac{\langle A B_\omega^{-t} x, B_\omega^{-t} x \rangle}{\langle x, x \rangle}.$$

Setting $x = B_\omega^t y$, we obtain

$$\lambda_{\max}(\tilde{A}_\omega) = \max_{x \neq 0} \frac{\langle \tilde{A}_\omega x, x \rangle}{\langle x, x \rangle} = \max_{y \neq 0} \frac{\langle Ay, y \rangle}{\langle C_\omega y, y \rangle}.$$

Similarly,

$$\lambda_{\min}(\tilde{A}_\omega) = \min_{y \neq 0} \frac{\langle Ay, y \rangle}{\langle C_\omega y, y \rangle}.$$

The goal is now to obtain the inequalities

$$0 < \alpha \leq \frac{\langle Ax, x \rangle}{\langle C_\omega x, x \rangle} \leq \beta, \qquad \forall x \neq 0, \tag{9.11}$$

from which we will deduce the bound $\mathrm{cond}_2(\tilde{A}_\omega) \leq \beta/\alpha$.

- To obtain the upper bound in (9.11) we decompose C_ω as

$$C_\omega = A + \frac{\omega}{2 - \omega} F_\omega D^{-1} F_\omega^t,$$

with $F_\omega = \frac{\omega - 1}{\omega} D - E$. For all $x \neq 0$, we have

$$\frac{2 - \omega}{\omega} \langle (A - C_\omega) x, x \rangle = -\langle F_\omega D^{-1} F_\omega^t x, x \rangle = -\langle D^{-1} F_\omega^t x, F_\omega^t x \rangle \leq 0,$$

since D^{-1} is positive definite. We can thus choose $\beta = 1$.
- To obtain the lower bound in (9.11), we write $(2 - \omega) C_\omega = A + aD + \omega G$ with

$$G = E D^{-1} E^t - \frac{1}{4} D \text{ and } a = \frac{(2 - \omega)^2}{4\omega}.$$

For $x \neq 0$, we compute

$$(2 - \omega) \frac{\langle C_\omega, x \rangle}{\langle Ax, x \rangle} = 1 + a \frac{\langle Dx, x \rangle}{\langle Ax, x \rangle} + \omega \frac{\langle Gx, x \rangle}{\langle Ax, x \rangle}.$$

Since $\langle Gx, x \rangle = -\frac{n^2}{2}|x_1|^2$, we have

$$(2 - \omega)\frac{\langle C_\omega, x \rangle}{\langle Ax, x \rangle} \leq 1 + a\frac{\langle Dx, x \rangle}{\langle Ax, x \rangle} = 1 + 2an^2\frac{\|x\|^2}{\langle Ax, x \rangle} \leq 1 + \frac{2an^2}{\lambda_{\min}(A)}.$$

We can therefore take $\alpha = (2 - \omega)\dfrac{1}{1 + \frac{2an^2}{\lambda_{\min}(A)}}$.

We now choose a value of ω that minimizes the function

$$f(\omega) = \frac{\beta}{\alpha} = \frac{1}{2 - \omega}\left(1 + \frac{2an^2}{\lambda_{\min}}\right) = \frac{1}{2 - \omega} + \gamma\frac{2 - \omega}{\omega},$$

where

$$\gamma = \frac{n^2}{2\lambda_{\min}} = \frac{1}{8\sin^2\frac{\pi}{2n}} \approx \frac{n^2}{2\pi^2}.$$

A simple computation shows that

$$\omega^2(2 - \omega)^2 f'(\omega) = \omega^2 - 2\gamma(2 - \omega)^2,$$

so the value ω_{opt} minimizing f is

$$\omega_{\mathrm{opt}} = \frac{2\sqrt{2\gamma}}{1 + \sqrt{2\gamma}} = \frac{2}{1 + 2\sin\frac{\pi}{2n}} \approx 2\left(1 - \frac{\pi}{n}\right).$$

For this optimal value of ω the conditioning of $\tilde{A}_\omega$ is bounded above as follows:

$$\mathrm{cond}_2(\tilde{A}_\omega) \leq f(\omega_{\mathrm{opt}}) = \frac{1}{2} + \frac{1}{2\sin\frac{\pi}{2n}}.$$

Thus for n large enough, $\mathrm{cond}_2(\tilde{A}_\omega) \leq \dfrac{n}{\pi}$, whereas $\mathrm{cond}_2(A) \approx \dfrac{4n^2}{\pi^2}$. We save one order of magnitude in n by preconditioning the initial linear system.

9.5.5 Chebyshev Polynomials

This section is devoted to some properties of Chebyshev polynomials used previously, in particular in the proof of Proposition 9.5.1. Chebyshev polynomials are first defined on the interval $[-1, 1]$, then extended to $\mathbb{R}$.

Study on $[-1, 1]$

On this interval, the Chebyshev polynomial T_n $(n \geq 1)$ is given by

$$T_n(t) = \cos(n\theta),$$

where $\theta \in [0, \pi]$ is defined by $\cos(\theta) = t$. We easily check that $T_0(t) = 1$, $T_1(t) = t$, and

$$T_{n+1}(t) = 2tT_n(t) - T_{n-1}(t), \qquad \forall n \geq 1, \tag{9.12}$$

which proves that T_n is a polynomial of degree n and can be defined on the whole of $\mathbb{R}$. We also check that T_n has exactly

- n zeros (called Chebyshev points), which are

$$t_k = \cos(\theta_k), \quad \theta_k = \frac{\pi}{2n} + k\frac{\pi}{n}, \quad 0 \le k \le n-1,$$

- $n+1$ extremal points where the polynomial takes its extremal values (-1 and 1), which are

$$\tilde{t}_k = \cos(k\pi/n), \quad 0 \le k \le n.$$

The first five Chebyshev polynomials are represented in Figure 9.3. Expressing

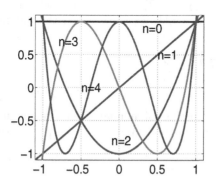

Fig. 9.3. Chebyshev polynomials T_n, for $n = 0, \ldots, 4$.

the cosine in exponential form, we have

$$2T_n(t) = e^{in\theta} + e^{-in\theta} = (\cos(\theta) + i\sin(\theta))^n + (\cos(\theta) - i\sin(\theta))^n.$$

Since $\theta \in [0, \pi]$, we have $\sin\theta = \sqrt{1-t^2}$, and T_n can be written explicitly in terms of t:

$$2T_n(t) = \left(t + i\sqrt{1-t^2}\right)^n + \left(t - i\sqrt{1-t^2}\right)^n. \tag{9.13}$$

For $\alpha \in \mathbb{R}$ with $|\alpha| > 1$, we denote by $\bar{\mathbb{P}}_n^\alpha$ the set of polynomials of $\mathbb{P}_n$ taking the value 1 at point α:

$$\bar{\mathbb{P}}_n^\alpha = \{p \in \mathbb{P}_n, \quad p(\alpha) = 1\}.$$

Proposition 9.5.2. *Chebyshev polynomials satisfy the following property:*

$$\min_{p \in \bar{\mathbb{P}}_n^\alpha} \max_{t \in [-1,1]} |p(t)| = \max_{t \in [-1,1]} |\bar{T}_n(t)| = \frac{1}{|T_n(\alpha)|}, \tag{9.14}$$

where we have set $\bar{T}_n = T_n/T_n(\alpha)$.

Proof. We know that $T_n(\alpha) \neq 0$, because all the zeros, or roots, of T_n belong to $]-1, 1[$. Hence $\bar{T}_n$ is well defined. On the one hand, if $\alpha > 1$, then $T_n(\alpha) > 0$, since the polynomials T_n do not change sign on $]1, +\infty[$ and $T_n(1) = 1$. On the other hand, if $\alpha < -1$, then $T_n(\alpha)$ has the same sign as $T_n(-1) = (-1)^n$. We shall therefore assume in the sequel, without loss of generality, that $\alpha > 1$, so that $T_n(\alpha) > 0$. To prove (9.14), we assume, to the contrary, that there exists $p \in \bar{\mathbb{P}}_n$ such that

$$\max_{t \in [-1,1]} |p(t)| < \frac{1}{T_n(\alpha)}. \tag{9.15}$$

We are going to show that $p = \bar{T}_n$, which contradicts our assumption (9.15). At each of the $n + 1$ extremal points $\tilde{t}_k$ of T_n, the polynomial $q = p - \bar{T}_n$ changes sign, since

- $q(\tilde{t}_0) = p(\tilde{t}_0) - \frac{1}{T_n(\alpha)} < 0$,
- $q(\tilde{t}_1) = p(\tilde{t}_1) + \frac{1}{T_n(\alpha)} > 0$,
- etc.

In addition, $q(\alpha) = 0$, so the polynomial $q \in \mathbb{P}_n$ has at least $n + 1$ distinct zeros, because the $\tilde{t}_k$ are distinct and $\alpha \notin [-1, 1]$. It must therefore vanish identically. $\qquad \square$

Study on $[a, b]$

The induction relation (9.12) shows that the Chebyshev polynomials can be defined on the whole real line $\mathbb{R}$. We can also define $T_n(x)$ for $x \in (-\infty, -1] \cup [1, +\infty)$ as follows: since $|x| \geq 1$, we have $\sqrt{1 - x^2} = i\sqrt{x^2 - 1}$, and by (9.13),

$$2T_n(x) = [x + \sqrt{x^2 - 1}]^n + [x - \sqrt{x^2 - 1}]^n. \tag{9.16}$$

To obtain the analogue of property (9.14) on any interval $[a, b]$, we define the linear function φ that maps $[-1, 1]$ onto $[a, b]$:

$$[-1, 1] \longrightarrow [a, b],$$
$$\varphi : t \longmapsto x = \frac{a+b}{2} + \frac{b-a}{2}t.$$

Proposition 9.5.3. *For any $\beta \notin [a, b]$, the Chebyshev polynomials satisfy*

$$\min_{q \in \mathbb{P}_n^\beta} \max_{x \in [a,b]} |q(x)| = \max_{x \in [a,b]} \frac{|T_n(\varphi^{-1}(x))|}{|T_n(\varphi^{-1}(\beta))|} = \frac{1}{|T_n(\varphi^{-1}(\beta))|}.$$

Proof. Since $\beta \notin [a, b]$, $\alpha = \varphi^{-1}(\beta) \notin [-1, 1]$, and the polynomial $p(t) = q(\varphi(t))$ belongs to $\mathbb{P}_n^\alpha$. By Proposition 9.5.2 we get

$$\min_{q \in \mathbb{P}_n^\beta} \max_{x \in [a,b]} |q(x)| = \min_{q \in \mathbb{P}_n^\beta} \max_{t \in [-1,1]} |q(\varphi(t))|$$

$$= \min_{p \in \mathbb{P}_n^\alpha} \max_{t \in [-1,1]} |p(t)|$$

$$= \max_{t \in [-1,1]} |\bar{T}_n(t)| = \frac{1}{|T_n(\varphi^{-1}(\beta))|},$$

with $\bar{T}_n(t) = T_n(t)/T_n(\varphi^{-1}(\beta))$, and the result is proved.   □

9.6 Exercises

In these exercises $A \in \mathcal{M}_n(\mathbb{R})$ is always a symmetric positive definite matrix. We recall that the following two problems are equivalent:

$$\min_{x \in \mathbb{R}^n} \left\{ f(x) = \frac{1}{2} \langle Ax, x \rangle - \langle b, x \rangle \right\}, \tag{9.17}$$

and solve the linear system

$$Ax = b. \tag{9.18}$$

9.1. The goal of this exercise is to program and study the constant step gradient algorithm.

1. Write a program `GradientS` that computes the minimizer of (9.17) by the gradient method; see Algorithm 9.1.
2. Take $n = 10$, `A=Laplacian1dD(n);xx=(1:n)'/(n+1);b=xx.*sin(xx)`. Compute the solution x_G of (9.18) obtained using this algorithm. Take $\alpha = 10^{-4}$ and limit the number of iterations to $N_{\text{iter}} = 10\,000$. The convergence criterion shall be written in terms of the residual norm which must be smaller than $\varepsilon = 10^{-4}$ times its initial value. How many iterations are necessary before convergence? Compare the obtained solution with that given by `Matlab`.
3. Take now $N_{\text{iter}} = 2000$ and $\varepsilon = 10^{-10}$. For α varying from $\alpha_{\min} = 32 \times 10^{-4}$ to $\alpha_{\max} = 42 \times 10^{-4}$, by steps of size 10^{-5}, plot a curve representing the number of iterations necessary to compute x_G. Determine numerically the value of α leading to the minimal number of iterations. Compare with the optimal value given by Theorem 9.1.1.

9.2. In order to improve the convergence of the gradient algorithm, we now program and study the variable step gradient algorithm. At each iteration, the step α is chosen equal to the value α_k that minimizes the norm of the residual $r_{k+1} = b - Ax_{k+1}$. In other words, α_k is defined by

$$\|\nabla f(x_k - \alpha_k \nabla f(x_k))\| = \inf_{\alpha \in \mathbb{R}} \|\nabla f(x_k - \alpha \nabla f(x_k))\|,$$

where $\|.\|$ denotes the Euclidean norm $\|.\|_2$.

1. Find an explicit formula for α_k.
2. Write a program `GradientV` that computes the solution of (9.17) by the variable step gradient method, see Algorithm 9.2.
3. Compare both algorithms (constant and variable step) and in particular the number of iterations and the total computational time for the same numerical accuracy. For this, take the data of Exercise 9.1 with, for the constant step gradient, α as the optimal value.

9.3 (∗). We now program and study the conjugate gradient algorithm.

1. Write a program `GradientC` that computes the solution of (9.17) using the conjugate gradient method (see Algorithm 9.3).
2. Compare this algorithm with the variable step gradient algorithm for the matrix defined in Exercise 9.1.
3. Let A and b be defined as follows:

$$A = \frac{1}{12} \begin{pmatrix} 5 & 4 & 3 & 2 & 1 \\ 4 & 5 & 4 & 3 & 2 \\ 3 & 4 & 5 & 4 & 3 \\ 2 & 3 & 4 & 5 & 4 \\ 1 & 2 & 3 & 4 & 5 \end{pmatrix}, \quad b = \begin{pmatrix} 1 \\ 2 \\ 3 \\ 4 \\ 5 \end{pmatrix}.$$

Solve the system $Ax = b$ using the program `GradientC` with the initial data $x_0 = (-2, 0, 0, 0, 10)^t$ and note down the number of iterations performed. Explain the result using the above script. Same question for $x_0 = (-1, 6, 12, 0, 17)^t$.

4. We no longer assume that the matrix A is positive definite, nor that it is symmetric, but only that it is nonsingular. Suggest a way of symmetrizing the linear system (9.18) so that one can apply the conjugate gradient method.

9.4. Write a program `GradientCP` that computes the solution of (9.17) by the preconditioned conjugate gradient algorithm (see Algorithm 9.4) with the SSOR preconditioning (see Section 9.5.4). Compare the programs `GradientC` and `GradientCP` by plotting on the same graph the errors of both methods in terms of the iteration number. Take $n = 50$ and $N_{\text{iter}} = 50$ iterations without a termination criterion on the residual and consider that the exact solution is `A\b`.

10

Methods for Computing Eigenvalues

10.1 Generalities

For a squared matrix $A \in \mathcal{M}_n(\mathbb{C})$, the problem that consists in finding the solutions $\lambda \in \mathbb{C}$ and nonzero $x \in \mathbb{C}^n$ of the algebraic equation

$$Ax = \lambda x \tag{10.1}$$

is called an eigenvalue problem. The scalar λ is called an "eigenvalue," and the vector x is an "eigenvector." If we consider real matrices $A \in \mathcal{M}_n(\mathbb{R})$, we can also study the eigenvalue problem (10.1), but the eigenvalue λ and eigenvector x may not be real and, in full generality, can belong to $\mathbb{C}$ or $\mathbb{C}^n$ respectively.

In Section 2.3 the existence and theoretical characterization of solutions of (10.1) were discussed. On the other hand, the present chapter is devoted to some numerical algorithms for solving (10.1) in practice.

We recall that the eigenvalues of a matrix A are the roots of its characteristic polynomial $\det(A - \lambda I_n)$. In order to compute these eigenvalues, one may naively think that it suffices to factorize its characteristic polynomial. Such a strategy is bound to fail. On the one hand, there is no explicit formula (using elementary operations such as addition, multiplication, and extraction of roots) for the zeros of most polynomials of degree greater than or equal to 5, as has been known since the work of Galois and Abel. On the other hand, there are no robust and efficient numerical algorithms for computing all the roots of large-degree polynomials. By the way, there is no special property of the characteristic polynomial, since any nth-degree polynomial of the type

$$p_A(\lambda) = (-1)^n \left(\lambda^n + a_1 \lambda^{n-1} + a_2 \lambda^{n-2} + \cdots + a_{n-1}\lambda + a_n \right)$$

is actually the characteristic polynomial (developed with respect to the last column) of the matrix (called the companion matrix of the polynomial)

$$A = \begin{pmatrix} -a_1 & -a_2 & \cdots & \cdots & -a_n \\ 1 & 0 & \cdots & \cdots & 0 \\ 0 & \ddots & \ddots & & \vdots \\ \vdots & & \ddots & \ddots & \vdots \\ 0 & \cdots & 0 & 1 & 0 \end{pmatrix}.$$

Accordingly, there cannot exist direct methods (that is, methods producing the result after a finite number of operations) for the determination of the eigenvalues! As a result, there are only iterative methods for computing eigenvalues (and eigenvectors).

It turns out that in practice, computing the eigenvalues and eigenvectors of a matrix is a much harder task than solving a linear system. Nevertheless, there exist numerically efficient algorithms for self-adjoint matrices or matrices having eigenvalues with distinct moduli. However, the general case is much more delicate to handle and will not be treated here. Finally, numerical stability issues are very complex. The possibility of having multiple eigenvalues seriously complicates their computation, especially in determining their corresponding eigenvectors.

10.2 Conditioning

The notion of conditioning of a matrix is essential for a good understanding of rounding errors in the computation of eigenvalues, but it is different from the one introduced for solving linear systems. Consider the following "ill-conditioned" example:

$$A = \begin{pmatrix} 0 & \cdots & \cdots & 0 & \varepsilon \\ 1 & 0 & \cdots & \cdots & 0 \\ 0 & \ddots & \ddots & & \vdots \\ \vdots & \ddots & \ddots & \ddots & \vdots \\ 0 & \cdots & 0 & 1 & 0 \end{pmatrix},$$

where $\varepsilon = 10^{-n}$. Noting that $p_A(\lambda) = (-1)^n(\lambda^n - \varepsilon)$, the eigenvalues of A are the nth roots of 10^{-n}, all of which are equal in modulus to 10^{-1}. However, had we taken $\varepsilon = 0$, then all eigenvalues would have been zero. Therefore, for large n, small variations in the matrix entries yield large variations in the eigenvalues.

Definition 10.2.1. *Let A be a diagonalizable matrix with eigenvalues $\lambda_1,...,\lambda_n$, and let $\| \cdot \|$ be a subordinate matrix norm. The real number defined by*

$$\Gamma(A) = \inf_{P^{-1}AP=\operatorname{diag}(\lambda_1,...,\lambda_n)} \operatorname{cond}(P),$$

where $\mathrm{cond}(P) = \|P\| \, \|P^{-1}\|$ *is the conditioning of the matrix* P, *is called the conditioning of the matrix* A, *relative to this norm, for computing its eigenvalues.*

For all diagonalizable matrices A, we have $\Gamma(A) \geq 1$. If A is normal, then A is diagonalizable in an orthonormal basis. In other words, there exists a unitary matrix U such that $U^{-1}AU$ is diagonal. For the induced matrix norm $\|\cdot\|_2$, we know that $\|U\|_2 = 1$, so $\mathrm{cond}_2(U) = 1$. Accordingly, if A is normal, then $\Gamma_2(A) = 1$. In particular, the 2-norm conditioning of self-adjoint matrices for computing eigenvalues is always equal to 1.

Theorem 10.2.1 (Bauer–Fike). *Let* A *be a diagonalizable matrix with eigenvalues* $(\lambda_1, \ldots, \lambda_n)$. *Let* $\|\cdot\|$ *be a subordinate matrix norm satisfying*

$$\|\operatorname{diag}(d_1, \ldots, d_n)\| = \max_{1 \leq i \leq n} |d_i|.$$

Then, for every matrix perturbation δA, *the eigenvalues of* $(A + \delta A)$ *are contained in the union of* n *disks of the complex plane* D_i *defined by*

$$D_i = \{ z \in \mathbb{C} \ \ |z - \lambda_i| \leq \Gamma(A) \|\delta A\| \}, \quad 1 \leq i \leq n.$$

Remark 10.2.1. The hypothesis on the subordinate matrix norm is satisfied by all norms $\|\cdot\|_p$ with $p \geq 1$; see Exercise 3.3.

Proof of Theorem 10.2.1. Let P be a nonsingular matrix satisfying

$$P^{-1}AP = \operatorname{diag}(\lambda_1, \ldots, \lambda_n).$$

Let us prove that the theorem is still valid if we substitute $\Gamma(A)$ by $\mathrm{cond}(P)$, which implies the desired result when P varies. Let λ be an eigenvalue of $(A + \delta A)$. If for some index i, we have $\lambda = \lambda_i$, then we are done. Otherwise, the matrix $\operatorname{diag}(\lambda_1 - \lambda, \ldots, \lambda_n - \lambda)$, denoted by $\operatorname{diag}(\lambda_i - \lambda)$, is nonsingular, so

$$P^{-1}(A + \delta A - \lambda I)P = \operatorname{diag}(\lambda_i - \lambda) + P^{-1}\delta A P \qquad (10.2)$$
$$= \operatorname{diag}(\lambda_i - \lambda)(I + B), \qquad (10.3)$$

where the matrix B is defined by

$$B = \operatorname{diag}(\lambda_i - \lambda)^{-1}P^{-1}\delta A P.$$

Assume that $\|B\| < 1$. Then, by Proposition 3.3.1 the matrix $(I + B)$ is invertible and equality (10.3) is thereby contradicted, since the matrix $(A + \delta A - \lambda I)$ is singular. Consequently, we infer that necessarily $\|B\| \geq 1$, and

$$1 \leq \|B\| \leq \|\operatorname{diag}(\lambda_i - \lambda)^{-1}\| \, \|P^{-1}\| \, \|\delta A\| \, \|P\|.$$

Thanks to the hypothesis on the norm of a diagonal matrix, we deduce

$$\min_{1 \leq i \leq n} |\lambda_i - \lambda| \leq \mathrm{cond}(P)\|\delta A\|,$$

which ends the proof. $\qquad\qquad\qquad\qquad\qquad\qquad\qquad\qquad\qquad\Box$

10.3 Power Method

The simplest method to compute eigenvalues and eigenvectors of a matrix is the power method. In practice, this method is confined to the computation of some (not all) extreme eigenvalues, provided that they are real and simple (their algebraic multiplicity is equal to 1). For the sake of simplicity we shall now restrict our attention to real matrices only. Let A be a real matrix of order n, not necessarily symmetric (unless otherwise mentioned). We call $(\lambda_1, \ldots, \lambda_n)$ its eigenvalues repeated with their algebraic multiplicity, and sorted in increasing order of their modulus $|\lambda_1| \leq |\lambda_2| \leq \cdots \leq |\lambda_n|$. Henceforth, we denote by $\| \cdot \|$ the Euclidean norm in $\mathbb{R}^n$.

The power method for computing the largest eigenvalue λ_n is defined by Algorithm 10.1. In the convergence test, ε is a small tolerance, for instance

---

Data: A . Output: $a \approx \lambda_n$ largest (in modulus) eigenvalue of A
$\qquad\qquad\qquad x_k \approx u_n$ eigenvector associated with λ_n

    Initialization:
        choose $x_0 \in \mathbb{R}^n$ such that $\|x_0\| = 1$.
    Iterations
        For $k \geq 1$ and $\|x_k - x_{k-1}\| \leq \varepsilon$
            $y_k = A x_{k-1}$
            $x_k = y_k / \|y_k\|$
        End
   $a = \|y_k\|$

Algorithm 10.1: Power method.

$\varepsilon = 10^{-6}$. If $\delta_k = x_k - x_{k-1}$ is small, then x_k is an approximate eigenvector of A associated with the approximate eigenvalue $\|y_k\|$ since

$$Ax_k - \|y_k\| x_k = A\delta_k.$$

Theorem 10.3.1. *Assume that A is diagonalizable, with real eigenvalues $(\lambda_1, \ldots, \lambda_n)$ associated to real eigenvectors $(e_1, \ldots, e_n)$, and that the eigenvalue with the largest modulus, denoted by λ_n, is simple and positive, i.e.,*

$$|\lambda_1| \leq \cdots \leq |\lambda_{n-1}| < \lambda_n.$$

Assume also that in the eigenvectors' basis, the initial vector reads $x_0 = \sum_{i=1}^n \beta_i e_i$ with $\beta_n \neq 0$. Then, the power method converges, that is,

$$\lim_{k \to +\infty} \|y_k\| = \lambda_n, \quad \lim_{k \to +\infty} x_k = x_\infty, \quad \text{where} \quad x_\infty = \pm e_n.$$

Moreover, the convergence speed is proportional to the ratio $|\lambda_{n-1}|/|\lambda_n|$:

$$\left| \|y_k\| - \lambda_n \right| \leq C \left(\frac{|\lambda_{n-1}|}{|\lambda_n|} \right)^k, \quad \text{and} \quad \|x_k - x_\infty\| \leq C \left(\frac{|\lambda_{n-1}|}{|\lambda_n|} \right)^k.$$

Remark 10.3.1. The assumption that the largest eigenvalue is real is crucial. However, if it is negative instead, we can apply Theorem 10.3.1 to $-A$. In practice, if we apply the above algorithm to a matrix whose largest-modulus eigenvalue is negative, it is the sequence $(-1)^k x_k$ that converges, while $-\|y_k\|$ converges to this eigenvalue. Let us recall that the simplicity condition on the eigenvalue λ_n expresses the fact that the associated generalized eigenspace is of dimension 1, or that λ_n is a simple root of the characteristic polynomial. Should there be several eigenvalues of maximal modulus, the sequences $\|y_k\|$ and x_k will not converge in general. Nevertheless, if the eigenvalue λ_n is multiple but is the only one with maximum modulus, then the sequence $\|y_k\|$ always converges to λ_n (but the sequence x_k may not converge).

Proof of Theorem 10.3.1.   The vector x_k is proportional to $A^k x_0$, and because of the form of the initial vector x_0, it is also proportional to

$$\sum_{i=1}^{n} \beta_i \lambda_i^k e_i,$$

which implies

$$x_k = \frac{\beta_n e_n + \sum_{i=1}^{n-1} \beta_i \left(\frac{\lambda_i}{\lambda_n}\right)^k e_i}{\left\| \beta_n e_n + \sum_{i=1}^{n-1} \beta_i \left(\frac{\lambda_i}{\lambda_n}\right)^k e_i \right\|}.$$

Note in passing that since $\beta_n \neq 0$, the norm $\|y_k\|$ never vanishes and there is no division by zero in Algorithm 10.1. From the assumption $|\lambda_i| < \lambda_n$ for $i \neq n$, we deduce that x_k converges to $\frac{\beta_n}{\|\beta_n e_n\|} e_n$. Likewise, we have

$$\|y_{k+1}\| = \lambda_n \frac{\left\| \beta_n e_n + \sum_{i=1}^{n-1} \beta_i \left(\frac{\lambda_i}{\lambda_n}\right)^{k+1} e_i \right\|}{\left\| \beta_n e_n + \sum_{i=1}^{n-1} \beta_i \left(\frac{\lambda_i}{\lambda_n}\right)^k e_i \right\|},$$

which therefore converges to λ_n.   □

It is possible to compute the smallest eigenvalue (in modulus) of A by applying the power method to A^{-1}: it is then called the inverse power method; see Algorithm 10.2 below. The computational cost of the inverse power method is higher, compared to the power method, because a linear system has to be solved at each iteration. If $\delta_k = x_k - x_{k-1}$ is small, then x_{k-1} is an approximate eigenvector associated to the approximate eigenvalue $1/\|y_k\|$, since

$$A x_{k-1} - \frac{x_{k-1}}{\|y_k\|} = -A \delta_k.$$

Theorem 10.3.2. *Assume that A is nonsingular and diagonalizable, with real eigenvalues $(\lambda_1, \ldots, \lambda_n)$ associated to real eigenvectors $(e_1, \ldots, e_n)$, and that*

Data: A . Output: $a \approx \lambda_1$ smallest (in modulus) eigenvalue of A
$$x_k \approx u_1 \text{ eigenvector associated with } \lambda_1$$
Initialization:
    choose $x_0 \in \mathbb{R}^n$ such that $\|x_0\| = 1$.
Iterations
    For $k \geq 1$ and $\|x_k - x_{k-1}\| \leq \varepsilon$,
        solve $Ay_k = x_{k-1}$
        $x_k = y_k / \|y_k\|$
    End
$a = 1/\|y_k\|$

Algorithm 10.2: Inverse power method.

the eigenvalue with the smallest modulus, denoted by λ_1, is simple and positive, i.e.,

$$0 < \lambda_1 < |\lambda_2| \leq \cdots \leq |\lambda_n|.$$

Assume also that in the basis of eigenvectors, the initial vector reads $x_0 = \sum_{i=1}^{n} \beta_i e_i$ with $\beta_1 \neq 0$. Then the inverse power method converges, that is,

$$\lim_{k \to +\infty} \frac{1}{\|y_k\|} = \lambda_1, \quad \lim_{k \to +\infty} x_k = x_\infty \quad \text{such that} \quad x_\infty = \pm e_1.$$

Moreover, the convergence speed is proportional to the ratio $|\lambda_1|/|\lambda_2|$:

$$\left| \|y_k\|^{-1} - \lambda_1 \right| \leq C \left(\frac{|\lambda_1|}{|\lambda_2|} \right)^k \quad \text{and} \quad \|x_k - x_\infty\| \leq C \left(\frac{|\lambda_1|}{|\lambda_2|} \right)^k.$$

Proof. The vector x_k is proportional to $A^{-k}x_0$ and thus to

$$\sum_{i=1}^{n} \beta_i \lambda_i^{-k} e_i,$$

which implies

$$x_k = \frac{\beta_1 e_1 + \sum_{i=2}^{n} \beta_i \left(\frac{\lambda_1}{\lambda_i} \right)^k e_i}{\left\| \beta_1 e_1 + \sum_{i=2}^{n} \beta_i \left(\frac{\lambda_1}{\lambda_i} \right)^k e_i \right\|}.$$

Note again that the norm $\|y_k\|$ never vanishes, so there is no division by zero in Algorithm 10.2. From $\lambda_1 < |\lambda_i|$ for $i \neq 1$, we deduce that x_k converges to $\frac{\beta_1}{\|\beta_1 e_1\|} e_1$, and that $\|y_k\|^{-1}$ converges to λ_1. □

When the matrix A is real symmetric, the statements of the previous theorems simplify a little, since A is automatically diagonalizable and the assumption on the initial vector x_0 simply requests that x_0 not be orthogonal to the eigenvector e_n (or e_1). Furthermore, the convergence toward the eigenvalue is faster in the real symmetric case. We only summarize, hereinafter, the power method; a similar result is also valid for the inverse power method.

Theorem 10.3.3. *Assume that the matrix A is real symmetric, with eigenvalues $(\lambda_1, \ldots, \lambda_n)$ associated to an orthonormal basis of eigenvectors $(e_1, \ldots, e_n)$, and that the eigenvalue with the largest modulus λ_n is simple and positive, i.e.,*

$$|\lambda_1| \leq \cdots \leq |\lambda_{n-1}| < \lambda_n.$$

Assume also that the initial vector x_0 is not orthogonal to e_n. Then the power method converges, and the sequence $\|y_k\|$ converges quadratically to the eigenvalue λ_n, that is,

$$\left| \|y_k\| - \lambda_n \right| \leq C \left(\frac{|\lambda_{n-1}|}{|\lambda_n|} \right)^{2k}.$$

Remark 10.3.2. The convergence speed of the approximate eigenvector x_k is unchanged compared to Theorem 10.3.1. However, the convergence of the approximate eigenvalue $\|y_k\|$ is faster, since $\left(\frac{|\lambda_{n-1}|}{|\lambda_n|} \right)^2 < \frac{|\lambda_{n-1}|}{|\lambda_n|}$.

Proof of Theorem 10.3.3.   In the orthonormal basis $(e_i)_{1 \leq i \leq n}$ we write the initial vector in the form $x_0 = \sum_{i=1}^{n} \beta_i e_i$. Thus, x_k reads

$$x_k = \frac{\beta_n e_n + \sum_{i=1}^{n-1} \beta_i \left(\frac{\lambda_i}{\lambda_n} \right)^k e_i}{\left(\beta_n^2 + \sum_{i=1}^{n-1} \beta_i^2 \left(\frac{\lambda_i}{\lambda_n} \right)^{2k} \right)^{1/2}}.$$

Then,

$$\|y_{k+1}\| = \|Ax_k\| = \lambda_n \frac{\left(\beta_n^2 + \sum_{i=1}^{n-1} \beta_i^2 \left(\frac{\lambda_i}{\lambda_n} \right)^{2(k+1)} \right)^{1/2}}{\left(\beta_n^2 + \sum_{i=1}^{n-1} \beta_i^2 \left(\frac{\lambda_i}{\lambda_n} \right)^{2k} \right)^{1/2}},$$

which yields the desired result.   □

Remark 10.3.3. To compute other eigenvalues of a real symmetric matrix (that is, not the smallest, nor the largest), we can use the so-called deflation method. For instance, if we are interested in the second-largest-modulus eigenvalue, namely λ_{n-1}, the deflation method amounts to the following process. We first compute the largest-modulus eigenvalue λ_n and an associated unitary eigenvector e_n such that $Ae_n = \lambda_n e_n$ with $\|e_n\| = 1$. Then, we apply again the power method to A starting with an initial vector x_0 orthogonal to e_n. This is equivalent to computing the largest eigenvalue of A, restricted to the subspace orthogonal to e_n (which is stable by A), that is, to evaluating λ_{n-1}. In practice, at each iteration the vector x_k is orthogonalized against e_n to be sure that it belongs to the subspace orthogonal to e_n. A similar idea works for computing the second-smallest-modulus eigenvalue λ_2 in the framework of the inverse power method. In practice, this technique is suitable only for the computation of some extreme eigenvalues of A (because of numerical loss of orthogonality). It is not recommended if all or only some intermediate eigenvalues have to be computed.

Let us conclude this section by mentioning that the convergence of the inverse power method can be accelerated by various means (see the next remark).

Remark 10.3.4. Let $\mu \notin \sigma(A)$ be a rough approximation of an eigenvalue λ of A. Since the eigenvectors of $(A - \mu I_n)$ are those of A and its spectrum is just shifted by μ, we can apply the inverse power method to the nonsingular matrix $(A - \mu I_n)$ in order to compute a better approximation of λ and a corresponding eigenvector. Note that the rate of convergence is improved, since if μ was a "not too bad" approximation of λ, the smallest-modulus eigenvalue of $(A - \mu I_n)$ is $(\lambda - \mu)$, which is small.

10.4 Jacobi Method

In this section we restrict our attention to the case of real symmetric matrices. The purpose of the Jacobi method is to compute all eigenvalues of a matrix A. Its principle is to iteratively multiply A by elementary rotation matrices, known as Givens matrices.

Definition 10.4.1. *Let $\theta \in \mathbb{R}$ be an angle. Let $p \neq q$ be two different integers between 1 and n. A Givens matrix is a rotation matrix $Q(p, q, \theta)$ defined by its entries $Q_{i,j}(p, q, \theta)$:*

$$
\begin{aligned}
Q_{p,p}(p, q, \theta) &= \cos \theta, \\
Q_{q,q}(p, q, \theta) &= \cos \theta, \\
Q_{p,q}(p, q, \theta) &= \sin \theta, \\
Q_{q,p}(p, q, \theta) &= -\sin \theta, \\
Q_{i,j}(p, q, \theta) &= \delta_{i,j} \text{ in all other cases.}
\end{aligned}
$$

The Givens matrix $Q(p, q, \theta)$ corresponds to a rotation of angle θ in the plane $span \{e_p, e_q\}$. It has the following shape:

$$
Q(p, q, \theta) = \begin{pmatrix}
1 & & & & & & & & & 0 \\
& \ddots & & & & & & & & \\
& & 1 & & & & & & & \\
& & & \cos \theta & & & \sin \theta & & & \\
& & & & 1 & & & & & \\
& & & & & \ddots & & & & \\
& & & & & & 1 & & & \\
& & & -\sin \theta & & & \cos \theta & & & \\
& & & & & & & 1 & & \\
& & & & & & & & \ddots & \\
0 & & & & & & & & & 1
\end{pmatrix}
\begin{matrix} \\ \\ \\ \leftarrow p \\ \\ \\ \\ \leftarrow q \\ \\ \\ \end{matrix}
$$

$$
\begin{matrix} \uparrow \\ p \end{matrix} \qquad \begin{matrix} \uparrow \\ q \end{matrix}
$$

Lemma 10.4.1. *The Givens rotation matrix* $Q(p, q, \theta)$ *is an orthogonal matrix that satisfies, for any real symmetric matrix* A,

$$\sum_{1 \le i,j \le n} b_{i,j}^2 = \sum_{1 \le i,j \le n} a_{i,j}^2,$$

where $B = Q^t(p, q, \theta) A Q(p, q, \theta)$. *Moreover, if* $a_{p,q} \ne 0$, *there exists a unique angle* $\theta \in \left]-\frac{\pi}{4}; 0\right[\cup \left]0; \frac{\pi}{4}\right]$, *defined by* $\cot(2\theta) = \frac{a_{q,q}-a_{p,p}}{2a_{p,q}}$, *such that*

$$b_{p,q} = 0 \quad and \quad \sum_{i=1}^n b_{ii}^2 = \sum_{i=1}^n a_{ii}^2 + 2a_{p,q}^2. \tag{10.4}$$

Proof. Since $Q(p, q, \theta)$ is a rotation matrix, it is orthogonal, i.e., $QQ^t = I_n$. A simple computation gives

$$\sum_{i,j=1}^n b_{i,j}^2 = \|B\|_F^2 = \operatorname{tr}(B^t B) = \operatorname{tr}(Q^t(p, q, \theta) A^t A Q(p, q, \theta)),$$

where $\| \cdot \|_F$ denotes the Frobenius norm. Since $\operatorname{tr}(MN) = \operatorname{tr}(NM)$, we deduce

$$\operatorname{tr}(Q^t(p, q, \theta) A^t A Q(p, q, \theta)) = \operatorname{tr}(AA^t),$$

and thus

$$\sum_{i,j=1}^n b_{i,j}^2 = \operatorname{tr}(A^t A) = \sum_{i,j=1}^n a_{i,j}^2.$$

We note that for any angle θ, only rows and columns of indices p and q of B change with respect to those of A. In addition, we have

$$\begin{pmatrix} b_{p,p} & b_{p,q} \\ b_{p,q} & b_{q,q} \end{pmatrix} = \begin{pmatrix} \cos\theta & -\sin\theta \\ \sin\theta & \cos\theta \end{pmatrix} \begin{pmatrix} a_{p,p} & a_{p,q} \\ a_{p,q} & a_{q,q} \end{pmatrix} \begin{pmatrix} \cos\theta & \sin\theta \\ -\sin\theta & \cos\theta \end{pmatrix}, \tag{10.5}$$

which leads to

$$\begin{cases} b_{p,p} = -2a_{p,q} \sin\theta \cos\theta + a_{p,p} \cos^2\theta + a_{q,q} \sin^2\theta, \\ b_{p,q} = a_{p,q}(\cos^2\theta - \sin^2\theta) + (a_{p,p} - a_{q,q}) \sin\theta \cos\theta, \\ b_{q,q} = 2a_{p,q} \sin\theta \cos\theta + a_{p,p} \sin^2\theta + a_{q,q} \cos^2\theta. \end{cases}$$

Consequently, $b_{p,q} = 0$ if and only if the angle θ satisfies

$$a_{p,q} \cos 2\theta + \frac{a_{p,p} - a_{q,q}}{2} \sin 2\theta = 0.$$

Such a choice of θ is always possible because $\cot 2\theta$ is onto on $\mathbb{R}$. For this precise value of θ, we deduce from (10.5) that $b_{p,q} = 0$ and

$$b_{p,p}^2 + b_{q,q}^2 = a_{p,p}^2 + a_{q,q}^2 + 2a_{p,q}^2.$$

Since all other diagonal entries of B are identical to those of A, this proves (10.4). $\qquad\square$

Definition 10.4.2. *The Jacobi method for computing the eigenvalues of a real symmetric matrix A amounts to building a sequence of matrices $A_k = (a_{i,j}^k)_{1 \le i,j \le n}$ defined by*

$$\begin{cases} A_1 = A, \\ A_{k+1} = Q^t(p_k, q_k, \theta_k) A_k Q(p_k, q_k, \theta_k), \end{cases}$$

where $Q(p_k, q_k, \theta_k)$ is a Givens rotation matrix with the following choice:

(i) (p_k, q_k) is a pair of indices such that $|a_{p_k,q_k}^k| = \max_{i \ne j} |a_{i,j}^k|$,
(ii) θ_k is the angle such that $a_{p_k,q_k}^{k+1} = 0$.

Remark 10.4.1. During the iterations, the previously zeroed off-diagonal entries do not remain so. In other words, although $a_{p_k,q_k}^{k+1} = 0$, for further iterations $l \ge k+2$ we have $a_{p_k,q_k}^l \ne 0$. Thus, we do not obtain a diagonal matrix after a finite number of iterations. However, (10.4) in Lemma 10.4.1 proves that the sum of all squared off-diagonal entries of A_k decreases as k increases.

Remark 10.4.2. From a numerical viewpoint, the angle θ_k is never computed explicitly. Indeed, trigonometric functions are computationally expensive, while only the cosine and sine of θ_k are required to compute A_{k+1}. Since we have an explicit formula for $\cot 2\theta_k$ and $2 \cot 2\theta_k = 1/\tan \theta_k - \tan \theta_k$, we deduce the value of $\tan \theta_k$ by computing the roots of a second-degree polynomial. Next, we obtain the values of $\cos \theta_k$ and $\sin \theta_k$ by computing again the roots of another second-degree polynomial.

Theorem 10.4.1. *Assume that the matrixe A is real symmetric with eigenvalues $(\lambda_1, \dots, \lambda_n)$. The sequence of matrices A_k of the Jacobi method converges, and we have*

$$\lim_{k \to +\infty} A_k = \text{diag}\,(\lambda_{\sigma(i)}),$$

where σ is a permutation of $\{1, 2, \dots, n\}$.

Theorem 10.4.2. *If, in addition, all eigenvalues of A are distinct, then the sequence of orthogonal matrices Q_k, defined by*

$$Q_k = Q(p_1, q_1, \theta_1) Q(p_2, q_2, \theta_2) \cdots Q(p_k, q_k, \theta_k),$$

converges to an orthogonal matrix whose column vectors are the eigenvectors of A arranged in the same order as the eigenvalues $\lambda_{\sigma(i)}$.

To prove these theorems, we need a technical lemma.

Lemma 10.4.2. *Consider a sequence of matrices M_k such that*

(i) $\lim_{k \to +\infty} \|M_{k+1} - M_k\| = 0$;
(ii) there exists C, independent of k, such that $\|M_k\| \le C$ for all $k \ge 1$;
(iii) the sequence M_k has a finite number of cluster points (limit points).

Then the sequence M_k converges.

Proof. Let $A_1, \ldots, A_p$ be all the cluster points of the sequence M_k. For every A_i, $1 \leq i \leq p$, there exists a subsequence $M_{k(i)}$ that converges to A_i. Let us show that for all $\varepsilon > 0$, there exists an integer $k(\varepsilon)$ such that

$$M_k \in \bigcup_{i=1}^{p} B(A_i, \varepsilon), \quad \forall k \geq k(\varepsilon),$$

where $B(A_i, \varepsilon)$ is the closed ball of center A_i and radius ε. If this were not true, there would exist $\varepsilon_0 > 0$ and an infinite subsequence k' such that

$$\|M_{k'} - A_i\| \geq \varepsilon_0, \quad 1 \leq i \leq p.$$

This new subsequence $M_{k'}$ has no A_i as a cluster point. However, it is bounded in a finite-dimensional space, so it must have at least one other cluster point, which contradicts the fact that there is no other cluster point of the sequence M_k but the A_i. Thus, for $\varepsilon = \frac{1}{4} \min_{i \neq i'} \|A_i - A_{i'}\|$, there exists $k(\varepsilon)$ such that

$$\|M_{k+1} - M_k\| \leq \varepsilon, \quad \text{and} \quad M_k \in \bigcup_{i=1}^{p} B(A_i, \varepsilon), \quad \forall k \geq k(\varepsilon).$$

In particular, there exists i_0 such that $M_{k(\varepsilon)} \in B(A_{i_0}, \varepsilon)$. Let us show that $M_k \in B(A_{i_0}, \varepsilon)$ for all $k \geq k(\varepsilon)$. Let k_1 be the largest integer greater than $k(\varepsilon)$ such that M_{k_1} belongs to $B(A_{i_0}, \varepsilon)$, but not M_{k_1+1}. In other words, k_1 satisfies

$$\|M_{k_1} - A_{i_0}\| \leq \varepsilon \text{ and } \|M_{k_1+1} - A_{i_0}\| > \varepsilon.$$

Accordingly, there exists i_1 such that $\|M_{k_1+1} - A_{i_1}\| \leq \varepsilon$. Therefore we deduce

$$\|A_{i_0} - A_{i_1}\| \leq \|M_{k_1} - A_{i_0}\| + \|M_{k_1+1} - M_{k_1}\| + \|M_{k_1+1} - A_{i_1}\|$$

$$\leq 3\varepsilon \leq \frac{3}{4} \min_{i \neq i'} \|A_i - A_{i'}\|,$$

which is not possible because $\min_{i \neq i'} \|A_i - A_{i'}\| > 0$. Therefore $k_1 = +\infty$.   □

Proof of Theorem 10.4.1.   We split the matrix $A_k = (a_{i,j}^k)_{1 \leq i,j \leq n}$ as follows:

$$A_k = D_k + B_k \quad \text{with} \quad D_k = \text{diag}(a_{i,i}^k).$$

Let $\varepsilon_k = \|B_k\|_F^2$, where $\|\cdot\|_F$ is the Frobenius norm. It satisfies

$$\varepsilon_k = \|A_k\|_F^2 - \|D_k\|_F^2.$$

From $\|A_k\|_F = \|A_{k+1}\|_F$ and $\|D_{k+1}\|_F^2 = \|D_k\|_F^2 - 2|a_{p_k q_k}^k|^2$ (by Lemma 10.4.1), we deduce that

$$\varepsilon_{k+1} = \varepsilon_k - 2|a_{p_k q_k}^k|^2.$$

Since $|a_{p_k q_k}^k| = \max_{i \neq j} |a_{i,j}^k|$, we have

$$\varepsilon_k \leq (n^2 - n)|a_{p_k q_k}^k|^2,$$

and so

$$\varepsilon_{k+1} \leq \left(1 - \frac{2}{n^2 - n}\right)\varepsilon_k.$$

As a result, the sequence ε_k tends to 0, and

$$\lim_{k \to +\infty} B_k = 0.$$

Since the off-diagonal part of A_k tends to 0, it remains to prove the convergence of its diagonal part D_k. For this purpose we check that D_k satisfies the assumptions of Lemma 10.4.2. The sequence D_k is bounded, because $\|D_k\|_F \leq \|A_k\|_F = \|A\|_F$. Let us show that D_k has a finite number of cluster points. Since D_k is a bounded sequence in a normed vector space of finite dimension, there exists a subsequence $D_{k'}$ that converges to a limit D. Hence, $A_{k'}$ converges to D. As a consequence,

$$\det (D - \lambda I) = \lim_{k' \to +\infty} \det (A_{k'} - \lambda I) = \lim_{k' \to +\infty} \det Q_{k'}^t (A - \lambda I) Q_{k'}$$
$$= \det (A - \lambda I),$$

where Q_k is the orthogonal matrix defined in Theorem 10.4.2. Since D is diagonal and has the same eigenvalues as A, D necessarily coincides with $\text{diag}\,(\lambda_{\sigma(i)})$ for some permutation σ. There exist $n!$ such permutations, therefore D_k has at most $n!$ cluster points. Finally, let us prove that $(D_{k+1} - D_k)$ converges to 0. By definition, $d_{i,i}^{k+1} = a_{i,i}^{k+1} = a_{i,i}^k = d_{i,i}^k$ if $i \neq p$ and $i \neq q$. It remains to compute $d_{p,p}^{k+1} - d_{p,p}^k = a_{p,p}^{k+1} - a_{p,p}^k$ (the case $a_{q,q}^{k+1} - a_{q,q}^k$ is symmetric):

$$a_{p,p}^{k+1} - a_{p,p}^k = -2a_{p,q}^k \sin \theta_k \cos \theta_k + (a_{q,q}^k - a_{p,p}^k) \sin^2 \theta_k.$$

By the formula $\cos 2\theta_k = 1 - 2\sin^2 \theta_k = \frac{a_{q,q}^k - a_{p,p}^k}{2a_{p,q}^k} \sin 2\theta_k$, we obtain

$$a_{q,q}^k - a_{p,p}^k = a_{p,q}^k \frac{1 - 2\sin^2 \theta_k}{\sin \theta_k \cos \theta_k},$$

from which we deduce that

$$a_{p,p}^{k+1} - a_{p,p}^k = -2a_{p,q}^k \sin \theta_k \cos \theta_k + a_{p,q}^k \frac{\sin \theta_k}{\cos \theta_k}(1 - 2\sin^2 \theta_k) = -a_{p,q}^k \tan \theta_k.$$

Since $\theta_k \in \left[-\frac{\pi}{4}; \frac{\pi}{4}\right]$, we have $|\tan \theta_k| \leq 1$. Thus

$$\|D^{k+1} - D^k\|_F^2 \leq 2|a_{p,q}^k|^2 \leq \|B_k\|_F^2 = \varepsilon_k,$$

which tends to 0 as k goes to infinity. Then we conclude by applying Lemma 10.4.2. □

Proof of Theorem 10.4.2.   We apply Lemma 10.4.2 to the sequence Q_k, which is bounded because $\|Q_k\|_F = 1$. Let us show that Q_k has a finite number of cluster points. Since it is bounded, there exists a subsequence $Q_{k'}$ that converges to a limit Q, and we have

$$\lim_{k' \to +\infty} A_{k'+1} = \operatorname{diag}(\lambda_{\sigma(i)}) = \lim_{k' \to +\infty} Q_{k'}^t A Q_{k'} = Q^t A Q,$$

which implies that the columns of Q are equal to $(\pm f_{\sigma(i)})$, where f_i is the normalized eigenvector corresponding to the eigenvalue λ_i of A (these eigenvectors are unique, up to a change of sign, because all eigenvalues are simple by assumption). Since there is a finite number of permutations and possible changes of sign, the cluster points, like Q, are finite in number. Finally, let us prove that $(Q_{k+1} - Q_k)$ converges to 0. We write

$$Q_{k+1} - Q_k = (Q(p_{k+1}, q_{k+1}, \theta_{k+1}) - I)Q_k.$$

By definition, $\tan 2\theta_k = 2a_{p_k q_k}^k / (a_{q_k q_k}^k - a_{p_k p_k}^k)$, and A_k converges to $\operatorname{diag}(\lambda_{\sigma(i)})$ with distinct eigenvalues. Consequently, for k large enough,

$$|a_{q_k q_k}^k - a_{p_k p_k}^k| \geq \frac{1}{2} \min_{i \neq j} |\lambda_i - \lambda_j| > 0.$$

Since $\lim_{k \to +\infty} a_{p_k q_k}^k = 0$, we deduce that θ_k tends to 0, and so $(Q_{k+1} - Q_k)$ converges to 0. Applying Lemma 10.4.2 finishes the proof. □

10.5 Givens–Householder Method

Once again we restrict ourselves to the case of real symmetric matrices. The main idea of the Givens–Householder method is to reduce a matrix to its tridiagonal form, the eigenvalues of which are easier to compute.

Definition 10.5.1. *The Givens–Householder method is decomposed into two successive steps:*

1. *By the Householder method, a symmetric matrix A is reduced to a tridiagonal matrix, that is, an orthogonal matrix Q is built such that $Q^t A Q$ is tridiagonal (this first step is executed in a finite number of operations).*
2. *The eigenvalues of this tridiagonal matrix are computed by a bisection (or dichotomy) method proposed by Givens (this second step is an iterative method).*

We begin with the first step, namely, the Householder method.

Proposition 10.5.1. *Let A be a real symmetric matrix. There exist $(n-2)$ orthogonal matrices H_k such that*

$$T = (H_1 H_2 \cdots H_{n-2})^t A (H_1 H_2 \cdots H_{n-2})$$

is tridiagonal. Note that A and T have the same eigenvalues.

Proof.   Starting with A, we build a sequence of matrices $(A_k)_{1 \le k \le n-1}$ such that $A_1 = A$, and $A_{k+1} = H_k^t A_k H_k$, where H_k is an orthogonal matrix chosen in such a way that A_k has the following block structure:

$$A_k = \begin{pmatrix} T_k & E_k^t \\ E_k & M_k \end{pmatrix}.$$

In the above, T_k is a tridiagonal square matrix of size k, M_k is a square matrix of size $n-k$, and E_k is a rectangular matrix with $(n-k)$ rows and k columns whose last column only, denoted by $a_k \in \mathbb{R}^{n-k}$, is nonzero:

$$T_k = \begin{pmatrix} \times & \times & & & \\ \times & \ddots & \ddots & & \\ & \ddots & \ddots & \times \\ & & \times & \times \end{pmatrix} \quad \text{and} \quad E_k = \begin{pmatrix} 0 \ldots 0 & a_{k,1} \\ \vdots & \vdots & a_{k,2} \\ \vdots & \vdots & \vdots \\ 0 \ldots 0 & a_{k,n-k} \end{pmatrix}.$$

Thus, it is clear that A_{n-1} is a tridiagonal matrix. We note that A is indeed of this form for $k = 1$. Let H_k be the matrix defined by

$$H_k = \begin{pmatrix} I_k & 0 \\ 0 & \tilde{H}_k \end{pmatrix},$$

where I_k is the identity matrix of order k, and $\tilde{H}_k$ is the Householder matrix (see Lemma 7.3.1) of order $n-k$ defined by

$$\tilde{H}_k = I_{n_k} - 2\frac{v_k v_k^t}{\|v_k\|^2}, \quad \text{with } v_k = a_k + \|a_k\| e_1, \tag{10.6}$$

where e_1 is the first vector of the canonical basis of $\mathbb{R}^{n-k}$. It satisfies $\tilde{H}_k a_k = -\|a_k\| e_1$. We observe that H_k is orthogonal and $H_k^t = H_k$. The definition (10.6) of the Householder matrix is valid only if a_k is not parallel to e_1; otherwise, the kth column of A_k is already of the desired form, so we take $H_k = I_{n-k}$. We compute $A_{k+1} = H_k^t A_k H_k$:

$$A_{k+1} = \begin{pmatrix} T_k & (\tilde{H}_k E_k)^t \\ \tilde{H}_k E_k & \tilde{H}_k M_k \tilde{H}_k \end{pmatrix},$$

where

$$\tilde{H}_k E_k = \begin{pmatrix} 0 \dots 0 & -\|a_k\| \\ \vdots & \vdots & 0 \\ \vdots & \vdots & \vdots \\ 0 \dots 0 & 0 \end{pmatrix}.$$

Accordingly, A_{k+1} takes the desired form:

$$A_{k+1} = \begin{pmatrix} T_{k+1} & E_{k+1}^t \\ E_{k+1} & M_{k+1} \end{pmatrix},$$

where T_{k+1} is a square tridiagonal matrix of size $k+1$, M_{k+1} is a square matrix of size $n - k - 1$, and E_{k+1} is a rectangular matrix of size $(n - k - 1, k + 1)$ whose only nonzero column is the last one. $\square$

Now we proceed to the second step of the algorithm, that is, the Givens bisection method.

Lemma 10.5.1. *Consider a real tridiagonal symmetric matrix*

$$A = \begin{pmatrix} b_1 & c_1 & & 0 \\ c_1 & \ddots & \ddots & \\ & \ddots & \ddots & c_{n-1} \\ 0 & & c_{n-1} & b_n \end{pmatrix}.$$

If there exists an index i such that $c_i = 0$, then

$$\det (A - \lambda I_n) = \det (A_i - \lambda I_i) \det (A_{n-i} - \lambda I_{n-i})$$

with

$$A_i = \begin{pmatrix} b_1 & c_1 & & 0 \\ c_1 & \ddots & \ddots & \\ & \ddots & \ddots & c_{i-1} \\ 0 & & c_{i-1} & b_i \end{pmatrix} \quad and \quad A_{n-i} = \begin{pmatrix} b_{i+1} & c_{i+1} & & 0 \\ c_{i+1} & \ddots & \ddots & \\ & \ddots & \ddots & c_{n-1} \\ 0 & & c_{n-1} & b_n \end{pmatrix}.$$

The proof of Lemma 10.5.1 is easy and left to the reader. It allows us to restrict our attention in the sequel to tridiagonal matrices with $c_i \neq 0$ for all $i \in \{1, \dots, n - 1\}$.

Proposition 10.5.2. *For $1 \le i \le n$, let A_i be the matrix of size i defined by*

$$A_i = \begin{pmatrix} b_1 & c_1 & & 0 \\ c_1 & \ddots & \ddots & \\ & \ddots & \ddots & c_{i-1} \\ 0 & & c_{i-1} & b_i \end{pmatrix},$$

with $c_i \neq 0$, and let $p_i(\lambda) = \det(A_i - \lambda I_i)$ be its characteristic polynomial. The sequence p_i satisfies the induction formula

$$p_i(\lambda) = (b_i - \lambda)p_{i-1}(\lambda) - c_{i-1}^2 p_{i-2}(\lambda), \qquad \forall i \geq 2,$$

with

$$p_0(\lambda) = 1 \quad and \quad p_1(\lambda) = b_1 - \lambda.$$

Moreover, for all $i \geq 1$, the polynomial p_i has the following properties:

1. $\lim_{\lambda \to -\infty} p_i(\lambda) = +\infty$;
2. if $p_i(\lambda_0) = 0$, then $p_{i-1}(\lambda_0)p_{i+1}(\lambda_0) < 0$;
3. the polynomial p_i has i real distinct roots that strictly separate the $(i+1)$ roots of p_{i+1}.

Remark 10.5.1. A consequence of Proposition 10.5.2 is that when all entries c_i are nonzero, the tridiagonal matrix A has only simple eigenvalues (in other words, they are all distinct).

Proof of Proposition 10.5.2.   Expanding $\det(A_i - \lambda I_i)$ with respect to the last row, we get the desired induction formula. The first property is obvious by definition of the characteristic polynomial. To prove the second property, we notice that if $p_i(\lambda_0) = 0$, then the induction formula implies that $p_{i+1}(\lambda_0) = -c_i^2 p_{i-1}(\lambda_0)$. Since $c_i \neq 0$, we deduce

$$p_{i-1}(\lambda_0)p_{i+1}(\lambda_0) \leq 0.$$

This inequality is actually strict; otherwise, if either $p_{i-1}(\lambda_0) = 0$ or $p_{i+1}(\lambda_0) = 0$, then the induction formula would imply that $p_k(\lambda_0) = 0$ for all $0 \leq k \leq i+1$, which is not possible because $p_0(\lambda_0) = 1$.

Concerning the third property, we first remark that $p_i(\lambda)$ has i real roots, denoted by $\lambda_1^i \leq \cdots \leq \lambda_i^i$, because A_i is real symmetric. Let us show by induction that these i roots of p_i are distinct and separated by those of p_{i-1}. First of all, this property is satisfied for $i = 2$. Indeed,

$$p_2(\lambda) = (b_2 - \lambda)(b_1 - \lambda) - c_1^2$$

has two roots $(\lambda_1^2, \lambda_2^2)$ that strictly bound the only root $\lambda_1^1 = b_1$ of $p_1(\lambda)$, i.e., $\lambda_1^2 < \lambda_1^1 < \lambda_2^2$. Assuming that $p_i(\lambda)$ has i real distinct roots separated by those of p_{i-1}, we now study the $i+1$ real roots of p_{i+1}. We define a polynomial q_i of degree $2i$ by

$$q_i(\lambda) = p_{i-1}(\lambda)p_{i+1}(\lambda).$$

We already know $i-1$ roots of q_i (those of p_{i-1}), and we also know that the i roots of p_i are such that $q_i(\lambda_k^i) < 0$. In other words,

$$q_i(\lambda_k^{i-1}) = 0, \text{ for } 1 \leq k \leq i-1, \quad q_i(\lambda_k^i) < 0, \text{ for } 1 \leq k \leq i,$$

with (see Figure 10.1)

$$\lambda_1^i < \lambda_1^{i-1} < \lambda_2^i < \cdots < \lambda_{i-1}^{i-1} < \lambda_i^i.$$

Between λ_k^i and λ_{k+1}^i, either q_i vanishes at another point $\gamma_k \neq \lambda_k^{i-1}$, in which case we have found another root of q_i, hence of p_{i+1}, or q_i vanishes only at λ_k^{i-1}, meaning it is at least a double root, since its derivative q_i' has to vanish at λ_k^{i-1} too. On the other hand, λ_k^{i-1} is a simple root of p_{i-1}, so λ_k^{i-1} is also a root of p_{i+1}. Because of the induction relation, this would prove that λ_k^{i-1} is a root for all polynomials p_j with $0 \leq j \leq i+1$, which is not possible because $p_0 = 1$ has no roots. As a consequence, we have proved that between each pair $\lambda_k^i, \lambda_{k+1}^i$ there exists another root $\gamma_k \neq \lambda_k^{i-1}$ of the polynomial q_i, thus of p_{i+1}. Overall, we have just found $(i-1)$ distinct roots of p_{i+1} that bound those of p_i. Moreover, $q_i(\lambda_1^i) < 0$ and $q_i(\lambda_i^i) < 0$, whereas

$$\lim_{\lambda \to \pm\infty} q_i(\lambda) = +\infty.$$

So we deduce the existence of two more distinct roots of q_i, hence of p_{i+1} (for a total number of $i+1$ roots), that intertwine those of p_i.   □

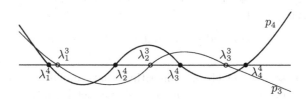

Fig. 10.1. Example of p_i polynomials in Proposition 10.5.2.

Proposition 10.5.3. *For all $\mu \in \mathbb{R}$, we define*

$$\operatorname{sgn} p_i(\mu) = \begin{cases} \text{sign of } p_i(\mu) & \text{if } p_i(\mu) \neq 0, \\ \text{sign of } p_{i-1}(\mu) & \text{if } p_i(\mu) = 0. \end{cases}$$

Let $N(i, \mu)$ be the number of sign changes between consecutive elements of the set $E(i, \mu) = \{+1, \operatorname{sgn} p_1(\mu), \operatorname{sgn} p_2(\mu), \ldots, \operatorname{sgn} p_i(\mu)\}$. Then $N(i, \mu)$ is the number of roots of p_i that are strictly less than μ.

Proof. First, we note that $\operatorname{sgn} p_i(\mu)$ is defined without ambiguity, since if $p_i(\mu) = 0$, then $p_{i-1}(\mu) \neq 0$ because of the second point in Proposition 10.5.2. We proceed by induction on i. For $i = 1$, we check the claim

$$\mu \leq b_1 \Rightarrow E(1, \mu) = \{+1, +1\} \Rightarrow N(1, \mu) = 0,$$

and

$$\mu > b_1 \Rightarrow E(1,\mu) = \{+1,-1\} \Rightarrow N(1,\mu) = 1.$$

We assume the claim to be true up till the ith order. Let $(\lambda_k^{i+1})_{1 \le k \le i+1}$ be the roots of p_{i+1} and $(\lambda_k^i)_{1 \le k \le i}$ be those of p_i, sorted in increasing order. By the induction assumption, we have

$$\lambda_1^i < \cdots < \lambda_{N(i,\mu)}^i < \mu \le \lambda_{N(i,\mu)+1}^i < \cdots < \lambda_i^i.$$

In addition,

$$\lambda_{N(i,\mu)}^i < \lambda_{N(i,\mu)+1}^{i+1} < \lambda_{N(i,\mu)+1}^i,$$

by virtue of the third point in Proposition 10.5.2. Therefore, there are three possible cases.

- ✗ First case. If $\lambda_{N(i,\mu)}^i < \mu \le \lambda_{N(i,\mu)+1}^{i+1}$, then $\operatorname{sgn} p_{i+1}(\mu) = \operatorname{sgn} p_i(\mu)$. Thus $N(i+1,\mu) = N(i,\mu)$.
- ✗ Second case. If $\lambda_{N(i,\mu)+1}^{i+1} < \mu < \lambda_{N(i,\mu)+1}^i$, then $\operatorname{sgn} p_{i+1}(\mu) = -\operatorname{sgn} p_i(\mu)$. Thus $N(i+1,\mu) = N(i,\mu) + 1$.
- ✗ Third case. If $\mu = \lambda_{N(i,\mu)+1}^i$, then $\operatorname{sgn} p_i(\mu) = \operatorname{sgn} p_{i-1}(\mu) = -\operatorname{sgn} p_{i+1}(\mu)$, according to the second point in Proposition 10.5.2. Therefore $N(i+1,\mu) = N(i,\mu) + 1$.

In all cases, $N(i+1,\mu)$ is indeed the number of roots of p_{i+1} that are strictly smaller than μ. □

We now describe the Givens algorithm that enables us to numerically compute some or all eigenvalues of the real symmetric matrix A. We denote by $\lambda_1 \le \cdots \le \lambda_n$ the eigenvalues of A arranged in increasing order.

Givens algorithm. In order to compute the ith eigenvalue λ_i of A, we consider an interval $[a_0, b_0]$ that we are sure λ_i belongs to (for instance, $-a_0 = b_0 = \|A\|_\infty$). Then we compute the number $N(n, \frac{a_0+b_0}{2})$ defined in Proposition 10.5.3 (the values of the sequence $p_j(\frac{a_0+b_0}{2})$, for $1 \le j \le n$, are computed by the induction formula of Proposition 10.5.2). If $N(n, \frac{a_0+b_0}{2}) \ge i$, then we conclude that λ_i belongs to the interval $[a_0, \frac{a_0+b_0}{2}[$. If, on the contrary, $N(n, \frac{a_0+b_0}{2}) < i$, then λ_i belongs to the other interval $[\frac{a_0+b_0}{2}, b_0]$. In both cases, we have divided by two the initial interval that contains λ_i. By dichotomy, that is, by repeating this procedure of dividing the interval containing λ_i, we approximate the exact value λ_i with the desired accuracy.

Remark 10.5.2. The Givens–Householder method allows us to compute one (or several) eigenvalue(s) of any rank i without having to compute all the eigenvalues (as is the case for the Jacobi method) or all the eigenvalues between the ith and the first or the last (as is the case for the power method with deflation).

10.6 QR Method

The QR method is the most-used algorithm to compute all the eigenvalues of a matrix. We restrict ourselves to the case of nonsingular real matrices whose eigenvalues have distinct moduli

$$0 < |\lambda_1| < \cdots < |\lambda_{n-1}| < |\lambda_n|. \tag{10.7}$$

The analysis of the general case is beyond the scope of this course (for more details we refer the reader to [3], [7], [13], and [15]). A real matrix A satisfying (10.7) is necessarily invertible, and diagonalizable, with distinct real eigenvalues, but we do not assume that it is symmetric.

Definition 10.6.1. *Let A be a real matrix satisfying (10.7). The QR method for computing its eigenvalues consists in building the sequence of matrices $(A_k)_{k \geq 1}$ with $A_1 = A$ and*

$$A_{k+1} = R_k Q_k,$$

where $Q_k R_k = A_k$ is the QR factorization of the nonsingular matrix A_k (see Section 6.4).

Recall that the matrices Q_k are orthogonal and the matrices R_k are upper triangular. We first prove a simple technical lemma.

Lemma 10.6.1. *1. The matrices A_k are all similar,*

$$A_{k+1} = Q_k^t A_k Q_k, \tag{10.8}$$
$$A_k = [Q^{(k)}]^t A Q^{(k)}, \tag{10.9}$$

with $Q^{(k)} = Q_1 \cdots Q_k$.
2. For all $k \geq 1$,

$$A Q^{(k)} = Q^{(k+1)} R_{k+1}. \tag{10.10}$$

3. The QR factorization of A^k, the kth power of A (not to be confused with A_k), is

$$A^k = Q^{(k)} R^{(k)} \tag{10.11}$$

with $R^{(k)} = R_k \cdots R_1$.

Proof.

1. By definition, $A_{k+1} = R_k Q_k = Q_k^t (Q_k R_k) Q_k = Q_k^t (A_k) Q_k$, and by induction,

$$A_{k+1} = Q_k^t (A_k) Q_k = Q_k^t Q_{k-1}^t (A_{k-1}) Q_{k-1} Q_k = \cdots = [Q^{(k+1)}]^t A Q^{(k+1)}.$$

2. We compute

$$Q^{(k)} R_k = Q_1 \cdots Q_{k-2} Q_{k-1} (Q_k R_k) = Q_1 \cdots Q_{k-2} Q_{k-1} (A_k)$$
$$= Q_1 \cdots Q_{k-2} Q_{k-1} (R_{k-1} Q_{k-1}) = Q_1 \cdots Q_{k-2} (A_{k-1}) Q_{k-1}$$
$$= Q_1 \cdots Q_{k-2} (R_{k-2} Q_{k-2}) Q_{k-1} = Q_1 \cdots (Q_{k-2} R_{k-2}) Q_{k-2} Q_{k-1}$$
$$= Q_1 \cdots (A_{k-2}) Q_{k-2} Q_{k-1} = A Q^{(k-1)}.$$

3. Using the previous identity, we get

$$Q^{(k)}R^{(k)} = (Q^{(k)}R_k)R^{(k-1)} = AQ^{(k-1)}R^{(k-1)},$$

and by induction, (10.11).

$\square$

Remark 10.6.1. Remarkably, the algorithm of Definition 10.6.1 features, as a special case, the power method and the inverse power method. Let $\alpha_1^{(k)} = (R_k)_{1,1}$, $\alpha_n^{(k)} = (R_k)_{n,n}$, and $u_1^{(k)}$ (resp. $u_n^{(k)}$) denote the first (resp. last) column of $Q^{(k)}$, i.e.,

$$Q^{(k)} = \left[u_1^{(k)} | \cdots | u_n^{(k)} \right].$$

Comparing the first column on both sides of identity (10.10), we get

$$Au_1^{(k)} = \alpha_1^{(k+1)} u_1^{(k+1)}. \tag{10.12}$$

Since $\|u_1^{(k+1)}\|_2 = 1$, we have $\alpha_1^{(k+1)} = \|Au_1^{(k)}\|_2$, and formula (10.12) is just the definition of the power method analyzed in Section 10.3. Under the hypothesis of Theorem 10.3.1 we know that the sequence α_1^k converges to $|\lambda_n|$ and $u_1^{(k)}$ converges to an eigenvector associated to λ_n.

On the other hand, (10.10) reads $A = Q^{(k+1)}R_{k+1}[Q^{(k)}]^t$, which implies by inversion and transposition (recall that by assumption, A and R_k are non-singular matrices)

$$A^{-t}Q^{(k)} = Q^{(k+1)}R_{k+1}^{-t}.$$

Comparing the last columns on both sides of this identity, we get

$$A^{-t}u_n^{(k)} = \frac{1}{\alpha_n^{(k+1)}} u_n^{(k+1)}. \tag{10.13}$$

Since $\|u_n^{(k+1)}\|_2 = 1$, we deduce $\alpha_n^{(k+1)} = \|A^{-t}u_1^{(k)}\|_2^{-1}$, and formula (10.13) is just the inverse power method. Under the hypothesis of Theorem 10.3.2 we know that the sequence $\alpha_n^{(k)}$ converges to $|\lambda_1|$ and $u_n^{(k)}$ converges to an eigenvector associated to λ_1.

The convergence of the QR algorithm is proved in the following theorem with some additional assumptions on top of (10.7).

Theorem 10.6.1. *Let A a be a real matrix satisfying (10.7). Assume further that P^{-1} admits an LU factorization, where P is the matrix of eigenvectors of A, i.e., $A = P \operatorname{diag}(\lambda_n, \ldots, \lambda_1)P^{-1}$. Then the sequence $(A_k)_{k \geq 1}$, generated by the QR method, converges to an upper triangular matrix whose diagonal entries are the eigenvalues of A.*

Proof. By assumption (10.7), the matrix A is diagonalizable, i.e., $A = PDP^{-1}$ with $D = \text{diag}(\lambda_n, \ldots, \lambda_1)$. Using the LU factorization of P^{-1} and the QR factorization of P we obtain

$$A^k = P(D^k)P^{-1} = (QR)(D^k)LU = (QR)(D^k LD^{-k})(D^k U).$$

The matrix $D^k LD^{-k}$ is lower triangular with entries

$$(D^k LD^{-k})_{i,j} = \begin{cases} 0 & \text{for } i < j, \\ 1 & \text{for } i = j, \\ \left(\frac{\lambda_{n-i+1}}{\lambda_{n-j+1}}\right)^k L_{i,j} & \text{for } i > j. \end{cases}$$

Hypothesis (10.7) implies that $\lim_{k \to +\infty} (D^k LD^{-k})_{i,j} = 0$ for $i > j$. Hence $D^k LD^{-k}$ tends to the identity matrix I_n as k goes to infinity, and we write

$$D^k LD^{-k} = I_n + E_k, \quad \text{with} \quad \lim_{k \to +\infty} E_k = 0_n$$

as well as

$$A^k = (QR)(I_n + E_k)(D^k U) = Q(I_n + RE_k R^{-1})RD^k U.$$

For k large enough, the matrix $I_n + RE_k R^{-1}$ is nonsingular and admits a QR factorization: $I_n + RE_k R^{-1} = \widetilde{Q}_k \widetilde{R}_k$. Since $\widetilde{R}_k RD^k U)$ is upper triangular as a product of upper triangular matrices (see Lemma 2.2.5), it yields a QR factorization of A^k:

$$A^k = (Q\widetilde{Q}_k)(\widetilde{R}_k RD^k U).$$

From (10.11) we already know another QR factorization of A^k. Thus, by uniqueness of the QR factorization of A^k (see Remark 6.4.2) there exists a diagonal matrix $\widetilde{D}_k$ such that

$$Q^{(k)} = Q\widetilde{Q}_k \widetilde{D}_k \quad \text{with} \quad |(\widetilde{D}_k)_{i,i}| = 1.$$

Plugging this into the expression for A_{k+1} given in (10.9), we obtain

$$A_{k+1} = [Q\widetilde{Q}_k \widetilde{D}_k]^* A[Q\widetilde{Q}_k \widetilde{D}_k] = \widetilde{D}_k^* [\widetilde{Q}_k^* Q^* AQ\widetilde{Q}_k] \widetilde{D}_k. \tag{10.14}$$

The entries of A_{k+1} are

$$(A_{k+1})_{i,j} = (\widetilde{D}_k^*)_{i,i} [\widetilde{Q}_k^* Q^* AQ\widetilde{Q}_k]_{i,j} (\widetilde{D}_k)_{j,j} = \pm[\widetilde{Q}_k^* Q^* AQ\widetilde{Q}_k]_{i,j}. \tag{10.15}$$

In particular, the diagonal entries of A_{k+1} are

$$(A_{k+1})_{i,i} = (\widetilde{D}_k^*)_{i,i} [\widetilde{Q}_k^* Q^* AQ\widetilde{Q}_k]_{i,i} (\widetilde{D}_k)_{i,i} = [\widetilde{Q}_k^* Q^* AQ\widetilde{Q}_k]_{i,i}.$$

Now we make two observations.

- Firstly, $Q^* AQ = Q^*(PDP^{-1})Q = Q^*(QRD(QR)^{-1})Q = RDR^{-1}$.

- Secondly, the sequence $(\widetilde{Q}_k)_k$ is bounded ($\|\widetilde{Q}_k\|_2 = 1$); it converges to a matrix $\widetilde{Q}$ (consider first a subsequence, then the entire sequence). As a consequence, the upper triangular matrices $\widetilde{R}_k = \widetilde{Q}_k^*(I_n + RE_kR^{-1})$ converge to the unitary matrix $\widetilde{Q}^*$, which is also triangular (as a limit of triangular matrices), hence it is diagonal (see the proof of Theorem 2.5.1), and its diagonal entries are ± 1.

Using these two remarks, we can pass to the limit in (10.15):

$$\lim_{k \to +\infty} (A_{k+1})_{i,j} = \pm(RDR^{-1})_{i,j},$$

which is equal to zero for $i > j$, and

$$\lim_{k \to +\infty} (A_{k+1})_{i,i} = (RDR^{-1})_{i,i} = D_{i,i} = \lambda_{n+1-i},$$

because of Lemma 2.2.5 on the product and inverse of upper triangular matrices (R is indeed upper triangular). In other words, the limit of A_k is upper triangular with the eigenvalues of A on its diagonal. $\square$

Remark 10.6.2. For a symmetric matrix A, the matrices A_k are symmetric too, by virtue of (10.9). From Theorem 10.6.1 we thus deduce that the limit of A_k is a diagonal matrix D. Since the sequence $(Q^{(k)})_k$ is bounded ($\|Q^{(k)}\|_2 = 1$), up to a subsequence it converges to a unitary matrix $Q^{(\infty)}$. Passing to the limit in (10.9) yields $D = [Q^{(\infty)}]^t A Q^{(\infty)}$, which implies that $Q^{(\infty)}$ is a matrix of eigenvectors of A.

Let us now study some practical aspects of the QR method. The computation of A_{k+1} from A_k requires the computation of a QR factorization and a matrix multiplication. The QR factorization should be computed by the Householder algorithm rather than by the Gram–Schmidt orthonormalization process (see Remark 7.3.3). A priori, the QR factorization of a matrix of order n requires on the order of $\mathcal{O}(n^3)$ operations. Such a complexity can be drastically reduced by first reducing the original matrix A to its upper Hessenberg form. An upper Hessenberg matrix is an "almost" upper triangular matrix as explained in the following definition.

Definition 10.6.2. *An $n \times n$ matrix T is called an upper Hessenberg matrix if $T_{i,j} = 0$, for all integers (i,j) such that $i > j + 1$.*

We admit the two following results (see [3], [7], [13], and [15] if necessary). The first one explains how to compute the upper Hessenberg form of a matrix (it is very similar to Proposition 10.5.1).

Proposition 10.6.1. *For any $n \times n$ matrix A, there exists a unitary matrix P, the product of $n - 2$ Householder matrices $H_1, \ldots, H_{n-2}$, such that the matrix P^*AP is an upper Hessenberg matrix.*

Note that the Hessenberg transformation $A \rightarrow P^* AP$ preserves the spectrum and the symmetry: if A is symmetric, so is $P^* AP$. Hence the latter matrix is tridiagonal (in which case Proposition 10.6.1 reduces to Proposition 10.5.1) and the Givens algorithm is an efficient way to compute its spectrum. The cost of the Hessenberg transformation is $\mathcal{O}(n^3)$, but it is done only once before starting the iterations of the QR method.

The second admitted result states that the structure of Hessenberg matrices is preserved during the iterations of the QR method.

Proposition 10.6.2. *If A is an upper Hessenberg matrix, then the matrices $(A_k)_{k \geq 1}$, defined by the QR method, are upper Hessenberg matrices too.*

The main practical interest of the upper Hessenberg form is that the QR factorization now requires only on the order of $\mathcal{O}(n^2)$ operations, instead of $\mathcal{O}(n^3)$ for a full matrix.

In order to implement the QR method, we have to define a termination criterion. A simple one consists in checking that the entries $(A_k)_{i,i-1}$ are very small (recall that the sequence A_k is upper Hessenberg). We proceed as follows: if $(A_k)_{n,n-1}$ is small, then $(A_k)_{n,n}$ is considered as a good approximation of an eigenvalue of A (more precisely, of the smallest one λ_1 according to Remark 10.6.1) and the algorithm continues with the $(n-1) \times (n-1)$ matrix obtained from A_k by removing the last row and column n. This is the so-called deflation algorithm. Actually, it can be proved that

$$(A_k)_{n,n-1} = \mathcal{O}\left(\left|\frac{\lambda_1}{\lambda_2}\right|^k\right), \tag{10.16}$$

which defines the speed of convergence of the QR method.

According to formula (10.16) for the speed of convergence, it is possible (and highly desirable) to speed up the QR algorithm by applying it to the "shifted" matrix $A - \sigma I_n$ instead of A. The spectrum of A is simply recovered by adding σ to the eigenvalues of $A - \sigma I_n$. Of course σ is chosen in such a way that

$$\mathcal{O}\left(\left|\frac{\lambda_1 - \sigma}{\lambda_2 - \sigma}\right|^k\right) \ll \mathcal{O}\left(\left|\frac{\lambda_1}{\lambda_2}\right|^k\right),$$

i.e., σ is a good approximation of the simple real eigenvalue λ_1. More precisely, the QR algorithm is modified as follows;

1. compute the QR factorization of the matrix $A_k - \sigma_k I_n$:

$$A_k - \sigma_k I_n = Q_k R_k;$$

2. define $A_{k+1} = R_k Q_k + \sigma_k I_n = Q_k^* A_k Q_k$.

Here the value of the shift is updated at each iteration k. A simple and efficient choice is

Data: matrix A and integer N (maximal number of iterations)
Output: v a vector containing the eigenvalues of A
      Initialization:
               $\varepsilon =, N =$, define the error tolerance and total number of iterations
               $m = n, k = 1$
               $a = hess(A)$. Compute the Hessenberg reduced form of A
      Iterations
             While $k = 1, \dots, N$ and $m > 1$
                    If $\|a_{m,m-1}\| < \varepsilon$
                        $v(m) = a_{m,m}$
                        $a(:, m) = [\,]$ delete column m of a
                        $a(m, :) = [\,]$ delete row m of a
                        $m = m - 1$
                  End
                  compute (Q, R) the QR factorization of a
                  $a = RQ$
                  $k = k + 1$
            End
            $v(1) = a(1, 1)$

Algorithm 10.3: QR method.

$$\sigma_k = (A_k)_{n,n}.$$

The case of real matrices with complex eigenvalues is more complicated (see, e.g., [13] and [15]). Note that these matrices do not fulfill the requirement (10.7) on the separation of the spectrum, since their complex eigenvalues come in pairs with equal modulus.

10.7 Lanczos Method

The Lanczos method computes the eigenvalues of a real symmetric matrix by using the notion of Krylov space, already introduced for the conjugate gradient method.

In the sequel, we denote by A a real symmetric matrix of order n, $r_0 \in \mathbb{R}^n$ some given nonzero vector, and K_k the Krylov space spanned by $\{r_0, Ar_0, \dots, A^k r_0\}$. Recall that there exists an integer $k_0 \leq n - 1$, called the Krylov critical dimension, which is characterized by $\dim K_k = k + 1$ if $k \leq k_0$, while $K_k = K_{k_0}$ if $k > k_0$.

The Lanczos algorithm builds a sequence of vectors $(v_j)_{1 \leq j \leq k_0 + 1}$ by the following induction formula:

$$v_0 = 0, \quad v_1 = \frac{r_0}{\|r_0\|}, \tag{10.17}$$

and for $2 \leq j \leq k_0 + 1$,

$$v_j = \frac{\hat{v}_j}{\|\hat{v}_j\|} \text{ with } \hat{v}_j = Av_{j-1} - \langle Av_{j-1}, v_{j-1}\rangle v_{j-1} - \|\hat{v}_{j-1}\|v_{j-2}. \quad (10.18)$$

We introduce some notation: for all integer $k \leq k_0 + 1$, we define an $n \times k$ matrix V_k whose columns are the vectors $(v_1, \ldots, v_k)$, as well as a tridiagonal symmetric matrix T_k of order k whose entries are

$$(T_k)_{i,i} = \langle Av_i, v_i \rangle, \ (T_k)_{i,i+1} = (T_k)_{i+1,i} = \|\hat{v}_{i+1}\|, \ (T_k)_{i,j} = 0 \text{ if } |i - j| \geq 2.$$

The Lanczos induction (10.17)–(10.18) satisfies remarkable properties.

Lemma 10.7.1. *The sequence* $(v_j)_{1 \leq j \leq k_0+1}$ *is well defined by (10.18), since* $\|\hat{v}_j\| \neq 0$ *for all* $1 \leq j \leq k_0 + 1$, *whereas* $\hat{v}_{k_0+2} = 0$. *For* $1 \leq k \leq k_0 + 1$, *the family* $(v_1, \ldots, v_{k+1})$ *coincides with the orthonormal basis of* K_k *built by application of the Gram–Schmidt procedure to the family* $(r_0, Ar_0, \ldots, A^k r_0)$. *Furthermore, for* $1 \leq k \leq k_0 + 1$, *we have*

$$AV_k = V_k T_k + \hat{v}_{k+1} e_k^t, \quad (10.19)$$

where e_k *is the kth vector of the canonical basis of* $\mathbb{R}^k$,

$$V_k^t AV_k = T_k, \quad and \quad V_k^t V_k = I_k, \quad (10.20)$$

where I_k *is the identity matrix of order* k.

Remark 10.7.1. Beware that the square matrices A and T_k are of different sizes, and the matrix V_k is rectangular, so it is not unitary (except if $k = n$).

Proof. Let us forget for the moment the definition (10.17)–(10.18) of the sequence (v_j) and substitute it with the new definition (which we will show to be equivalent to (10.18)) $v_0 = 0$, $v_1 = \frac{r_0}{\|r_0\|}$, and for $j \geq 2$,

$$v_j = \frac{\hat{v}_j}{\|\hat{v}_j\|}, \quad \text{where} \quad \hat{v}_j = Av_{j-1} - \sum_{i=1}^{j-1} \langle Av_{j-1}, v_i \rangle v_i. \quad (10.21)$$

Of course, (10.21) is meaningless unless $\|\hat{v}_j\| \neq 0$. If $\|\hat{v}_j\| = 0$, we shall say that the algorithm stops at index j. By definition, v_j is orthogonal to all v_i for $1 \leq i \leq j - 1$. By induction we easily check that $v_j \in K_{j-1}$. The sequence of Krylov spaces K_j is strictly increasing for $j \leq k_0 + 1$, that is, $K_{j-1} \subset K_j$ and $\dim K_{j-1} = j - 1 < \dim K_j = j$. Therefore, as long as the algorithm has not stopped (i.e., $\|\hat{v}_j\| \neq 0$), the vectors $(v_1, \ldots, v_j)$ form an orthonormal basis of K_{j-1}. Consequently, v_j, being orthogonal to $(v_1, \ldots, v_{j-1})$, is also orthogonal to K_{j-2}. Hence, according to the uniqueness result of Lemma 9.5.1, we have just proved that the family $(v_1, \ldots, v_j)$, defined by (10.21), coincides with the orthonormal basis of K_{j-1} built by the Gram–Schmidt procedure applied to the family $(r_0, Ar_0, \ldots, A^{j-1} r_0)$. In particular, this proves that the only possibility for the algorithm to stop is that the family $(r_0, Ar_0, \ldots, A^{j-1} r_0)$ is linearly dependent, i.e., that $j - 1$ is larger than the Krylov critical dimension k_0. Hence, $\|\hat{v}_j\| \neq 0$ as long as $j \leq k_0 + 1$, and $\hat{v}_{k_0+2} = 0$.

Now, let us show that definitions (10.18) and (10.21) of the sequence (v_j) are identical. Since A is symmetric, we have

$$\langle Av_{j-1}, v_i \rangle = \langle v_{j-1}, Av_i \rangle = \langle v_{j-1}, \hat{v}_{i+1} \rangle + \sum_{k=1}^{i} \langle Av_i, v_k \rangle \langle v_{j-1}, v_k \rangle.$$

Thanks to the orthogonality properties of (v_k), we deduce that $\langle Av_{j-1}, v_i \rangle = 0$ if $1 \leq i \leq j - 3$ and $\langle Av_{j-1}, v_{j-2} \rangle = \|\hat{v}_{j-1}\|$. Therefore, definitions (10.18) and (10.21) coincide.

Finally, the matrix equality (10.19), taken column by column, is nothing else than (10.18) rewritten, for $2 \leq j \leq k$,

$$Av_{j-1} = \|\hat{v}_j\| v_j + \langle Av_{j-1}, v_{j-1} \rangle v_{j-1} + \|\hat{v}_{j-1}\| v_{j-2},$$

and

$$Av_k = \hat{v}_{k+1} + \langle Av_k, v_k \rangle v_k + \|\hat{v}_k\| v_{k-1}.$$

The property that $V_k^t V_k = I_k$ is due to the orthonormality properties of $(v_1, \ldots, v_k)$, whereas the relation $V_k^t A V_k = T_k$ is obtained by multiplying (10.19) on the left by V_k^t and taking into account that $V_k^t \hat{v}_{k+1} = 0$.   □

Remark 10.7.2. The computational cost of the Lanczos algorithm (10.18) is obviously much less than that of the Gram–Schmidt algorithm applied to the family $(r_0, Ar_0, \ldots, A^k r_0)$, which yields the same result. The point is that the sum in (10.18) contains only two terms, while the corresponding sum in the Gram–Schmidt algorithm contains all previous terms (see Theorem 2.1.1). When the Krylov critical dimension is maximal, i.e., $k_0 = n - 1$, relation (10.19) or (10.20) for $k = k_0 + 1$ shows that the matrices A and T_{k_0+1} are similar (since V_{k_0+1} is a square nonsingular matrix if $k_0 = n - 1$). In other words, the Lanczos algorithm can be seen as a tridiagonal reduction method, like the Householder algorithm of Section 10.5. Nevertheless, the Lanczos algorithm is not used in practice as a tridiagonalization method. In effect, for n large, the rounding errors partially destroy the orthogonality of the last vectors v_j with respect to the first ones (a shortcoming already observed for the Gram–Schmidt algorithm).

We now compare the eigenvalues and eigenvectors of A and T_{k_0+1}. Let us recall right away that these matrices are usually not of the same size (except when $k_0 + 1 = n$). Consider $\lambda_1 < \lambda_2 < \cdots < \lambda_m$, the distinct eigenvalues of A (with $1 \leq m \leq n$), and $P_1, \ldots, P_m$, the orthogonal projection matrices on the corresponding eigensubspaces of A. We recall that

$$A = \sum_{i=1}^{m} \lambda_i P_i, \quad I = \sum_{i=1}^{m} P_i, \quad \text{and} \quad P_i P_j = 0 \text{ if } i \neq j. \tag{10.22}$$

Lemma 10.7.2. *The eigenvalues of T_{k_0+1} are simple and are eigenvalues of A too. Conversely, if r_0 satisfies $P_i r_0 \neq 0$ for all $1 \leq i \leq m$, then $k_0 + 1 = m$, and all eigenvalues of A are also eigenvalues of T_{k_0+1}.*

Remark 10.7.3. When $P_i r_0 \neq 0$ for all i, A and T_{k_0+1} have exactly the same eigenvalues, albeit with possibly different multiplicities. The condition imposed on r_0 for the converse part of this lemma is indeed necessary. Indeed, if r_0 is an eigenvector of A, then $k_0 = 0$ and the matrix T_{k_0+1} has a unique eigenvalue, the one associated with r_0.

Proof of Lemma 10.7.2.   Let λ and $y \in \mathbb{R}^{k_0+1}$ be an eigenvalue and eigenvector of T_{k_0+1}, i.e., $T_{k_0+1} y = \lambda y$. Since $\hat{v}_{k_0+2} = 0$, (10.19) becomes for $k = k_0+1$

$$AV_{k_0+1} = V_{k_0+1} T_{k_0+1}.$$

Multiplication by the vector y yields $A\left(V_{k_0+1} y\right) = \lambda\left(V_{k_0+1} y\right)$. The vector $V_{k_0+1} y$ is nonzero, since $y \neq 0$ and the columns of V_{k_0+1} are linearly independent. Consequently, $V_{k_0+1} y$ is an eigenvector of A associated with the eigenvalue λ, which is therefore an eigenvalue of A too.

Conversely, we introduce a vector subspace E_m of $\mathbb{R}^n$, spanned by the vectors $(P_1 r_0, \ldots, P_m r_0)$, which are assumed to be nonzero, $P_i r_0 \neq 0$, for all $1 \leq i \leq m$. These vectors are linearly independent, since projections on P_i are mutually orthogonal. Accordingly, the dimension of E_m is exactly m. Let us show under this assumption that $m = k_0 + 1$. By (10.22) we have

$$A^k r_0 = \sum_{i=1}^{m} \lambda_i^k P_i r_0,$$

i.e., $A^k r_0 \in E_m$. Hence the Krylov spaces satisfy $K_k \subset E_m$ for all $k \geq 0$. In particular, this implies that $\dim K_{k_0} = k_0 + 1 \leq m$. On the other hand, in the basis $(P_1 r_0, \ldots, P_m r_0)$ of E_m, the coordinates of $A^k r_0$ are $(\lambda_1^k, \ldots, \lambda_m^k)$. Writing the coordinates of the family $(r_0, A r_0, \ldots, A^{m-1} r_0)$ in the basis $(P_1 r_0, \ldots, P_m r_0)$ yields a matrix representation M:

$$M = \begin{pmatrix} 1 & \lambda_1 & \lambda_1^2 & \ldots & \lambda_1^{m-1} \\ \vdots & & & & \vdots \\ 1 & \lambda_m & \lambda_m^2 & \ldots & \lambda_m^{m-1} \end{pmatrix},$$

which is just a Vandermonde matrix of order m. It is nonsingular, since

$$\det(M) = \prod_{i=1}^{m-1} \prod_{j>i} (\lambda_j - \lambda_i)$$

and the eigenvalues λ_i are distinct. As a result, the family $(r_0, A r_0, \ldots, A^{m-1} r_0)$ is linearly independent, which implies that $\dim K_{m-1} = m$, and accordingly

$m - 1 \leq k_0$. We thus conclude that $m = k_0 + 1$ and $E_m = K_{k_0}$. On the other hand, multiplying (10.22) by $P_i r_0$ yields

$$A(P_i r_0) = \lambda_i (P_i r_0).$$

Since $P_i r_0$ is nonzero, it is indeed an eigenvector of A associated with the eigenvalue λ_i. Because $E_m = K_{k_0}$ and the columns of V_{k_0+1} are a basis of K_{k_0}, we deduce the existence of a nonzero vector $y_i \in \mathbb{R}^m$ such that

$$P_i r_0 = V_{k_0+1} y_i.$$

We multiply the first equality of (10.20) by y_i to obtain

$$
\begin{aligned}
T_{k_0+1} y_i = V_{k_0+1}^t A V_{k_0+1} y_i &= V_{k_0+1}^t A P_i r_0 \\
&= \lambda_i V_{k_0+1}^t P_i r_0 = \lambda_i V_{k_0+1}^t V_{k_0+1} y_i = \lambda_i y_i,
\end{aligned}
$$

which proves that y_i is an eigenvector of T_{k_0+1} for the eigenvalue λ_i.   □

In view of Lemma 10.7.2 one may believe that the Lanczos algorithm has to be carried up to the maximal iteration number $k_0 + 1$, before computing the eigenvalues of T_{k_0+1} in order to deduce the eigenvalues of A. Such a practice makes the Lanczos method comparable to the Givens–Householder algorithm, since usually k_0 is of order n. Moreover, if n or k_0 is large, the Lanczos algorithm will be numerically unstable because of orthogonality losses for the vectors v_j (caused by unavoidable rounding errors; see Remark 10.7.2). However, as for the conjugate gradient method, it is not necessary to perform as many iterations as $k_0 + 1$ to obtain good approximate results. Indeed, in numerical practice one usually stops the algorithm after k iterations (with k much smaller than k_0 or n) and computes the eigenvalues of T_k, which turn out to be good approximations of those of A, according to the following lemma.

Lemma 10.7.3. *Fix the iteration number $1 \leq k \leq k_0 + 1$. For any eigenvalue λ of T_k, there exists an eigenvalue λ_i of A satisfying*

$$|\lambda - \lambda_i| \leq \|\hat{v}_{k+1}\|. \tag{10.23}$$

Furthermore, if $y \in \mathbb{R}^k$ is a nonzero eigenvector of T_k associated to the eigenvalue λ, then there exists an eigenvalue λ_i of A such that

$$|\lambda - \lambda_i| \leq \|\hat{v}_{k+1}\| \frac{|\langle e_k, y \rangle|}{\|y\|}, \tag{10.24}$$

where e_k is the kth vector of the canonical basis of $\mathbb{R}^k$.

Remark 10.7.4. The first conclusion (10.23) of Lemma 10.7.3 states that the eigenvalues of T_k are good approximations of some eigenvalues of A, provided that $\|\hat{v}_{k+1}\|$ is small. The second conclusion (10.24) is the most valuable one:

if the last entry of y is small, then λ is a good approximation of an eigenvalue of A even if $\|\hat{v}_{k+1}\|$ is not small. In practice, we test the magnitude of the last entry of an eigenvector of T_k to know whether the corresponding eigenvalue is a good approximation of an eigenvalue of A. For details of the numerical implementation of this method we refer to more advanced monographs, e.g., [2]. The Lanczos method is very efficient for large n, since it gives good results for a total number of iterations k much smaller than n, and it is at the root of many fruitful generalizations.

Proof of Lemma 10.7.3.   Consider a nonzero eigenvector $y \in \mathbb{R}^k$ such that $T_k y = \lambda y$. Multiplying (10.19) by y, we get

$$AV_k y = V_k T_k y + \langle e_k, y \rangle \hat{v}_{k+1},$$

from which we deduce

$$A(V_k y) - \lambda(V_k y) = \langle e_k, y \rangle \hat{v}_{k+1}. \tag{10.25}$$

Then, we expand $V_k y$ in the eigenvectors basis of A, $V_k y = \sum_{i=1}^m P_i(V_k y)$. Taking the scalar product of (10.25) with $V_k y$, and using relations (10.22) leads to

$$\sum_{i=1}^m (\lambda_i - \lambda) |P_i(V_k y)|^2 = \langle e_k, y \rangle \langle \hat{v}_{k+1}, V_k y \rangle. \tag{10.26}$$

Applying the Cauchy–Schwarz inequality to the right-hand side of (10.26) yields

$$\min_{1 \le i \le m} |\lambda_i - \lambda| \|V_k y\|^2 \le \|y\| \|\hat{v}_{k+1}\| \|V_k y\|. \tag{10.27}$$

Since the columns of V_k are orthonormal, we have $\|V_k y\| = \|y\|$, and simplifying (10.27), we obtain the first result (10.23). This conclusion can be improved if we do not apply Cauchy–Schwarz to the term $\langle e_k, y \rangle$ in (10.26). In this case, we directly obtain (10.24).   □

10.8 Exercises

10.1. What is the spectrum of the following bidiagonal matrix (called a Wilkinson matrix)?

$$W(n) = \begin{pmatrix} n & n & 0 & \cdots & 0 \\ 0 & n-1 & n & 0 & \vdots \\ \vdots & & \ddots & \ddots & \vdots \\ \vdots & & & 2 & n \\ 0 & \cdots & \cdots & 0 & 1 \end{pmatrix} \in \mathcal{M}_n(\mathbb{R}).$$

For $n = 20$, compare (using `Matlab`) the spectrum of $W(n)$ with that of the matrix $\tilde{W}(n)$ obtained from $W(n)$ by modifying the single entry in row n and column 1, $\tilde{W}(n,1) = 10^{-10}$. Comment.

10.2. Define

$$A = \begin{pmatrix} 7.94 & 5.61 & 4.29 \\ 5.61 & -3.28 & -2.97 \\ 4.29 & -2.97 & -2.62 \end{pmatrix}, \quad T = \begin{pmatrix} 1 & 1 & 1 \\ 1 & 0 & 1 \\ 0 & 0 & 1 \end{pmatrix}, \quad \text{and} \quad b = \begin{pmatrix} 1 \\ 1 \\ 1 \end{pmatrix}.$$

1. Compute the spectrum of A and the solution x of $Ax = b$.
2. Define $A_1 = A + 0.01T$. Compute the spectrum of A_1 and the solution x_1 of $A_1 x = b$.
3. Explain the results.

10.3. The goal of this exercise is to study the notion of left eigenvectors for a matrix $A \in \mathcal{M}_n(\mathbb{C})$.

1. Prove the equivalence $\lambda \in \sigma(A) \Longleftrightarrow \bar{\lambda} \in \sigma(A^*)$.
2. Let $\lambda \in \sigma(A)$ be an eigenvalue of A. Show that there exists (at least) one nonzero vector $y \in \mathbb{C}^n$ such that $y^* A = \lambda y^*$. Such a vector is called a left eigenvector of A associated with the eigenvalue λ.
3. Prove that the left eigenvectors of a Hermitian matrix are eigenvectors.
4. Let λ and μ be two distinct eigenvalues of A. Show that all left eigenvectors associated with λ are orthogonal to all left eigenvectors associated with μ.
5. Use `Matlab` to compute the left eigenvectors of the matrix

$$A = \begin{pmatrix} 1 & -2 & -2 & -2 \\ -4 & 0 & -2 & -4 \\ 1 & 2 & 4 & 2 \\ 3 & 1 & 1 & 5 \end{pmatrix}.$$

10.4. Let $A \in \mathcal{M}_n(\mathbb{C})$ be a diagonalizable matrix, i.e., there exists an invertible matrix P such that $P^{-1}AP = \mathrm{diag}(\lambda_1, \ldots, \lambda_n)$, where the λ_i are the eigenvalues of A. Denote by x_i an eigenvector associated to λ_i, and y_i a left eigenvector associated to λ_i (see Exercise 10.3).

1. Define $Q^* = P^{-1}$. Prove that the columns of Q are left eigenvectors of the matrix A.
2. Deduce that if the eigenvalue λ_i is simple, then $y_i^* x_i \neq 0$.

10.5. We study the conditioning of an eigenvalue.

1. Define a matrix

$$A = \begin{pmatrix} -97 & 100 & 98 \\ 1 & 2 & -1 \\ -100 & 100 & 101 \end{pmatrix}.$$

Determine the spectrum of A. We define random perturbations of the matrix A by the command `B=A+0.01*rand(3,3)`. Determine the spectra of B for several realizations. Which eigenvalues of A have been modified the most, and which ones have been the least?

2. The goal is now to understand why some eigenvalues are more sensitive than others to variations of the matrix entries. Let λ_0 be a simple eigenvalue of a diagonalizable matrix A_0. We denote by x_0 (respectively, y_0) an eigenvector (respectively, a left eigenvector) associated to λ_0 (we assume that $\|x_0\|_2 = \|y_0\|_2 = 1$). For ε small, we define the matrix $A_\varepsilon = A_0 + \varepsilon E$, where E is some given matrix such that $\|E\|_2 = 1$. We define λ_ε to be an eigenvalue of A_ε that is the closest to λ_0 (it is well defined for ε small enough), and let x_ε be an eigenvector of A_ε associated to λ_ε.

   (a) Show that the mapping $\varepsilon \mapsto \lambda_\varepsilon$ is continuous and that $\lim_{\varepsilon \to 0} \lambda_\varepsilon = \lambda_0$. We admit that $\lim_{\varepsilon \to 0} x_\varepsilon = x_0$.

   (b) We denote by $\delta\lambda = \lambda_\varepsilon - \lambda_0$ the variation of λ, and by $\delta x = x_\varepsilon - x_0$ the variation of x. Prove that $A_0(\delta x) + \varepsilon E x_\varepsilon = (\delta\lambda)x_\varepsilon + \lambda_0(\delta x)$.

   (c) Deduce that the mapping $\lambda : \varepsilon \mapsto \lambda_\varepsilon$ is differentiable and that

$$\lambda'(0) = \frac{y_0^* E x_0}{y_0^* x_0}.$$

   (d) Explain why $\mathrm{cond}(A, \lambda_0) = 1/|y_0^* x_0|$ is called "conditioning of the eigenvalue λ_0."

3. Compute the conditioning of each eigenvalue of the matrix A. Explain the results observed in question 1.

10.6 (∗). Write a function $[\mathbf{l}, \mathbf{u}] = \mathtt{PowerD(A)}$ that computes by the power method (Algorithm 10.1) the approximations l and u of the largest eigenvalue (in modulus) and corresponding eigenvector of A. Initialize the algorithm with a unit vector x_0 with equal entries. Test the program with each of the following symmetric matrices and comment the obtained results.

$$A = \begin{pmatrix} 2 & 2 & 1 & 0 \\ 2 & 0 & 0 & 0 \\ 1 & 0 & 0 & 2 \\ 0 & 0 & 2 & -2 \end{pmatrix}, \quad B = \begin{pmatrix} 15 & 0 & 9 & 0 \\ 0 & 24 & 0 & 0 \\ 9 & 0 & 15 & 0 \\ 0 & 0 & 0 & 16 \end{pmatrix}, \quad C = \begin{pmatrix} 1 & 2 & -3 & 4 \\ 2 & 1 & 4 & -3 \\ -3 & 4 & 1 & 2 \\ 4 & -3 & 2 & 1 \end{pmatrix}.$$

10.7. Compute by the power method and the deflation technique (see Remark 10.3.3) the two largest (in modulus) eigenvalues of the following matrix

$$A = \begin{pmatrix} 2 & 1 & 2 & 2 & 2 \\ 1 & 2 & 1 & 2 & 2 \\ 2 & 1 & 2 & 1 & 2 \\ 2 & 2 & 1 & 2 & 1 \\ 2 & 2 & 2 & 1 & 2 \end{pmatrix}.$$

Modify the function $\mathtt{PowerD}$ into a function $\mathbf{l} = \mathtt{PowerDef(A,u)}$, where the iteration vector x_k is orthogonal to a given vector u.

10.8. Program a function $[\mathbf{l}, \mathbf{u}] = \mathtt{PowerI(A)}$ that implements the inverse power method (Algorithm 10.2) to compute approximations, l and u, of the smallest eigenvalue (in modulus) of A and its associated eigenvector. Test the program on the matrices defined by $\mathtt{A = Laplacian1dD(n)}$, for different values of n.

10.9. Program a function T=HouseholderTri(A) that implements the House-holder algorithm to reduce a symmetric matrix A to a tridiagonal matrix T (following the proof of Proposition 10.5.1). Check with various examples that

- the matrix T is tridiagonal (write a function for this purpose);
- the spectra of the matrices A and T are the same.

10.10. Program a function Givens(T,i) that computes the eigenvalue λ_i (labeled in increasing order) of a symmetric tridiagonal matrix T using the Givens method. Test the program by computing the eigenvalues of matrices obtained by the instructions

u=rand(n,1);v=rand(n-1,1);T=diag(u)+diag(v,1)+diag(v,-1)

for various values of n.

10.11. By gathering the routines of the two previous exercises, program the Givens–Householder algorithm to compute the eigenvalues of a real symmetric matrix. Run this program to obtain the eigenvalues of the matrix A defined in Exercise 10.7.

10.12. The goal is to numerically compute the eigenvalues of the matrix

$$A = \begin{pmatrix} 5 & 3 & 4 & 3 & 3 \\ 3 & 5 & 2 & 3 & 3 \\ 4 & 2 & 4 & 2 & 4 \\ 3 & 3 & 2 & 5 & 3 \\ 3 & 3 & 4 & 3 & 5 \end{pmatrix}$$

by the Lanczos method. The notation is that of Lemma 10.7.1.

1. Let $r_0 = (1, 2, 3, 4, 5)^t$. Compute the sequence of vectors v_j. Deduce k_0, the critical dimension of the Krylov space associated with A and r_0.
2. For $k = k_0 + 1$, compute the matrix T_k, as well as its eigenvalues and eigenvectors. Which eigenvalues of A do you find in the spectrum of T?
3. Answer the same questions for $r_0 = (1, 1, 1, 1, 1)^t$, then for $r_0 = (1, -1, 0, 1, -1)^t$.

10.13. Let A=Laplacian2dD(n) be the matrix of the discretized Laplacian on an $n \times n$ square mesh of the unit square; see Exercise 6.7. The notation is that of the previous exercise. We fix $n = 7$.

1. Starting with $r_0 = (1, \ldots, 1)^t \in \mathbb{R}^{n^2}$, compute the matrix T_n. Plot the eigenvalues of A (by a symbol) and those of T_n (by another symbol) on the same graph. Are the eigenvalues of T_n good approximations of some eigenvalues of A?
2. Same questions starting with the vector $r_0 = (1, 2, \ldots, n^2)^t$. Comment on your observations.

11

Solutions and Programs

11.1 Exercises of Chapter 2

Solution of Exercise 2.1

1. u is the ith vector of the canonical basis of $\mathbb{R}^n$.
2. The scalar product $\langle v, u \rangle = u^t v$ is computed by u'*v (or by the Matlab function dot).

   ```
   >> n=10;u=rand(n,1);v=rand(n,1);
   >> w=v-u'*v*u/(u'*u);
   >> fprintf(' the scalar product <w,u> = %f \n',u'*w)
      the scalar product <w,u> = 0.000000
   ```

   The two vectors are orthogonal. We recognize the Gram–Schmidt ortho-normalization process applied to (u, v).
3. The scalar product $\langle Cx, x \rangle = x^t C x$ is computed by the Matlab instruction x'*C*x.

   (a) The matrix C being antisymmetric, i.e., $C^t = -C$, we have

   $$\langle Cx, x \rangle = \langle x, C^t x \rangle = -\langle x, Cx \rangle = -\langle Cx, x \rangle \implies \langle Cx, x \rangle = 0.$$

   (b) Since $A = B + C$, we have $\langle Ax, x \rangle = \langle Bx, x \rangle$. The matrix B is called the symmetric part of A, and the matrix C its antisymmetric (or skew-symmetric) part.

Solution of Exercise 2.2 The following functions are possible (nonunique) solutions.

1. ```
   function A=SymmetricMat(n)
   A=rand(n,n);
   A=A+A';
   ```
2. ```
   function A=NonsingularMat(n)
   A=rand(n,n);
   A=A+norm(A,'inf');   % make the diagonal of A dominant
   ```

```
3. function A=LowNonsingularMat(n)
   A=tril(rand(n,n));
   A=A+norm(A,'inf')*eye(size(A));% make the diagonal of A dominant
4. function A=UpNonsingularMat(n)
   A=triu(rand(n,n));
   A=A+norm(A,'inf')*eye(size(A));% make the diagonal of A dominant
5. function A=ChanceMat(m,n,p)
   %nargin  = number of input arguments of the function
   switch  nargin                    % arguments of the function
      case 1
         m=n;A=rand(m,n);            % The entries of A
      case 2                         % take values
         A=rand(m,n);               % between 0 and 1
      else
         A=rand(m,n);A=p*(2*A-1);  % affine transformation
   end;
```

We may call this function with 1, 2 or 3 arguments.

```
6. function A=BinChanceMat(m,n)
   A=rand(m,n);
   A(A<0.5)=0;A(A>=0.5)=1;
7. function H = HilbertMat(n,m)
   % this function returns a Hilbert matrix
   %nargin  = number of input arguments of the function
   if nargin==1, m=n; end;
   H=zeros(n,m);
   for i=1:n
      for j=1:m
         H(i,j)=1/(i+j-1);
      end;
   end;
```

For square Hilbert matrices, use the Matlab function hilb.

Solution of Exercise 2.7 Matrix of fixed rank.

```
function A= MatRank(m,n,r)
s=min(m,n);S=max(m,n);
if r>min(m,n)
   fprintf('The rank cannot be greater than %i ',s)
   error('Error in function MatRank.')
else
   A=NonsingularMat(s);
   if m>=n
      A=[A; ones(S-s,s)];             % m × n matrix of rank r
      for k=r+1:s,A(:,k)=rand()*A(:,r);end;
   else
```

```
        A=[A   ones(s,S-s)];                % m × n matrix of rank r
        for k=r+1:s,A(k,:)=rand()*A(r,:);end;
    end;
end
```

Solution of Exercise 2.10 The Gram–Schmidt algorithm.

1. To determine the vector u_p, we compute $(p-1)$ scalar products of vectors of size m and $(p-1)$ multiplications of a scalar by a vector (we do not take into account the test if), that is, $2m(p-1)$ operations. The total number of operations is then $\sum_{p=1}^{n} 2m(p-1) \approx mn^2$.

2.
```
function B = GramSchmidt(A)
pres=1.e-12;
[m,n]=size(A);
B=zeros(m,n);
for i=1:n
    s=zeros(m,1);
    for k=1:i-1
        s=s+(A(:,i)'*B(:,k))*B(:,k);
    end;
    s=A(:,i)-s;
    if norm(s) > pres
        B(:,i) =s/norm(s);
    end;
end;
```

For the matrix defined by

```
>> n=5;u=1:n; u=u'; c2=cos(2*u); c=cos(u); s=sin(u);
>> A=[u c2 ones(n,1) rand()*c.*c exp(u) s.*s];
```

we get

```
>> U=GramSchmidt(A)
U =
       0.1348   -0.2471    0.7456        0    0.4173        0
       0.2697   -0.3683    0.4490        0   -0.4791        0
       0.4045    0.8167    0.2561        0    0.1772        0
       0.5394    0.0831   -0.0891        0   -0.5860        0
       0.6742   -0.3598   -0.4111        0    0.4707        0
```

(a) We have
```
>> U'*U
ans =
       1.0000    0.0000    0.0000        0    0.0000        0
       0.0000    1.0000   -0.0000        0   -0.0000        0
       0.0000   -0.0000    1.0000        0   -0.0000        0
            0         0         0        0         0        0
       0.0000   -0.0000   -0.0000        0    1.0000        0
```

$$0 \qquad 0 \qquad 0 \quad 0 \qquad 0 \quad 0$$

Explanation: the (nonzero) columns of U being orthonormal, it is clear that $(U^tU)_{i,j} = \langle u_i, u_j \rangle = \delta_{i,j}$ if u_i and u_j are nonzero. If one of the vectors is zero, then $(U^tU)_{i,j} = 0$. The situation is different for UU^t:

```
>> U*U'
ans =
```

0.8092	0.2622	0.1175	-0.2587	0.0697
0.2622	0.6395	-0.1616	0.3556	-0.0958
0.1175	-0.1616	0.9276	0.1594	-0.0430
-0.2587	0.3556	0.1594	0.6491	0.0945
0.0697	-0.0958	-0.0430	0.0945	0.9745

Note: If all columns of U are orthonormal, then U is nonsingular and $U^t = U^{-1}$. In this case, we also obtain $UU^t = I$ (U is an orthogonal matrix).

(b) The algorithm is stable in the sense that applied to an orthogonal matrix U, it provides the same matrix, up to a small remainder term.

```
>> V=GramSchmidt(U);norm(V-U)
ans =
    7.8400e-16
```

3. The following function answers the question:

```
function B = GramSchmidt1(A)
pres=1.e-12;
colnn=0;   % points out the current nonzero column
[m,n]=size(A);
B=zeros(m,n);
for i=1:n
    s=zeros(m,1);
    for k=1:i-1   % we can describe k=1:colnn
        s=s+(A(:,i)'*B(:,k))*B(:,k);
    end;
    s=A(:,i)-s;
    if norm(s) > pres
        colnn=colnn+1;
        B(:,colnn) =s/norm(s);
    end;
end;
>> W=GramSchmidt1(A)
W =
```

0.1348	-0.2471	0.7456	0.4173	0	0
0.2697	-0.3683	0.4490	-0.4791	0	0
0.4045	0.8167	0.2561	0.1772	0	0
0.5394	0.0831	-0.0891	-0.5860	0	0
0.6742	-0.3598	-0.4111	0.4707	0	0

Solution of Exercise 2.11 The modified Gram–Schmidt algorithm.

1.
$$\text{For } p = 1 \nearrow n$$
$$\text{If } \|a_p\| \neq 0 \text{ then}$$
$$u_p = a_p / \|a_p\|$$
$$\text{Otherwise}$$
$$u_p = 0$$
$$\text{End}$$
$$\text{For } k = p + 1 \nearrow n$$
$$a_k = a_k - \langle a_k, u_p \rangle u_p$$
$$\text{End}$$

$$\text{End}$$

Modified Gram-Schmidt algorithm

Note that both algorithms have the same algorithmic complexity. Here is a Matlab programming of the modified Gram–Schmidt algorithm.

```
function B = MGramSchmidt(A)
pres=1.e-12;
[m,n]=size(A);
B=zeros(m,n);
for i=1:n
    s=A(:,i);
    if norm(s) > pres
        B(:,i) =s/norm(s);
        for k=i+1:n
            A(:,k)=A(:,k)-(A(:,k)'*B(:,i))*B(:,i);
        end;
    else
        error('linearly dependent vectors')
    end;
end;
```

2. Comparison of the two algorithms. For a randomly chosen matrix (rand), we do not observe a noteworthy difference between the two algorithms. For a Hilbert matrix, we get

```
>> n=10;A=hilb(n);
>> U=GramSchmidt1(A);V=MGramSchmidt(A);I=eye(n,n);
>> norm(U'*U-I), norm(V'*V-I)
ans =
    2.9969
ans =
    2.3033e-04
```

We remark on this example that the modified Gram–Schmidt algorithm is more accurate than the standard Gram–Schmidt algorithm.

3. We improve the Gram–Schmidt algorithm by iterating it several times:

```
>> U=GramSchmidt1(A);norm(U'*U-I)
ans =
     2.9969
>> U=GramSchmidt1(U);norm(U'*U-I)
ans =
     1.3377e-07
>> U=GramSchmidt1(U);norm(U'*U-I)
ans =
     3.4155e-16
```

Solution of Exercise 2.12 Warning: the running times obtained depend on what computer is used. However, their ordering on each computer is always the same.

```
>> t1,t2
t1 =
     1.6910
t2 =
     0.7932
```

Explanation: it is clearly more efficient to declare beforehand the (large-sized) matrices by initializing them. This prevents Matlab from resizing the matrix at each creation of a new entry $A_{i,j}$.

```
>> t2,t3
t2 =
     0.7932
t3 =
     0.7887
```

Explanation: since matrices being stored column by column, it is slightly better to "span" a matrix in this order and not row by row, since the entries $A_{i,j}$ and $A_{i+1,j}$ are stored in neighboring "slots" in the computer's memory (for $1 \leq i \leq n-1$)). In this way, we reduce the access time to these slots.

```
>> t3,t4
t3 =
     0.7887
t4 =
     0.0097
```

Explanation: a tremendous speedup is obtained by using a "vectorizable" definition of the matrices. Efficient Matlab programs will always be written this way.

Solution of Exercise 2.18 The following script produces Figure 11.1.

```
>> A=[10 2; 2 4];
>> t=0:0.1:2*pi;x=cos(t)';y=sin(t)';
>> x=[x;x(1)];y=[y;y(1)];
```

```
>> for i=1:length(x)
>>   z(i)=[x(i),y(i)]*A*[x(i);y(i)];
>> end;
>>plot3(x,y,z,x,y,zeros(size(x))),'MarkerSize',10,'LineWidth',3);
     box
```

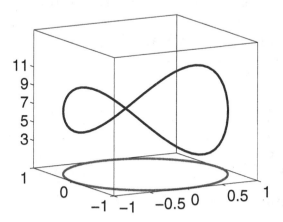

Fig. 11.1. Computation of the Rayleigh quotient on the unit circle.

We deduce from Figure 11.1 that the minimal value is close to 3.4 and the maximal value is close to 10.6. The spectrum of A is

```
>> eig(A)
ans =
    3.3944
   10.6056
```

The eigenvalues are therefore very close to the minimal and maximal values of $x^t A x$ on the unit circle. Since A is symmetric, we know from Theorem 2.6.1 and Remark 2.6.1 that the minimum of the Rayleigh quotient is equal to the minimal eigenvalue of A and its maximum is equal to the maximal eigenvalue of A.

Solution of Exercise 2.20 Let us comment on the main instructions in the definition of A:

```
A=PdSMat(n)                    % symmetric matrix
[P,D]=eig(A);D=abs(D);         % diagonalization: PDP⁻¹ = A
D=D+norm(D)*eye(size(D))       % the eigenvalues of D are > 0
A=P*D*inv(P)                   % positive definite symmetric matrix
```

with the comments rendered as: *symmetric matrix*; *diagonalization: $PDP^{-1} = A$*; *the eigenvalues of D are > 0*; *positive definite symmetric matrix*.

1. We always have $\det(A) > 0$, because the determinant of a matrix is equal to the product of its eigenvalues, which are positive here.

2. We easily check that all the main subdeterminants are strictly positive. For $x \in \mathbb{C}^k$ a nonzero vector, we denote by $\tilde{x}$ the vector in $\mathbb{C}^n$ obtained by adding $n - k$ zero entries to x. We have

$$x^t A_k x = \tilde{x}^t A \tilde{x} > 0,$$

since A is positive definite on $\mathbb{R}^n$.

3. The spectrum of A_k is not (always) included in that of A, as illustrated by the following example:

```
>> n=5;A=PdSMat(n);Ak=A(1:3,1:3);
>> eig(A)
ans =
   10.7565
    5.5563
    6.6192
    6.0613
    6.2220
>> eig(Ak)
ans =
    9.0230
    5.6343
    6.1977
```

Solution of Exercise 2.23   We fix $n = 5$. For A=rand(n,n), we have the following results:

1. ```
>> sp=eig(A)
sp =

    2.7208
   -0.5774 + 0.1228i
   -0.5774 - 0.1228i
    0.3839 + 0.1570i
    0.3839 - 0.1570i
```

2. The command sum(abs(X),2) returns a column vector whose entry $\ell$ is equal to the sum of the entries of row $\ell$ of matrix $X$.

   (a) ```
>> Gamma=sum(abs(A),2)-diag(abs(A))
Gamma =
    2.5085
    3.5037
    1.2874
    1.8653
    2.1864
```

   (b) ```
n=length(sp);
for lambda=sp
```
```

```
ItIsTrue=0;
for k=1:n                              % = length(Gamma)
    if abs(lambda-A(k,k))<=Gamma(k)
        ItIsTrue=1;break;
    end;
end;
if ~ItIsTrue
    fprintf('Error with the eigenvalue %f\n',lambda);
end;
end;
```

Thus the spectrum seems to be included in the union of the Gershgorin disks.

(c) We prove the previous remark, known as the Gershgorin–Hadamard theorem. Let $\lambda \in \sigma(A)$ be an eigenvalue of matrix A and $u(\neq 0) \in \mathbb{R}^n$ a corresponding eigenvector. We have

$$(\lambda - a_{i,i})u_i = \sum_{j \neq i} a_{i,j}u_j.$$

In particular, if i is such that $|u_i| = \max_j |u_j|$, we have

$$|(\lambda - a_{i,i})u_i| \leq \sum_{j \neq i} |a_{i,j}| \, |u_j|,$$
$$|\lambda - a_{i,i}| \leq \sum_{j \neq i} |a_{i,j}| \frac{|u_j|}{|u_i|},$$
$$|\lambda - a_{i,i}| \leq \sum_{j \neq i} |a_{i,j}| = \gamma_i.$$

We deduce that any $\lambda \in \sigma(A)$ belongs to at least one disk D_i, and so $\sigma(A) \subset \bigcup_{i=1}^{n} D_i$.

3. (a) `function A=DiagDomMat(n,dom)`
   % *returns a square strictly diagonally dominant matrix*
   % **dom** *determines the extent of the dominance of the diagonal*
   `A=rand(n,n);`
   `%nargin` = *number of input arguments of the function*
   `if nargin==1   dom=1, end;`   % *default value*
   `A=A-diag(diag(A));`
   `A=A+diag(sum(abs(A),2))+dom*eye(size(A))`

   (b) We note that the matrix `A=DiagDomMat(n)` is always nonsingular.

   (c) Assume that A is singular. Then 0 is an eigenvalue of A, and by the previous calculation, there would exist an index i such that $0 \in D_i$, i.e.,

$$|a_{i,i}| \leq \sum_{j \neq i} |a_{i,j}| = \gamma_i,$$

which contradicts the fact that matrix A is diagonally dominant. Thus A is nonsingular.

4. ```
function PlotGersh(a)
% plots the Gershgorin-Hadamard circles
[m,n]=size(a);
if m~=n, error('the matrix is not square'), end;
d=diag(a);
radii=sum(abs(a),2);
radii=radii-abs(d);
% define a large rectangle containing all the circles
%   determine the ''lower-left'' corner
cornerx=real(diag(a))-radii;cornery=imag(diag(a))-radii;
mx=min(cornerx);my=min(cornery);mx=round(mx-1);my=round(my-1);
%   we determine the ''upper-right'' corners
cornerx=real(diag(a))+radii;cornery=imag(diag(a))+radii;
Mx=max(cornerx);My=max(cornery);Mx=round(Mx+1);My=round(My+1);
% specify eigenvalues by symbol +
eigA=eig(a);
plot(real(eigA),imag(eigA),'+','MarkerSize',10,'LineWidth',3)
axis([mx Mx my My]);
set(gca,'XTick',-5:2:10,'YTick',-5:2:5,'FontSize',24);
grid on;
%
Theta = linspace(0,2*pi);
for i=1:n
   X=real(a(i,i))+radii(i)*cos(Theta);
   Y=imag(a(i,i))+radii(i)*sin(Theta);
   hold on;plot(X,Y)   %circle i
end
```

The eigenvalues of $A$ and $A^t$ being the same, we have at our disposal two bounds on the spectrum of $A$ by applying the Gershgorin–Hadamard theorem to both matrices. We display in Figure 11.2 the Gershgorin circles of $A$ (left) and those of $A^t$ (right). Of course, we obtain a better bound of the spectrum of $A$ by superposing the two figures.

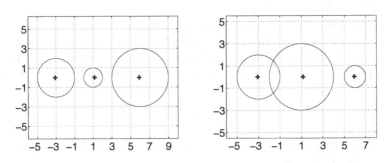

**Fig. 11.2.** Gershgorin disks for a matrix (left) and its transpose (right).
```

Solution of Exercise 2.29 Another definition of the pseudoinverse matrix.

1. ```
>> m=10,n=7;A=MatRank(m,n,5);
>> % determination of P
>> X=orth(A');   % note: Im (A^t) = (Ker A)^⊥
>> P=X*X';
```
2. ```
>> X=orth(A);           % image of A
>> Q=X*X';
```
3. We note that
```
>> norm(pinv(A)*A-P)
ans =
   2.7981e-14
>> norm(A*pinv(A)-Q)
ans =
   1.5443e-14
```
Let us show that indeed $P = A^\dagger A$ and $Q = AA^\dagger$.
(a) We first check that $X = A^\dagger A$ is an orthogonal projection, i.e., that $X = X^*$ and $X^2 = X$. This is a consequence of the Moore–Penrose relations
   i. $X^* = (A^\dagger A)^* = A^\dagger A = X$,
   ii. $X^2 = (A^\dagger A)(A^\dagger A) = (A^\dagger AA^\dagger)A = A^\dagger A = X$.
   To prove that $P = A^\dagger A$, it remains to show that $\operatorname{Im} X = (\operatorname{Ker} A)^\perp$. We know that $(\operatorname{Ker} A)^\perp = \operatorname{Im}(A^*)$. Let us prove that $\operatorname{Im} X = \operatorname{Im}(A^*)$.
   - For any $y \in \operatorname{Im}(A^\dagger A)$, there exists $x \in \mathbb{C}^n$ such that
   $$y = A^\dagger Ax = (A^\dagger A)^* x = A^*(A^\dagger)^* x \Longrightarrow y \in \operatorname{Im} A^*.$$
   - For any $y \in \operatorname{Im}(A^*)$, there exists $x \in \mathbb{C}^m$ such that $y = A^* x$
   $$y = (AA^\dagger A)^* x = (A^\dagger A)^* A^* x = A^\dagger AA^* x \Longrightarrow y \in \operatorname{Im}(A^\dagger A).$$
(b) Similarly, we now check that $Y = AA^\dagger$ is an orthogonal projection:
   i. $Y^* = (AA^\dagger)^* = AA^\dagger = Y$,
   ii. $Y^2 = (AA^\dagger)(AA^\dagger) = (AA^\dagger A)A^\dagger = AA^\dagger = Y$.
   It remains to show that $\operatorname{Im} Y = \operatorname{Im} A$.
   - $y \in \operatorname{Im}(AA^\dagger) \Longrightarrow \exists x \in \mathbb{C}^m, y = AA^\dagger x \Longrightarrow y \in \operatorname{Im} A$.
   - $y \in \operatorname{Im} A \Longrightarrow \exists x \in \mathbb{C}^n, y = Ax = AA^\dagger Ax \Longrightarrow y \in \operatorname{Im}(AA^\dagger)$.
4. First, there exists at least one x such that $Ax = Qy$, since $Q = AA^\dagger$, and thus the existence of at least one $x_1 = Px$ is obvious. Let us prove the uniqueness of x_1. For x and $\tilde{x}$ in $\mathbb{C}^n$ such that $Qy = Ax = A\tilde{x}$, since $P = A^\dagger A$, we have

$$Ax = A\tilde{x} \Longleftrightarrow x - \tilde{x} \in \operatorname{Ker} A \Longrightarrow x - \tilde{x} \in \operatorname{Ker} P \Longleftrightarrow Px = P\tilde{x}.$$

By definition, $Ax = Qy = AA^\dagger y$, and multiplying by $A^\dagger$ and using the relation $A^\dagger = A^\dagger AA^\dagger$, we deduce that $A^\dagger Ax = A^\dagger y$. Since $A^\dagger A = P$, we conclude that $A^\dagger y = Px = x_1$, which is the desired result.

11.2 Exercises of Chapter 3

Solution of Exercise 3.3 We observe that all these norms coincide for a diagonal matrix A, and are equal to $\max_i |a_{i,i}|$. Let us prove this result. We start with the case $p = \infty$. Let x be such that $\|x\|_\infty = 1$ we have $\|Ax\|_\infty = \max_i |A_{i,i}x_i|$, so that $\|A\|_\infty \leq \max_i |A_{i,i}|$. Let I be an index for which $|A_{I,I}| = \max_i |A_{i,i}|$. We have $\|Ae_I\|_\infty = \max_i |A_{i,i}|$, which implies that $\|A\|_\infty = \max_i |A_{i,i}|$.

For $p \in [1, \infty)$ and x such that $\|x\|_p = 1$, we have $\|Dx\|_p^p = \sum_i |D_{i,i}x_i|^p$, whence $\|Dx\|_p^p \leq (\max_i |D_{i,i}|^p) \|x\|_p^p$, and so $\|D\|_p \leq \max_i |D_{i,i}|$. We end the proof as in the case $p = \infty$.

Solution of Exercise 3.7

1. We recall the bounds

$$\lambda_{\min} \|x\|_2^2 \leq \langle Ax, x \rangle \leq \lambda_{\max} \|x\|_2^2,$$

   from which we easily infer the result.
2. Plot of S_A for $n = 2$.
   (a) $\langle Ax, x \rangle = \left\langle A\begin{pmatrix} x_1 \\ px_1 \end{pmatrix}, \begin{pmatrix} x_1 \\ px_1 \end{pmatrix} \right\rangle = |x_1|^2 A_p$ setting $A_p = \left\langle A\begin{pmatrix} 1 \\ p \end{pmatrix}, \begin{pmatrix} 1 \\ p \end{pmatrix} \right\rangle$.
       Then, $x = (x_1, x_2)^t$ belongs to the intersection of Γ_p and S_A if and only if $|x_1| = 1/\sqrt{A_p}$ and $x_2 = p/\sqrt{A_p}$.
   (b) We rotate the half-line $x_2 = px_1$ around the origin and determine for each value of p the intersection of Γ_p and S_A.
   ```
   function [x,y]=UnitCircle(A,n)
   % add a test to check that the matrix
   % is symmetric positive definite
   >> x=[];y=[];h=2*pi/n;
   >> for alpha=0:h:2*pi
   >>     p=tan(alpha);
   >>     ap=[1 p]*A*[1;p];
   >>     if cos(alpha)>0
   >>         x=[x 1/sqrt(ap)];
   >>         y=[y p/sqrt(ap)];
   >>     else
   >>         x=[x -1/sqrt(ap)];
   >>         y=[y -p/sqrt(ap)];
   >>     end;
   >> end;
   ```
   (c)
   ```
   >> n=100;A= [7 5; 5 7];
   >> [x1,y1]=UnitCircle(A,n);
   >> [x2,y2]=UnitCircle(eye(2,2),n);
   >> plot(x1,y1,x2,y2,'.','MarkerSize',10,'LineWidth',3)
   >> axis([-1.2 1.2 -1.2 1.2]);grid on; axis equal;
   ```

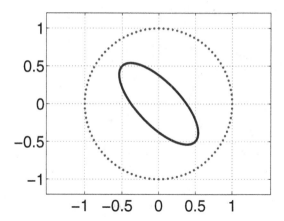

Fig. 11.3. Unit circles for the Euclidean and the norm induced by the matrix A; see (3.12).

```
>> set(gca,'XTick',-1:.5:1,'YTick',-1:.5:1,'FontSize',24);
```
In view of Figure 11.3, it seems that S_A is an ellipse. Let us prove that it is indeed so. Let A be a symmetric positive definite matrix of size $n \times n$. It is diagonalizable in an orthonormal basis of eigenvectors, $A = PDP^t$:

$$\langle Ax, x \rangle = 1 \iff \langle DP^t x, P^t x \rangle = 1 \iff \sum_{i=1}^{n} \lambda_i y_i^2 = 1,$$

where $y = P^t x$ and λ_i are the eigenvalues of A. The last equation may be written

$$\sum_{i=1}^{n} \left(\frac{y_i}{\frac{1}{\sqrt{\lambda_i}}} \right)^2 = 1,$$

since the eigenvalues are positive. We recognize the equation, in the basis of eigenvectors y_i, of an ellipsoid of semiaxes $1/\sqrt{\lambda_i}$.

(d) The results of the following instructions (with $n = 100$) are shown in Figure 11.4.
```
>> [x1,y1]=UnitCircle(A,n);
>> [x2,y2]=UnitCircle(B,n);
>> [x3,y3]=UnitCircle(C,n);
>> plot(x1,y1,x2,y2,'.',x3,y3,'+',...
>> 'MarkerSize',10,'LineWidth',3)
>> grid on; axis equal;
>> set(gca,'XTick',-.8:.4:.8,'YTick',-.8:.4:.8,...
>> 'FontSize',24);
```
The most elongated ellipse corresponds to the matrix C, and the least elongated one to A. Explanation (see previous question): the three

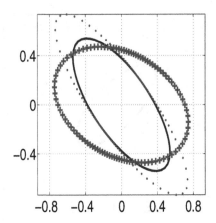

Fig. 11.4. Unit circles for the norms defined by matrices defined in (3.12).

matrices have a common first eigenvalue, equal to 12. The second eigenvalue is equal to 2 for A, 1 for B, and $1/2$ for C.

Solution of Exercise 3.8 We fix the dimension n. The execution of the following instructions,

```
>> A=rand(n,n); s=eigs(A,1)
>> for i=1:10
>>     k=i*10;
>>     abs((norm(A^k))^(1/k)-s)
>> end
```

shows that $\|A^k\|^{1/k}$ seemingly tends to $\varrho(A)$, whatever the chosen norm. Let us prove that it is actually the case. Consider a square matrix A and a matrix norm $\|.\|$.

1. Since $\lambda \in \sigma(A) \implies \lambda^k \in \sigma(A^k)$, we have $|\lambda|^k \leq \varrho(A^k) \leq \|A^k\|$, and in particular, $\varrho(A)^k \leq \varrho(A^k) \leq \|A^k\|$ so $\varrho(A) \leq \|A^k\|^{1/k}$.
2. By construction of the matrix A_ε, we indeed have $\varrho(A_\varepsilon) < 1$ and thus $\lim_{k\to+\infty} \|A_\varepsilon^k\| = 0$. We deduce that there exists an index k_0 such that for all $k \geq k_0$, we have $\|A_\varepsilon^k\| \leq 1$, or even $\varrho(A) + \varepsilon \geq \|A^k\|^{1/k}$.
3. Conclusion: for all $\varepsilon > 0$, there exists then k_0 such that $k \geq k_0$ implies that $\varrho(A) \leq \|A^k\|^{1/k} \leq \varrho(A) + \varepsilon$, and therefore

$$\varrho(A) = \lim_{k\to+\infty} \|A^k\|^{1/k}.$$

11.3 Exercises of Chapter 4

Solution of Exercise 4.1

1. The computation of a scalar product $\langle u, v \rangle = \sum_{i=1}^{n} u_i v_i$ is carried out in n multiplications (and $n - 1$ additions, but we do not take into account additions in operation counts). The computation of $\|u\|_2 = \sqrt{\langle u, u \rangle}$ is carried out in n multiplications and one square root extraction (which is negligible when n is large). Computing the rank-one matrix uv^t requires n^2 operations, since $(uv^t)_{i,j} = u_i v_j$.

2. We denote by a_i the rows of A. Each entry $(Au)_i$ being equal to the scalar product $\langle a_i, u \rangle$, the total cost of the computation of Au is n^2 operations. We call b_j the columns of matrix B. Each element $(AB)_{i,j}$ being equal to the scalar product $\langle a_i, b_j \rangle$, the total cost for computing AB is n^3 operations.

3. The result of the instructions below is displayed in Figure 11.5.

```
>> n1=500;in=1:5;
>> for k=in;
>>    d=k*n1;a=rand(d,d);b=rand(d,d);
>>    tic;x=a*b;time(k)=toc;
>> end
>> n=n1*in';plot(n,time,'-+','MarkerSize',10,'LineWidth',3)
>> text(2200,.5,'n','FontSize',24);
>> text(600,5.2,'T(n)','FontSize',24)
```

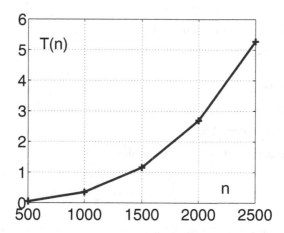

Fig. 11.5. Running time for computing a product of two matrices.

4. The assumption $T(n) \approx Cn^s$ defines an affine relation between logarithms

$$\ln T(n)) \approx s \ln n + D, \qquad (11.1)$$

with $D = \ln C$. We display in Figure 11.6 the result of the following instructions:

```
>> x=log(n);y=log(time);
>> plot(x,y,'-+','MarkerSize',10,'LineWidth',3)
>> text(7.5,-3.5,'log(n)','FontSize',24);
>> text(6.2,1.5,'log(T(n))','FontSize',24)
```

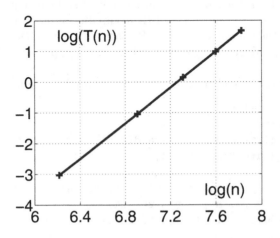

Fig. 11.6. Running time for computing a product of two matrices (log-log scale).

The curve is close to a line of slope s:

```
>> s=(y(5)-y(1))/(x(5)-x(1))
s =
      2.9234
```

The practical running time is close to the theoretical computational time.

Solution of Exercise 4.2

1. Since the matrix A is lower triangular, we have

$$c_{i,j} = \sum_{k=1}^{n} a_{i,k} b_{k,j} = \sum_{k=1}^{i} a_{i,k} b_{k,j}.$$

So $c_{i,j}$ is computed in i multiplications; moreover, there are n elements to be computed per row i. The total cost is

$$\sum_{i=1}^{n} ni = n \frac{n(n+1)}{2} \approx \frac{n^3}{2}.$$

2.   For $i = 1 \nearrow n$ *computing row i of matrix C*
                For $j = 1 \nearrow n$
                    $s = 0$
                    For $k = 1 \nearrow i$
                        $s = s + a_{i,k} b_{k,j}$
                    End For k
                    $c_{i,j} = s$
                End For j
            End For i

3. If both matrices A and B are lower triangular, so is the product $C = AB$. We have, for $j \le i$,

$$c_{i,j} = \sum_{k=j}^{i} a_{i,k} b_{k,j}.$$

Hence, $c_{i,j}$ is computed in $i - j + 1$ multiplications. Furthermore, there are only i elements left to be computed per row i. The total cost is

$$\sum_{i=1}^{n} \sum_{j=1}^{i} (i - j + 1) = \sum_{i=1}^{n} \sum_{j=1}^{i} j = \sum_{i=1}^{n} \frac{i(i+1)}{2} \approx \frac{n^3}{6}.$$

4. The function `LowTriMatMult` defined below turns out to be slower than the usual matrix computation by `Matlab` (which does not take into account the triangular character of the matrices).

```
function C=LowTriMatMult(A,B)
% Multplication of two lower triangular matrices
[m,n]=size(A);[n1,p]=size(B);
if n~=n1
    error('Wrong dimensions of the matrices')
end;
C=zeros(m,p);
for i=1:m
    for j=1:i
        s=0;
        for k=j:i
            s=s+A(i,k)*B(k,j);
        end;
        C(i,j)=s;
    end;
end;
```

This is not a surprise, since the above function `LowTriMatMult` is not vectorized and optimized as are standard `Matlab` operations, such as the product of two matrices. In order to see the computational gain in taking into account the triangular structure of matrices, one has to compare

LowTriMatMult with a function programmed in Matlab language that executes the product of two matrices without vectorization.

5. Now we compare LowTriMatMult with the function MatMult defined below:

```
function C=MatMult(A,B)
% Multiplication of two matrices
% tests whether the dimensions are compatible
% as in function LowTriMatMult
[m,n]=size(A);p=size(B,2);
C=zeros(m,p);
for i=1:m
    for j=1:p
        s=0;
        for k=1:n
            s=s+A(i,k)*B(k,j);
        end;
        C(i,j)=s;
    end;
end;
```

The comparison gives the advantage to LowTriMatMult: the speedup is approximately a factor 6:

```
>> n=1000;a=tril(rand(n,n));b=tril(rand(n,n));
>> tic;c=MatMult(a,b);t1=toc;
>> tic;d=LowTriMatMult(a,b);t2=toc;
>> t1, t2
t1 =
    23.5513
t2 =
     4.1414
```

6. With the instruction sparse, Matlab takes into account the sparse structure of the matrix:

```
>> n=300;
>> a=triu(rand(n,n));b=triu(rand(n,n));
>> tic;a*b;t1=toc
>> sa=sparse(a);sb=sparse(b);
>> tic;sa*sb;t2=toc
t1 =
    0.0161
t2 =
    0.0291
```

There is no gain in computational time. Actually, the advantage of sparse relies on the storage

```
>> n=300;a=triu(rand(n,n));spa=sparse(a);
>> whos a
  Name       Size                      Bytes  Class
   a        300x300                   720000  double array
Grand total is 90000 elements using 720000 bytes
>> whos spa
  Name       Size                   Bytes  Class
   spa      300x300                 543004  double array (sparse)
Grand total is 45150 elements using 543004 bytes
```

11.4 Exercises of Chapter 5

Solution of Exercise 5.2

1. ```
   function x=ForwSub(A,b)
   % Computes the solution of system Ax = b
   % A is nonsingular lower triangular
   [m,n]=size(A);o=length(b);
   if m~=n | o~=n, error('dimension problem'), end;
   small=1.e-12;
   if norm(A-tril(A),'inf')>small
       error('non lower triangular matrix')
   end;
   x=zeros(n,1);
   if abs(A(1,1))<small, error('noninvertible matrix'), end;
   x(1)=b(1)/A(1,1);
   for i=2:n
       if abs(A(i,i))<small
           error('noninvertible matrix')
       end;
       x(i)=(b(i)-A(i,1:i-1)*x(1:i-1))/A(i,i);
   end;
   ```

2. ```
   function x=BackSub(A,b)
   % Computes the solution of system Ax = b
   % A is nonsingular upper triangular
   [m,n]=size(A);o=length(b);
   if m~=n | o~=n, error('dimension problem'), end;
   small=1.e-12;
   if norm(A-triu(A),'inf')>small
       error('non upper triangular matrix')
   end;
   x=zeros(n,1);
   if abs(A(n,n))<small, error('noninvertible matrix'), end;
   x(n)=b(n)/A(n,n);
   ```

```
    for i=n-1:-1:1
        if abs(A(i,i))<small
            error('noninvertible matrix')
        end;
        x(i)=(b(i)-A(i,i+1:n)*x(i+1:n))/A(i,i);
    end;
```

Solution of Exercise 5.3 We store in a vector u the lower triangular matrix A row by row, starting with the first one. The mapping between indices is $A_{i,j} = u_{j+i(i-1)/2}$. Here are the programs:

1.
```
function aL=StoreL()
fprintf('Storage of a lower triangular matrix')
fprintf('the matrix is stored row by row')
n=input('enter the dimension n of the square matrix')
for i=1:n
    fprintf('row %i \n',i)
    ii=i*(i-1)/2;
    for j=1:i
        fprintf('enter element (%i,%i) of the matrix',i,j)
        aL(j+ii)=input(' ');
    end;
end;
```

2.
```
function y=StoreLpv(a,b)
% a is a lower triangular matrix stored by  StoreL
[m,n]=size(b);
if n~=1
        error('b is not a vector')
end;
if m*(m+1)/2~=length(a)
        error('incompatible dimensions')
end;
for i=1:m
    ii=i*(i-1)/2;
    s=0;
    for j=1:i
        s=s+a(j+ii)*b(j);
    end;
    y(i)=s;
end;
```
The inner loop could be replaced by the more compact and efficient instruction

```
y(i)=a(ii+1:i)*b(1:i);
```

3.
```
function x=ForwSubL(a,b)
% a is a lower triangular matrix stored by  StoreL
```

```
% add the compatibility tests of function StoreLpv(a,b)
m=length(b);
for i=1:m
    ii=i*(i-1)/2;
    s=0;
    for j=1:i-1
        s=s+a(j+ii)*x(j);
    end;
    x(i)=(b(i)-s)/a(i+ii); % check if a(i+ii) is zero
end;
```

Solution of Exercise 5.13 Hager algorithm.

1. Proposition 3.1.2 furnishes a formula for $\|A\|_1$ (at a negligible computational cost):

$$\|A\|_1 = \max_{1 \le j \le n} \left(\sum_{i=1}^{n} |a_{i,j}| \right).$$

2. From the previous formula we deduce that there exists an index j_0 such that

$$\|A^{-1}\|_1 = \sum_{i=1}^{n} |(A^{-1})_{i,j_0}| = \|(A^{-1})_{j_0}\|_1,$$

where $(A^{-1})_{j_0}$ is the j_0 column of A^{-1}. Hence

$$\|A^{-1}\|_1 = \|A^{-1}e_{j_0}\|_1 = f(e_{j_0}).$$

3. We write

$$f(x) = \sum_{i=1}^{n} |(A^{-1}x)_i| = \sum_{i=1}^{n} |\tilde{x}_i| = \sum_{i=1}^{n} s_i \tilde{x}_i = \langle \tilde{x}, s \rangle.$$

4. We compute $\tilde{x}^t(a - x)$:

$$\tilde{x}^t(a - x) = (A^{-t}s)^t(a - x) = s^t A^{-1}(a - x) = \langle A^{-1}a, s \rangle - \langle \tilde{x}, s \rangle.$$

According to question 3,

$$f(x) + \tilde{x}^t(a - x) = \langle A^{-1}a, s \rangle = \sum_{j=1}^{n} (A^{-1}a)_j s_j.$$

Each term in this sum is bounded from above by $|(A^{-1}a)_j|$; we therefore deduce

$$f(x) + \tilde{x}^t(a - x) \le \sum_{j=1}^{n} |(A^{-1}a)_j| = \|A^{-1}a\|_1 = f(a),$$

from which we get the result.

5. According to question 4, we have

$$f(e_j) - f(x) \geq \bar{x}^t(e_j - x) = \langle \bar{x}, e_j \rangle - \langle x, \bar{x} \rangle = \bar{x}_j - \langle x, \bar{x} \rangle > 0,$$

which yields the result.

6. (a) According to question 3, we have $f(y) = \langle \tilde{y}, s' \rangle$, where s' is the sign vector of $\tilde{y}$. However, for y close enough to x, the sign of $\tilde{y} = A^{-1}y$ is equal to the sign of $\tilde{x}$ (we have assumed that $\tilde{x}_i \neq 0$ for all i). Thus we have

$$f(y) = \langle \tilde{y}, s \rangle = f(x) + \langle \widetilde{y - x}, s \rangle = f(x) + \langle A^{-1}(y - x), s \rangle$$
$$= f(x) + s^t A^{-1}(y - x),$$

which proves the result.

(b) x is a local maximum of f if and only if for all y close enough to x, we have $s^t A^{-1}(y - x) \leq 0$. We write

$$s^t A^{-1}(y - x) = \langle A^{-1}(y - x), s \rangle = \langle y - x, A^{-t}s \rangle = \langle y - x, \bar{x} \rangle$$
$$= \langle y, \bar{x} \rangle - \langle x, \bar{x} \rangle.$$

We infer that if $\|\bar{x}\|_\infty \leq \langle x, \bar{x} \rangle$, we have

$$s^t A^{-1}(y - x) \leq \langle y, \bar{x} \rangle - \|\bar{x}\|_\infty \leq (\|y\|_1 - 1)\|\bar{x}\|_\infty = 0,$$

for $y \in S$.

7. The Hager algorithm in pseudolanguage:

```
choose x of norm ||x||₁ = 1
compute
    x̃ by Ax̃ = x
    s by sᵢ = sign(x̃ᵢ)
    compute x̄ by Aᵗx̄ = s
While ||x̄||∞ > ⟨x, x̄⟩
    compute j such that x̄ⱼ = ||x̄||∞
    set x = eⱼ
    compute x̃, s and x̄
    If ||x̄||∞ ≤ ⟨x, x̄⟩
        cond₁(A) ≈ ||A||₁ ||x̄||∞
    End If
End While
```

8. The Hager algorithm in Matlab.

```
function conda=Cond1(a)
% computes an approximation of the 1-norm
% condition number of a square matrix
% by optimization criteria
%
```

```
% initialization
n=size(a,1);
x=ones(n,1)/n;                            % x_i = 1/n
xt=a\x;ksi=sign(xt);xb=a'\ksi;
notgood=norm(xb,'inf')>xb'*x;
conda=norm(xt,1)*norm(a,1);
% loop
while notgood
   [maxx,j]=max(abs(xb));
   x=zeros(n,1);x(j)=1;
   xt=a\x;ksi=sign(xt);xb=a'\ksi;
   if norm(xb,'xinf')<=xb'*x
      conda=norm(xt,1)*norm(a,1);
      notgood=0;
   else
      [maxx,j]=max(abs(xb));
      x=zeros(n,1);x(j)=1;
   end;
end;
```

This algorithm gives remarkably precise results:

```
>> n=5;a=NonsingularMat(n);
>> [Cond1(a), norm(a,1)*norm(inv(a),1)]
ans =
   238.2540   238.2540
>> n=10;a=NonsingularMat(n);
>> [Cond1(a), norm(a,1)*norm(inv(a),1)]
ans =
   900.5285   900.5285
```

Solution of Exercise 5.15 Polynomial preconditioner $C^{-1} = p(A)$.

```
function PA=PrecondP(A,k)
% Polynomial preconditioning
% check the condition  ||I - A|| < 1
PA=eye(size(A));
if k~=0
   C=eye(size(A))-A;
   if max(abs(eig(C)))>=.99                % compute ϱ(C)
       error('look for another preconditioner')
   end;
   for i=1:k
       PA=PA+C^i;
   end;
end;
```

Some examples:

```
>> n=10;a=PdSMat(n);a=a/(2*norm(a));
>> cond(a)
ans =
    1.9565
>> ap=PrecondP(a,1);cond(ap*a)
ans =
    1.6824
>> ap=PrecondP(a,10);cond(ap*a)
ans =
    1.0400
>> ap=PrecondP(a,20);cond(ap*a)
ans =
    1.0020
```

The larger k is, the better is the conditioning; however, its computational cost increases.

Solution of Exercise 5.16 Finite difference approximation of Laplace equation.

1. Computation of the matrix and right-hand side of (5.19).
   (a) The following function gives the value of the matrix A_n.
   ```
   function A=Laplacian1dD(n)
   % computes the 1D Laplacian matrix
   % discretized by centered finite differences
   % with Dirichlet boundary conditions.
   A=zeros(n,n);
   for i=1:n-1
       A(i,i)=2;
       A(i,i+1)=-1;
       A(i+1,i)=-1;
   end;
   A(n,n)=2;
   A=A*(n+1)^2;
   ```
   There are other ways of constructing A_n. For instance,
   i. Function `toeplitz(u)`, which returns a symmetric Toeplitz matrix whose first row is the vector u. The instruction
   ```
   Ah=toeplitz([2, -1, zeros(1, n-2)])*(n+1)^2
   ```
   then defines A_n.
   ii. The following instructions define also A_n.
   ```
   u=ones(n-1,1);v=[1;u];a=(n+1)^2;
   Aht=2*a.*diag(v,0)-a.*diag(u,-1)-a.*diag(u,1);
   ```
   iii. The reader can check that the instructions
   ```
   A=[eye(n,n) zeros(n,1)];A=eye(n,n)-A(:,2:n+1);
   A=A+A';A=A*(n+1)^2;
   ```
   also define the matrix A_n.

(b) The following function gives the value of the right-hand side $b^{(n)}$.

```
function v=InitRHS(n)
x=1:n;h=1./(n+1);
x=h*x';v=f(x);
```

It calls a function f that corresponds to the right-hand side $f(x)$ of the differential equation.

2. Validation.
   (a) When $f(x) \equiv 1$, $u^e(x) = x(1-x)/2$ is the solution of problem (5.18).

```
function [exa,sm]=Check(n)
x=1:n;h=1./(n+1);x=h*x';
exa=x.*(1-x)/2;
sm=ones(n,1);
```

Examples:

```
>> n=10;Ah=Laplacian1dD(n);
>> [exasol,sm]=Check(n);sol=Ah\sm;
>> norm(sol-exasol)
ans =
    2.6304e-16
```

We note that no error was made; the solution is exactly computed with the accuracy of the machine. This was predictable, since the discretization of u'' performed in (5.19) is exact for polynomials of degree less than or equal to 3.

   (b) Convergence of the method. We use here the function

```
function [exa,sm]=Check2(n)
x=1:n;h=1./(n+1);x=h*x';
exa=(x-1).*sin(10*x);
sm=-20*cos(10*x)+100*(x-1).*sin(10*x);
```

and the approximation error is given by the script

```
for i=1:10
    n=10*i;[exasol,sm]=Check2(n);
    Ah=Laplacian1dD(n);sol=Ah\sm;
    y(i)=log(norm(sol-exasol,'inf'));
    x(i)=log(n);
end
plot(x,y,'.-','MarkerSize',20,'LineWidth',3)
grid on;
set(gca,'XTick',2:1:5,'YTick',-7.5:1:-3,'FontSize',24);
```

In view of Figure 11.7, the logarithm of the error is an affine function of the logarithm of n; the slope of this straight line is about -2. Thus, we deduce that $\|\text{sol} - \text{exasol}\|_\infty \approx C^{ste} \times n^{-2}$, which was predicted by Theorem 1.1.1.

3. Eigenvalues of the matrix A_n.
   (a) Solving the differential equation $u'' + \lambda u = 0$ shows that the eigenvalues of the operator $u \mapsto -u''$ endowed with homogeneous Dirichlet

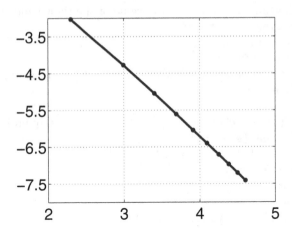

Fig. 11.7. Logarithm of the error in terms of logarithm of n (log-log scale).

boundary conditions are exactly $\lambda_k = k^2\pi^2$, for $k > 0$, with corresponding eigenfunctions $\varphi_k(x) = \sin(k\pi x)$. When n tends to $+\infty$ (i.e., h tends to 0) and for a fixed k, we obtain the limit

$$\lambda_{h,k} = \frac{4}{h^2}\sin^2\left(\frac{k\pi h}{2}\right) \approx k^2\pi^2 = \lambda_k.$$

In other words, the "first" eigenvalues of A_n are close to the eigenvalues of the continuous operator; see Figure 11.8.

```
n=20;x=(1:n)';h=1./(n+1);
Ahev=(1-cos(pi.*h.*x)).*2./h./h;
contev=pi.*pi.*x.*x;
plot(x,contev,x,Ahev,'.-','MarkerSize',20,'LineWidth',3)
```

(b) The function `eig(X)` returns a vector containing the eigenvalues of matrix X:

```
n=500;Ah=Laplacian1dD(n);
x=(1:n)';h=1./(n+1);
exacev=(1-cos(pi.*h.*x)).*2./h./h;
matlabev=eig(Ah);
plot(x,matlabev-exacev)
```

The eigenvalues of A_n are accurately computed by Matlab: for example, for $n = 500$ the maximal error committed is less than 10^{-9}.

(c) We plot on a log-log scale the condition number of A_n:

```
for i=1:7
    n=50*i;Ah=Laplacian1dD(n);
    en(i)=log(n);condn(i)=log(cond(Ah));
end
plot(en,condn)
```

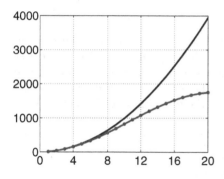 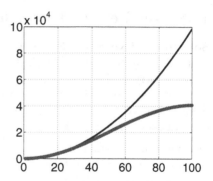

Fig. 11.8. Comparison of the eigenvalues of the continuous and discrete problems for $n = 20$ (left) and $n = 100$ (right).

We observe that the 2-norm conditioning of A_n behaves as $C^{ste} \times n^2$. This result agrees with the theory, since, A_n being symmetric, its 2-norm conditioning is equal to the ratio of its extremal eigenvalues

$$cond_2(A_n) = \frac{\max_k |\lambda_{h,k}|}{\min_k |\lambda_{h,k}|} \approx \frac{4}{\pi^2} n^2.$$

4. (a) We have, of course, $\tilde{A}_n = A_n + cI_n$, and the eigenvalues of $\tilde{A}_n$ are $\tilde{\lambda}_{h,k} = \lambda_{h,k} + c$. In particular, the first eigenvalue is

$$\tilde{\lambda}_{h,1} = \frac{4}{h^2} \sin^2\left(\frac{\pi h}{2}\right) + c.$$

(b) We solve the linear system for $n = 100$, $f = 1$.
```
>> n=100;Ah=Laplacian1dD(n);
>> s=eig(Ah);c=-s(1);        % Warning: it may be necessary
>> At=Ah+c*eye(n,n);         % to sort out vector s with
>> b=InitRHS(n);sol=At\b;    % command sort
```
To check the reliability of the result we compute the norm of the solution `sol` given by Matlab:
```
>> norm(sol)
ans =
    1.0470e+12
```
which is a clear indication that something is going wrong .... Since c is an eigenvalue of A_n, the matrix $\tilde{A}_n$ is singular. Thanks to the rounding errors, Matlab nevertheless find a solution without any warning.

11.5 Exercises of Chapter 6

Solution of Exercise 6.3 The function Gauss(A,b) computes the solution of the linear system $Ax = b$ by the Gauss method; the pivot in step k is the first nonzero entry in the set $(A_{j,k})_{j \geq k}$. In the function GaussWithoutPivot there is no pivot strategy; the program stops if the entry $A_{k,k}$ is too small. Partial pivoting and complete pivoting are used in the functions GaussPartialPivot and GaussCompletePivot.

1.
```
function x=Gauss(A,b)
% solve system Ax = b by
% the Gauss method with partial pivoting
[m,n]=size(A);o=length(b);
if m~=n | o~=n, error('dimension problem'), end;
% Initialization
small=1.e-16;
for k=1:n-1
% Search for the pivot
    u=A(k,k:n);pivot=A(k,k);i0=k;
    while abs(pivot)<small
        i0=i0+1;pivot=A(i0,k);
    end;
    if abs(pivot)<small
        error('singular matrix')
    end;
% Exchange rows for A and b
    if i0~=k
        u=A(i0,k:n);A(i0,k:n)=A(k,k:n);A(k,k:n)=u;
        s=b(i0);b(i0)=b(k);b(k)=s;
    end
    for j=k+1:n
        s=A(j,k)/pivot;v=A(j,k:n);
        A(j,k:n)=v-s*u;b(j)=b(j)-s*b(k);
    end;
end;
% A = An is an upper triangular matrix
% we solve Anx = bn by back substitution
x=zeros(n,1);
if abs(A(n,n))>=small
    x(n)=b(n)/A(n,n);
else
    error('singular matrix')
end;
for i=n-1:-1:1
    x(i)=(b(i)-A(i,i+1:n)*x(i+1:n))/A(i,i);
end;
```

2.
```
function x=GaussWithoutPivot(A,b)
% solve system Ax = b by
% the Gauss method without pivoting
[m,n]=size(A);o=length(b);
if m~=n | o~=n, error('dimension problem'), end;
% Initialization
small=1.e-16;
for k=1:n-1
    u=A(k,k:n);pivot=A(k,k);i0=k;
    if abs(pivot)<small, error('stop: zero pivot'), end;
    for j=k+1:n
        s=A(j,k)/pivot;v=A(j,k:n);
        A(j,k:n)=v-s*u;b(j)=b(j)-s*b(k);
    end;
end;
% A = A_n is an upper triangular matrix
% we solve A_n x = b_n by the back substitution method
x=zeros(n,1);
if abs(A(n,n))>=small
    x(n)=b(n)/A(n,n);
else
    error('singular matrix')
end;
for i=n-1:-1:1
    x(i)=(b(i)-A(i,i+1:n)*x(i+1:n))/A(i,i);
end;
```
3.
```
function x=GaussPartialPivot(A,b)
% solve system Ax = b by
% the partial pivoting Gauss method
[m,n]=size(A);o=length(b);
if m~=n | o~=n, error('dimension problem'), end;
% Initialization
small=1.e-16;
for k=1:n-1
    B=A(k:n,k);
% We determine i_0
    [pivot,index]=max(abs(B));
    if abs(pivot)<small
        error('singular matrix')
    end;
    i0=k-1+index(1,1);;;
% Exchange rows for A and b
    if i0~=k
        u=A(i0,k:n);A(i0,k:n)=A(k,k:n);A(k,k:n)=u;
        s=b(i0);b(i0)=b(k);b(k)=s;
```

```
          end
    % We carry out the Gauss elimination
        u=A(k,k:n);pivot=A(k,k);
        for j=k+1:n
            s=A(j,k)/pivot;v=A(j,k:n);
            A(j,k:n)=v-s*u;b(j)=b(j)-s*b(k);
        end;
    end;
    % A = A_n is an upper triangular matrix
    % we solve A_n x = b_n by the back substitution method
    x=zeros(n,1);
    if abs(A(n,n))>=small
        x(n)=b(n)/A(n,n);
    else
        error('singular matrix')
    end;
    for i=n-1:-1:1
        x(i)=(b(i)-A(i,i+1:n)*x(i+1:n))/A(i,i);
    end;
4. function x=GaussCompletePivot(A,b)
    % solve system Ax = b by
    % the complete pivoting Gauss method
    [m,n]=size(A);o=length(b);
    if m~=n | o~=n, error('dimension problem'), end;
    % Initialization
    small=1.e-8;
    ix=1:n;
    for k=1:n-1
        B=A(k:n,k:n);
    % We determine i0,j0
        [P,I]=max(abs(B));
        [p,index]=max(P); pivot=B(I(index),index);
        if abs(pivot)<small, error('singular matrix'), end;
        i0=k-1+I(index);j0=k-1+index;
    % Exchange rows for A and b
        if i0~=k
            u=A(i0,k:n);A(i0,k:n)=A(k,k:n);A(k,k:n)=u;
            s=b(i0);b(i0)=b(k);b(k)=s;
        end
    % Exchange columns of A and rows of x
        if j0~=k
            u=A(:,k);A(:,k)=A(:,j0);A(:,j0)=u;
            s=ix(j0);ix(j0)=ix(k);ix(k)=s;
        end
    % We carry out the Gauss elimination
```

```
    u=A(k,k:n);pivot=A(k,k);
    for j=k+1:n
        s=A(j,k)/pivot;
        v=A(j,k:n);
        A(j,k:n)=v-s*u;
        b(j)=b(j)-s*b(k);
    end;
end;
% A = A_n is an upper triangular matrix
% we solve A_n x_n = b_n by the back substitution method
y=zeros(n,1);
if abs(A(n,n))>=small
    y(n)=b(n)/A(n,n);
else
    error('singular matrix')
end;
for i=n-1:-1:1
    y(i)=(b(i)-A(i,i+1:n)*y(i+1:n))/A(i,i);
end;
% we rearrange the entries of x
x=zeros(n,1);x(ix)=y;
```

5. (a) Dividing by a small pivot yields bad numerical results because of rounding errors:

```
>> e=1.E-15;
>> A=[e 1 1;1 1 -1;1 1 2];x=[1 -1 1]';b=A*x;
>> norm(Gauss(A,b)-x)
ans =
   7.9928e-04
>> norm(GaussPartialPivot(A,b)-x)
ans =
   2.2204e-16
```

(b) Now we compare the strategies complete pivoting/partial pivoting.

    i. We modify the headings of the programs by adding a new output argument. For instance, we now define the function Gauss, function [x,g]=Gauss(A,b). The rate g is computed by adding the instruction g=max(max(abs(A)))/a0 just after the computation of the triangular matrix. The variable a0=max(max(abs(A))) is computed at the beginning of the program.

    ii. For diagonally dominant matrices, we note that the rates are all close to 1.

```
>> n=40;b=rand(n,1);A=DiagDomMat(n);
>> [x gwp]=GaussWithoutPivot(A,b);[x,g]=Gauss(A,b);
>> [x gpp]=GaussPartialPivot(A,b);
>> [x,gcp]=GaussCompletePivot(A,b);
>> [gwp, g, gpp, gcp]
```

```
ans =
        0.9900      0.9900      0.9900      1.0000
```
The same holds true for positive definite symmetric matrices.
```
>> n=40;b=rand(n,1);A=PdSMat(n);
>> [x gwp]=GaussWithoutPivot(A,b);[x,g]=Gauss(A,b);
>> [x gpp]=GaussPartialPivot(A,b);
>> [x,gcp]=GaussCompletePivot(A,b);
>> n=40;b=rand(n,1);A=PdSMat(n);
>> [gwp, g, gpp, gcp]
ans =
        0.9969      0.9969      0.9969      1.0000
```
We can therefore apply, without bad surprises, the Gauss method to these matrix classes. For "random" matrices, the result is different:
```
>> n=40;b=rand(n,1);A=rand(n,n);
>> [x gwp]=GaussWithoutPivot(A,b);[x,g]=Gauss(A,b);
>> [x gpp]=GaussPartialPivot(A,b);
>> [x,gcp]=GaussCompletePivot(A,b);
>> [gwp, g, gpp, gcp]
ans =
      128.3570   128.3570      3.0585      1.8927
```
For these matrices, it is better to use the Gauss methods with partial or complete pivoting.

iii. Comparison of ϱ_{GPP} and ϱ_{GCP}.

A. The following instructions produce Figure 11.9, which clearly shows that complete pivoting is more stable. The drawback of this method is its slow speed, since it performs $(n - k)^2$ comparisons (or logical tests) at each step of the algorithm, whereas partial pivoting requires only $n - k$ comparisons.

```
>> for k=1:10
>>     n=10*k;b=ones(n,1);
>>     A=n*NonSingularMat(n);
>>     [x g]=GaussPartialPivot(A,b);Pp1(k)=g;
>>     [x g]=GaussCompletePivot(A,b);Cp1(k)=g;
>>     A=n*NonSingularMat(n);
>>     [x g]=GaussPartialPivot(A,b);Pp2(k)=g;
>>     [x g]=GaussCompletePivot(A,b);Cp2(k)=g;
>>     A=n*NonSingularMat(n);
>>     [x g]=GaussPartialPivot(A,b);Pp3(k)=g;
>>     [x g]=GaussCompletePivot(A,b);Cp3(k)=g;
>> end;
>> n=10*(1:10);
>> plot(n,Pp1,'+',n,Pp2,'+',n,Pp3,'+',n,Cp1,...
>> 'x',n,Cp2,'x',n,Cp3,'x','MarkerSize',10,...
>> 'LineWidth',3)
```

```
>> set(gca,'XTick',0:20:100,'YTick',0:3:6,...
>> 'FontSize',24);
```

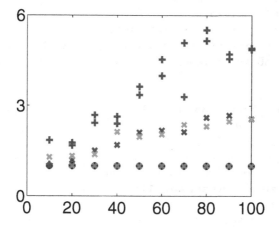

Fig. 11.9. Growth rate, in terms of n, for several runs of Gaussian elimination with partial pivoting $(+)$ and complete pivoting (x).

B. The following matrix has been cooked up so that $\varrho_{\text{GPP}} = 2^{n-1}$, while, as for any matrix, ϱ_{GCP} does not grow much more quickly than n. However, in usual practice, the partial pivoting strategy is as efficient as the complete pivoting Gauss method.

```
>> for k=1:5
>>     n=2*k;b=ones(n,1);
>>     A=-tril(ones(n,n))+2*diag(ones(n,1));
>>     A=A+[zeros(n,n-1) [ones(n-1,1);0]];
>>     [x g]=GaussPartialPivot(A,b);Pp(k)=g;
>>     [x g]=GaussCompletePivot(A,b);Cp(k)=g;
>> end;
>> dim=2*(1:5);
[dim; Pp;Cp]
ans =
       2     4     6     8    10
       2     8    32   128   512
       2     2     2     2     2
```

Solution of Exercise 6.5 Storage of a band matrix.

```
1. function aB=StoreB(p)
   fprintf('Storage of a triangular band matrix\n')
   fprintf('of bandwidth 2*p+1\n')
```

```
        fprintf('the matrix is stored row by row \n')
        n=input('enter the dimension n of the square matrix ')
        for i=1:n
            fprintf('row %i \n',i)
            ip=(2*i-1)*p;
            for j=max(1,i-p):min(n,i+p)
                fprintf('enter entry (%i,%i) of the matrix',i,j)
                aB(j+ip)=input(' ');
            end;
        end;
2. function aBb=StoreBpv(a,p,b)
   % a is a band matrix
   % stored by  StockB
   % Warning: execute all compatibility tests
   % here we assume that a and b are compatible
   [m,n]=size(b);aBb=zeros(m,1);
   for i=1:m
       ip=(2*i-1)*p;
       s=0;
       for j=max(1,i-p):min(m,i+p)
           s=s+a(j+ip)*b(j);
       end;
       aBb(i)=s;
   end;
```

Solution of Exercise 6.6

```
function aB=LUBand(aB,p)
% Compute the LU factorization of
% a band matrix A with half-bandwidth p.
% A (in aB), L and U (also in aB)
% are in the form computed by program StoreB
n=round(length(aB)/(2*p+1));zero=1.e-16;
for k=1:n-1
  if abs(aB((2*k-1)*p+k))<zero, error('error : zero pivot'),end;
  Ind=min(n,k+p);
  for i=(k+1):Ind
    aB((2*i-1)*p+k)=aB((2*i-1)*p+k)/aB((2*k-1)*p+k);
    IndR=(k+1):Ind;
    aB((2*i-1)*p+IndR)=aB((2*i-1)*p+IndR)- ...
    aB((2*k-1)*p+IndR)*aB((2*i-1)*p+k);
  end;
end;
```

11.6 Exercises of Chapter 7

Solution of Exercise 7.5 The following function computes the QR factorization of a square matrix by the Householder algorithm.

```
function [Q,R]=Householder(A)
% QR decomposition computed by
% the Householder algorithm
% Matrix A is square
[m,n]=size(A);
R=A;Q=eye(size(A));
for k=1:m-1
    i=k-1;j=n-i;v=R(k:n,k);w=v+norm(v)*[1;zeros(j-1,1)];
    Hw=House(w);
    Hk=[eye(i,i) zeros(i,j); zeros(j,i) Hw];
    Q=Hk*Q;
    R(k:n,k:n)=Hw*R(k:n,k:n);
end;
Q=Q';
```

The function House computes the Householder matrix of a vector.

```
function H=House(v)
% Elementary Householder matrix
[n,m]=size(v);
if m~=1
    error('enter a vector')
else
    H=eye(n,n);
    n=norm(v);
    if n>1.e-10
        H=H -2*v*v'/n/n;
    end;
end;
```

We check the results.

```
>> n=20;A=rand(n,n);[Q,R]=Householder(A);
% we check that QR = A
>> norm(Q*R-A)
ans =
    3.3120e-15
% we check that Q is unitary
>> norm(Q*Q'-eye(size(A))), norm(Q'*Q-eye(size(A)))
ans =
    1.8994e-15
ans =
    1.8193e-15
```

```
% we check that R is upper triangular
>> norm(R-triu(R))
ans =
   4.3890e-16
```

Finally, we compare the results with the modified Gram–Schmidt method of Exercise 2.11.

```
>>n=10;A=HilbertMat(n);B=MGramSchmidt(A);norm(B*B'-eye(size(A)))
ans =
  2.3033e-04
>> [Q,R]=Householder(A);norm(Q*Q'-eye(size(A)))
ans =
  1.0050e-15
```

For larger values of n, we have

```
>>n=20;A=HilbertMat(n);B=MGramSchmidt(A);norm(B*B'-eye(size(A)))
?? Error using ==> MGramSchmidt
linearly dependent vectors
>> [Q,R]=Householder(A);norm(Q*Q'-eye(size(A)))
ans =
  1.6434e-15
```

The Householder method is more robust and provides an orthogonal matrix, whereas the Gram–Schmidt algorithm finds zero vectors (more precisely of norm less than 10^{-12}).

11.7 Exercises of Chapter 8

Solution of Exercise 8.6 Program for the relaxation method.

```
function [x, iter]=Relax(A,b,w,tol,MaxIter,x)
% Computes by the relaxation method the solution of system Ax=b
% w =  relaxation parameter
% tol =  ε of the termination criterion
% MaxIter = maximum number of iterations
%  x = x0 initial vector
[m,n]=size(A);
if m~=n, error('the matrix is not square'), end;
if abs(det(A)) < 1.e-12
    error('the matrix is singular')
end;
if ~w, error('omega = zero');end;
% nargin  = number of input arguments of the function
% Default values of the arguments
if nargin==5 , x=zeros(size(b));end;
```

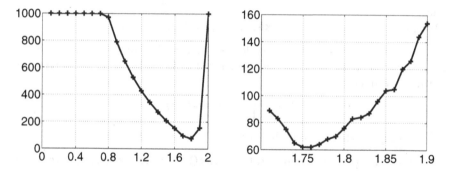

Fig. 11.10. Relaxation method: number of iterations in terms of ω.

```
if nargin==4 , x=zeros(size(b));MaxIter=200;end;
if nargin==3 , x=zeros(size(b));MaxIter=200;tol=1.e-4;end;
if nargin==2 , x=zeros(size(b));MaxIter=200;tol=1.e-4;w=1;end;
M=diag((1-w)*diag(A)/w)+tril(A);
% Initialization
iter=0;r=b-A*x;
% Iterations
while (norm(r)>tol)&(iter<MaxIter)
    y=M\r;
    x=x+y;
    r=r-A*y;
    iter=iter+1;
end;
```

We run the program for $\omega \in (0.1, 2)$.

```
>> n=20;A=Laplacian1dD(n);b=sin((1:n)/(n+1))';sol=A\b;
>> pas=0.1;
>> for i=1:20
>>      omega(i)=i*pas;
>>      [x, iter]=Relax(A,b,omega(i),1.e-6,1000,zeros(size(b)));
>>      itera(i)=iter;
>> end;
>> plot(omega,itera,'-+','MarkerSize',10,'LineWidth',3)
```

In view of the results in Figure 11.10 (left), it turns out that the optimal parameter is between 1.7 and 1.9. Hence, we zoom on this region.

```
pas=0.01;
for i=1:20
    omega(i)=1.7+i*pas;
    [x, iter]=Relax(A,b,omega(i),1.e-6,1000,zeros(size(b)));
    itera(i)=iter;
```

```
end;
plot(omega,itera,'-+','MarkerSize',10,'LineWidth',3)
```

The optimal parameter seems be close to 1.75 in Figure 11.10 (right). Since A is tridiagonal, symmetric, and positive definite, Theorem 8.3.2 gives the optimal parameter

$$\omega_{opt} = \frac{2}{1 + \sqrt{1 - \varrho(\mathcal{J})^2}}.$$

Let us compute this value with Matlab

```
>> D=diag(diag(A));J=eye(size(A))-inv(D)*A;rhoJ= max(abs(eig(J)))
rhoJ =
   0.9888
>> wopt=2/(1+sqrt(1-rhoJ^2))
wopt =
   1.7406
```

which is indeed close to 1.75.

11.8 Exercises of Chapter 9

Solution of Exercise 9.3

```
1. function [x, iter]=GradientC(A,b,tol,MaxIter,x)
   % Computes by the conjugate gradient method
   % the solution of system Ax = b
   %  tol = ε termination criterion
   %  MaxIter = maximal number of iterations
   %  x =x₀
   % nargin  = number of input arguments of the function
   % Default values of the arguments
   if nargin==4 , x=zeros(size(b));end;
   if nargin==3 , x=zeros(size(b));MaxIter=2000;end;
   if nargin==2 , x=zeros(size(b));MaxIter=2000;tol=1.e-4;end;
   % Initialization
   iter=0;r=b-A*x;tol2=tol*tol;normr2=r'*r;p=r;
   % Iterations
   while (normr2>tol2)&(iter<MaxIter)
       Ap=A*p;
       alpha=normr2/(p'*Ap);
       x=x+alpha*p;
       r=r-alpha*Ap;
       beta=r'*r/normr2;
       p=r+beta*p;
       normr2=r'*r;
```

```
      iter=iter+1;
   end;
```

2. The number of required iterations is always much smaller for the conjugate gradient method.

```
>> n=5;A=Laplacian1dD(n);xx=(1:n)'/(n+1);b=xx.*sin(xx);
>> [xVG, iterVG]=GradientV(A,b,1.e-4,10000);
>> [xCG, iterCG]=GradientC(A,b,1.e-4,10000);
>> [iterVG, iterCG]
ans =
      60     5
>> n=10;A=Laplacian1dD(n);xx=(1:n)'/(n+1);b=xx.*sin(xx);
>> [xVG, iterVG]=GradientV(A,b,1.e-4,10000);
>> [xCG, iterCG]=GradientC(A,b,1.e-4,10000);
>> [iterVG, iterCG]
 ans =
     220    10
>> n=20;A=Laplacian1dD(n);xx=(1:n)'/(n+1);b=xx.*sin(xx);
>> [xVG, iterVG]=GradientV(A,b,1.e-4,10000);
>> [xCG, iterCG]=GradientC(A,b,1.e-4,10000);
>> [iterVG, iterCG]
 ans =
     848    20
>> n=30;A=Laplacian1dD(n);xx=(1:n)'/(n+1);b=xx.*sin(xx);
>> [xVG, iterVG]=GradientV(A,b,1.e-4,10000);
>> [xCG, iterCG]=GradientC(A,b,1.e-4,10000);
>> [iterVG, iterCG]
 ans =
          1902            30
```

3.
```
>> n=5;A=toeplitz(n:-1:1)/12;b=(1:n)';
>> x0=[-2,0,0,0,10]';
>> [xCG, iter0]=GradientC(A,b,1.e-10,50,x0);
>> x1=[-1,6,12,0,17]';
>> [xCG, iter1]=GradientC(A,b,1.e-10,50,x1);
>> [iter0 iter1]
ans =
      3     5
```

For the initial guess x_0, the conjugate gradient algorithm has converged in three iterations. Let us check that the Krylov critical dimension corresponding to the residual $b - Ax_0$ is equal to 2.

```
>> r=b-A*x0;
>> X=[];x=r;
>> for k=1:n
>>     X=[X x];[k-1 rank(X)]
```

```
>>      x=A*x;
>> end;
ans =
        0    1
ans =
        1    2
ans =
        2    3
ans =
        3    3
ans =
        4    3
```

For the second initial guess, the Krylov critical dimension turns out to be 4.

4. System $Ax = b$ is equivalent to $A^t Ax = A^t b$, whose matrix $A^t A$ is symmetric, positive definite. The conjugate gradient algorithm can be applied to the latter system. However, observe that computing $A^t A$ costs $n^3/2$ operations, which is very expensive, and therefore it will not be done if we want to minimize the computational time. Remember that the explicit form of $A^t A$ is not required: it is enough to know how to multiply a vector by this matrix.

11.9 Exercises of Chapter 10

Solution of Exercise 10.6 Here is a a program for the power method.

```
function [l,u]=PowerD(A)
% Computes by the power method
%  l = approximation of |λn|
%  u = a corresponding eigenvector
% Initialization
n=size(A,1);x0=ones(n,1)/sqrt(n); % x0
converge=0;eps=1.e-6;
iter=0;MaxIter=100;
% beginning of iterations
while (iter<MaxIter)&(~converge)
  u=A*x0;
  x=u/norm(u);
  converge=norm(x-x0)<eps;
  x0=x;iter=iter+1;
end
l=norm(u);
```

Application to the three suggested matrices:

✗ For matrix A, everything seems to work as expected. The algorithm returns the maximal eigenvalue that is simple.

```
>> A=[2 2  1 0;2 0 0 0;  1 0 0 2;0 0 2 -2];
[l,u]=PowerD(A); fprintf('l =%f',l)
l =3.502384>>
>> A=[2 2  1 0;2 0 0 0;  1 0 0 2;0 0 2 -2];
[l,u]=PowerD(A); fprintf('l =%f \n',l)
l =3.502384
>> disp(eig(A)')
    -3.3063   -1.2776   1.0815   3.5024
```

✗ For matrix B, we also get the maximal eigenvalue, although it is double; see Remark 10.3.1.

```
>> B=[15 0 9 0;0 24 0 0;9 0 15 0;0 0 0 16];
>> [l,u]=PowerD(B);
>> fprintf('l =%f \',l)
>> l =24.000000
>> disp(eig(B)')
     6    16    24    24
```

✗ For matrix C, the algorithm does not compute the eigenvalue of maximal modulus, even though all eigenvalues are simple.

```
>> C=[1 2 -3 4;2 1 4 -3;-3 4 1 2;4 -3 2 1];
>> [l,u]=PowerD(C);
>> fprintf('l =%f \',l)
l =4.000000
>> disp(eig(C)')
    -8.0000    2.0000    4.0000    6.0000
disp(eig(C)')
```

Explanation: let us calculate the eigenvectors of matrix D, then compare them with the initial data of the algorithm $x_0 = \left(\frac{1}{2}, \frac{1}{2}, \frac{1}{2}, \frac{1}{2}\right)^t$.

```
>> [P,X]=eig(C); P, X
P =
    0.5000   -0.5000   -0.5000   -0.5000
   -0.5000   -0.5000   -0.5000    0.5000
    0.5000    0.5000   -0.5000    0.5000
   -0.5000    0.5000   -0.5000   -0.5000
X =
   -8.0000         0         0         0
         0    2.0000         0         0
         0         0    4.0000         0
         0         0         0    6.0000
```

We see that x_0 is an eigenvector of C corresponding to the eigenvalue $\lambda = 4$. In this case, it is easy to see that the sequence of approximated eigenvectors, generated by the power method, is stationary (equal to x_0). The sequence of eigenvalues is stationary too.

References

1. BRASSARD, G., BRATLEY, P. *Fundamentals of Algorithmics*, Prentice-Hall, Inc. Upper Saddle River, NJ, USA (1996).
2. CHATELIN, F. *Eigenvalues of Matrices*, John Wiley & Sons, Ltd., Chichester (1993).
3. CIARLET, P.G. *Introduction to Numerical Linear Algebra and Optimisation*, Cambridge Texts in Applied Mathematics. Cambridge University Press, Cambridge (1989).
4. DEMMEL, W.J. *Applied Numerical Linear Algebra*. Siam, Philadelphia (1997).
5. DUFF, I.S., ERISMAN, A.M., REID, J.K. *Direct Methods for Sparse Matrices*, Clarendon Press, Oxford (1986).
6. GENTLE, J.E. *Numerical Linear Algebra for Applications in Statistics*, Springer-Verlag, New York (1998).
7. GOLUB, G., VAN LOAN, C. *Matrix Computations*. The John Hopkins University Press, Baltimore (1984).
8. LASCAUX P., THEODOR R. *Analyse numérique matricielle appliquée à l'art de l'ingénieur*. Masson, Paris (1985).
9. LANG S. *Linear Algebra*. Undergraduate Texts in Mathematics, Springer-Verlag, New York (1989).
10. LAX P. *Linear algebra*. John Wiley, New York (1997).
11. LUCQUIN B., PIRONNEAU, O. *Introduction to Scientific Computing*, John Wiley & Sons, Ltd., Chichester (1998).
12. ORTEGA, J. *Introduction to Parallel and Vector Solution of Linear Systems*. Frontiers of computer Science. Plenum Press, New York (1988).
13. QUARTERONI, A., SACCO, R. AND SALERI, F. *Numerical Mathematics*, Texts in Applied Mathematics, Springer-Verlag, Berlin, (2000).
14. SAAD, Y. *Iterative Methods for Sparse Linear Systems*. PWS Publishing, Boston (1996).
15. STOER, J., BULIRSCH, R. *Introduction to Numerical Analysis*, Texts in Applied Mathematics, Springer-Verlag, New York, (2002).
16. STRANG, G. *Linear Algebra and its Applications*. Academic Press, New York (1980).
17. TREFETHEN L., BAU D. *Numerical Linear Algebra*, Society for Industrial and Applied Mathematics (SIAM), Philadelphia, PA (1997).
18. WILKINSON, J.H. *The Algebraic Eigenvalue Problem*. Clarendon Press, Oxford (1965).

References

Index

Index of Programs

Texts in Applied Mathematics

(*continued from page ii*)